Henry Reid

A practical Treatise on natural and artificial Concrete

Its Varieties and constructive Adaptations

Henry Reid

A practical Treatise on natural and artificial Concrete
Its Varieties and constructive Adaptations

ISBN/EAN: 9783337106447

Printed in Europe, USA, Canada, Australia, Japan

Cover: Foto ©ninafisch / pixelio.de

More available books at **www.hansebooks.com**

A PRACTICAL TREATISE
ON
NATURAL AND ARTIFICIAL
CONCRETE:
ITS
VARIETIES AND CONSTRUCTIVE ADAPTATIONS.

BY

HENRY REID.

AUTHOR OF 'THE SCIENCE AND ART OF THE MANUFACTURE OF PORTLAND CEMENT,' ETC.

NEW EDITION.

LONDON:
E. & F. N. SPON, 46, CHARING CROSS.

NEW YORK:
446, BROOME STREET.

1879.

PREFACE.

Owing to the favourable reception accorded to my Concrete book of 1869, which has passed through a second edition, I have been induced to prepare the following work, treating more fully on the subject, and I trust more worthily of this important question.

The subject of concrete and its varieties now receives much attention, more especially in consequence of the numerous novel adaptations of comparatively recent origin. In its more extensive application for important engineering works there has not been so much attention given to the accurate scientific rules according to which this material should be made. For minor purposes, however, and in what may be regarded as its domestic direction, great efforts are being made to keep pace with the most recent knowledge on the subject of concrete-making. Of the paving and pipe-making industries, together with a variety of others of probably less importance, because not yet sufficiently established, I have in these pages given such information as will at all events lead to a further consideration of their peculiar features and advantages.

Marine architecture has during many years been beneficially assisted by that system of concrete-making which received its greatest impetus when Portland cement was first made good enough to warrant its acceptance by engineers without fear of dangerous results. In too many cases, however, the acquisition of a faultless matrix has been regarded by engineers as the only desideratum, and a degree of indifference has in consequence resulted, to the quality and character of the aggregates. In the best examples of concrete of established quality I have shown that much of their excellence is due to the care bestowed on their fabrication and a just and intelligent knowledge of the matrices and aggregates of which they are composed. There has always been much difficulty in advancing good reasons why concrete made from Portland cement was superior to that produced from other matrices having a more ancient and established reputation. The apparently strong and indisputable reasons which antiquity furnished to the advocates of lime and puzzolana tended much to retard the use of Portland cement long after its excellent qualities had been well known and approved. Roman buildings and others of more recent origin were cited as evidence against the adoption of the new cement, and perhaps its greatest and most remarkable success is due to the fact that through it constructive feats have been accomplished, which the most sanguine of ancient constructors could not have supposed possible even in their dreamiest of moods. Ponderosity and bulk were not unknown to ancient architects and builders, as the Pyramids and Roman ruins fully show. Practical science had not, however, in those remote ages furnished the means required to enable the engineers of antiquity to lift a River Liffey concrete block weighing nearly 400 tons, or float a Tay Bridge pier to its required site and deposit its weight of 200 tons safely and unharmed in deep water.

Notwithstanding these engineering achievements, now regarded as of comparatively easy accomplishment, considerable harbour and dock works are still progressing regardless of the lessons which such able and advanced engineering teaches and illustrates. Less attention is attracted to these important works because the public have, or at least take, but little interest in their progress. It is otherwise, however, in matters of purely domestic importance, where the poisonous sewer-gas lays low the helpless victim of centralized constructive authority, or where the walls of the dwellings are saturated with moisture, and worse, also contribute their quota to the continuous and never-ceasing sacrifice demanded by the Juggernauts of unhealthy homes.

In the matter referring to the paving and pipes I have endeavoured to show that by the use of concrete of the right quality great advantages are secured; while the various references to dwelling construction illustrate the benefits and advantages which are easily commanded in that department.

Satisfactory progress in the realization of any of the advantages sought from the use of concrete in the numerous directions in which it is applicable, is impossible if the question of its accurate scientific preparation is disregarded.

I have shown very clearly that concretes differ in quality according to the care bestowed in the selection of the materials used and the methods by which they are compounded. The various illustrations which I have prepared of natural and artificial concretes show that when careful rules are adopted and intelligently followed even more than has yet been accomplished is capable of realization. The reference to natural rocks and their uncertain and dangerous peculiarities affords additional reasons for the still more earnest prosecution of an industry which can only be yet regarded as in a condition of comparative childhood.

The gradual introduction of machinery in producing concrete forms will tend greatly to the advancement of the art, for however satisfactory the results of hand treatment may be considered, the final and much-to-be-desired excellence is only attainable through the agency of carefully devised mechanical appliances. The reference to *impact* concrete is intended to illustrate the desirability, and indeed necessity, of bringing into as close relationship as possible the matrix and aggregates without the hurtful agency of water.

Although there are many novel and perhaps useful cement combinations before the constructive world, I have purposely avoided direct reference to them, or by suggestion even recommending their use. I have too much regard for, and feel too highly, the importance of limiting the attention of concrete-makers to the one fully tried and excellent Portland cement, which in its best form is undoubtedly the most valuable constructive agent at our command. Many suggestions and *inventions* are obtruded on our notice, some to remedy the defects of faulty Portland cement or supersede its use altogether; and from the danger which such teaching is likely to engender, I have discussed the subject generally in Chapter XVI., treating of "Bastard" mortars.

In my Portland cement book, and indeed whenever by other means I could advance the *cause* of this valuable material, I have not flinched from such observations as I honestly thought would assist in concentrating the attention of the makers of that article to a clearer understanding of their duty to the public. The natural growth of the Portland cement industry has not, I am sorry to say, been entirely due to the efforts of the manufacturers, for they have, as it were, been dragged by the consumer into a prosperous condition which some of them hardly deserved.

When Sir Joseph Bazalgette, I might also say, timidly imposed a moderate test great opposition to its acceptance was offered. The consequences which resulted from that step are now too well known to require to be more particularly referred to in these pages.

Foreign engineers, while solely dependent on English supplies of cement, were much troubled in the obtainment of the quality which their scientific researches had considered to be essentially necessary. While, however, the watchfulness and knowledge of external intelligence insisted upon quality of such a standard as was of comparatively easy accomplishment, internal vigilance had not been sufficiently aroused to the importance of a thoroughly good testable cement capable of being guaranteed.

French and German engineers may be said to have laid the foundation of the reputation of English Portland cement, and now, being no longer entirely dependent on supplies from England, have the satisfaction of obtaining more reliable qualities of home manufacture. It is from a full knowledge of Portland cement progress in other countries that I feel called upon to speak unreservedly, and it may be said captiously, of the indifference of English Portland cement-makers, more especially in the London district, to the quality of their manufactures.

Concrete-makers—I should say scientific concrete-makers —are in a continual state of anxiety as to the quality of the cements which the exigencies of their industry demand, and they are therefore in frequent conflict with the cement-makers about the character and quality of the required tests. The want of uniformity in the method of testing in this country, and the variety of machines or apparatus employed in its performance, continue to surround the question with much that is erroneous and misleading. This difficulty is only capable of removal when manufacturers of cement think proper to discontinue the practice of making a variety of qualities of an article, which experience elsewhere proves is capable of being uniform and unvarying.

I shall in all probability be accused by many with want of patriotism in holding up the German and other cement-makers as guides for our home manufacturers to follow. By those, however, who have experienced the difficulties of obtaining first-class cement in England, my motives in thus alluding so pointedly to this subject, in these days of depressed trade and foreign competition, will be favourably appreciated. The qualities of foreign cements which I have considered it necessary to point out, will at least show that the German cement-makers to whom we imparted the rudiments of cement-making knowledge, are now outrunning us in the race of competition, and not only prepared to supplant us in distant markets, but also bid for customers for their cements at our own doors.

Good concrete means good cement, and if that cannot be found at home it must be brought from abroad. The "Stern" cement of Stettin, so luckily put in my hands to assist in the discussion of my argument, proves that the quality of uniformity, high tests, and their guarantee are of facile accomplishment. I purposely produce all the facts of the case for the information of concrete-makers, to show that their present demands on cement-makers are not only moderate in character, but that elsewhere a higher standard is supplied in quite a voluntary manner.

In the introductory chapter, I have endeavoured to illustrate the necessity of substituting an artificial material, which by its excellence is competent to supply a want and assist in removing the distrust and anxiety attending the use of our best building stones. I have by the best means at my command, pointed out the varieties of suitable aggregates, and also the manner of their conversion to render them useful agents in concrete mixtures. I consider these observations of much importance, because they point out in a practical manner the necessity of giving intelligent attention to the ingredients on which so much depends, and which if unsuitable would render the best concrete valueless, however intelligently prepared.

When I undertook the preparation of this book, I did not fully realize the extent of the field of observation which I have I fear but imperfectly surveyed. It was impossible for me, within the limits to which I was necessarily prescribed, to refer to all the numerous adaptations of concrete art and industry, now being prosecuted in this and other countries. Those referred to have been selected either from the prominence they occupy, or the merits which to my mind they possess, and from what has been said and illustrated, those interested will be at no loss to select that which they regard as most suitable to their wants or desires.

Building by frames has been for a long time in fashion, and although not implicitly committing myself to that principle, I have referred to a simple and useful apparatus which may be adopted in ordinary cases, where elaborate or ambitious designs are undesirable or not required. For moderate structures in country districts, frames will no doubt be found useful; and for many of the subordinate buildings of large establishments, such as stables, outhouses, retaining walls, or other cognate purposes, such an apparatus as I have referred to has many advantages and may within reasonable limits be successfully applied.

In tracing the progress of scientific concrete-making, which may be said to have originated with Mr. Buckwell, and following its several developments through the phases of Victoria stone paving and silicated pipe making, I have endeavoured to show that the success achieved by these prosperous industries is due to the steady adherence of their several adapters to the accurate rules which should guide all concrete-makers.

Comparatively recent applications of a novel character of cement concrete to more ambitious and less substantial purposes, as in the case of the Lascelles system of construction, indicate that for ornamental and highly decorative designs, the dependence on Portland cement has not been in vain. That and other compounds, outcomes of concrete, as in the case of Hornblower's fireproof construction, may be regarded as only pioneers of other adaptations of it to more extended purposes.

Lest I may be accused of ambitiously parading my knowledge of the subject on which this book has, I fear, tediously dwelt, I freely confess my indebtedness to many sources of information, without access to which reference to many of the interesting facts which I have collected would have been impossible. From General Gillmore's 'Treatise on Béton Coignet' I have derived valuable information regarding the concrete industries of the United States of America. To those other authorities to which I have had recourse, my acknowledgments have been made at the places of their occurrence in these pages.

I cannot, however, launch these lines before the public, without acknowledging my obligations to all and every one to whose industries or processes I have referred, for their ready and willing response to my inquiries, and the assistance they have rendered me in the performance of my task.

HENRY REID.

21, ARUNDEL STREET,
   THAMES EMBANKMENT, LONDON.
      July, 1879.

CONTENTS.

CHAPTER I.

INTRODUCTION.

Definition of concrete, 1 — Dangers of its indiscriminate use, 2 — Ancient buildings defaced by age, 3 — Terms, béton and concrete, 5 — Granitic breccia, 5 — Béton aggloméré, 6 — New World ruins, 7 — Ancient buildings in Grecian Archipelago, 9 — Cob dwellings, 10 — Great Pyramid, 11 — British granites, 12 — Waterloo Bridge, 13 — Cornwall and Devon granites, 15 — Lundy Island, 16 — Portland stone, 17 — Houses of Parliament, 19 — Masonry of Jerusalem, 21 — Mediæval buildings, 22 — Metamorphic rocks, 23 — Sedimentary rocks, 24 — Artificial stones, 26 — Silicated stones, 29 — Coral formations, 30.

CHAPTER II.

WHAT IS CONCRETE?

Conditions requiring observance, 34 — Italian and Roman concretes, 35 — Divisions of concrete-making, 36 — Modern (carelessly made) concrete, 39 — Victoria stone, 42 — Road-making, 44 — Size of road metal, 46.

CHAPTER III.

MATRICES.

Portland cement: properties of cementing agents, 48 — Portland cement: sources of raw materials from which it is made,. 49 — Characteristics of Portland cement, 50 — Process of its manufacture, 52 — Testing Portland cement, various methods, 53 — Deacon's extemporized tester, 54 — Bailey's testing machine, 56 — Pallant's, 57 — Briquette press, 58 — German testing machine, its description and method of using it, 59 — Portland cement analyses, 62 — Carbonic acid in Portland cement, 63 — Analyses of faulty Portland cement, 64 — Analyses of varieties of Portland cement and their average, 65 — Difference in value of briquettes under varied exposures, 66 — Experiments on Portland cement of different finenesses, 67 — Mixtures with different sands, 68 — Indifference of Portland cement makers as to its quality, 70 — Reason why consumers should exert their authority, 71 — Lieutenant Innes's experiments with heavy cement and different sands, 72-75 — Dr. Heintzel's experiments as to the effect of light on cement, 74 — Tests of exceptionally heavy cement, 77 — United States of America experiences, 78 — General Gillmore's views on cement, 79 — Mr. Maclay's experiments and observations, 79 — Clinker sent to New York, 85 — Limes: hydraulic limes from the lias formation, their analyses, 87 — Properties of hydraulic limes, 88 — Thiel limestone and lime, their analyses, 89 — Slaking and grinding lime, 90 — "Calp" limestone, 91 — Scotch limestone, 91 — Roman cement, its properties and derivation, 92 — Analyses of various sorts, 93 — Experiments on its strength, Pasley and others, 95 — Its value when mixed with sands, 96 — Comparative experiments of building stones, 96 — Mr. Ranger's artificial stone, 98 — Gypsum, 99 — Keene's and Parian cements, 100.

CHAPTER IV.

AGGREGATES.

Definition of the term, 101 — Peculiarities of various rocks, 102 — Granite, 103 — Felspar and mica analyses, 104 — Granite sources, 105 — Experiments as to its durability, 107 — Large granite block, 109 — Kentish ragstone, 109 — Its analysis, 110 — Source and quality, 111 — Portland stone, its source, peculiarities, and analysis, 112 — Carboniferous limestone, its source and analysis, 113 — Magnesian limestone, analyses and characteristics, 114 — Sandstones, their varieties, 115 — Craigleith stone analysis, 116 — Basaltic rocks, 116 — Gravels, 117 — Sands, 118 — Flints, 119 — Burnt shale, 119 — Burnt ballast, 120 — Broken pottery, 120 — Iron and other slags, 121 — Osnabruck bricks, 122 — Siegen bricks, 122 — Experiments as to their permeability, by Pettenkofer, 123.

CHAPTER V.

MACHINERY FOR REDUCING THE AGGREGATES.

Classification of stone-crushers, 125 — Crushers, rollers, &c., 126 — Squeezers, Blake's, 127 — Hall's multiple Blake, 128 — Combined crusher and roller, 129 — Portable Blake, 130 — Broadbent's and Gimson's stone-breakers, 131 — Goodman's crushers, 133 — Review of their several peculiarities and fitness, 135 — Impact reducers, Patterson's stamper, 136 — Its description, 138 — Experiments as to results, 139 — Sholl's stamps, 140 — Abraders, millstones, &c., 143 — Motte's grinder, 143 — Experiments and results, 146 — Hall's disintegrator, 147 — Horizontal millstones, 149.

CHAPTER VI.

TREATMENT OF THE AGGREGATES.

Necessity of effective cleansing, 150 — Size of aggregates, 150 — Sands and gravels, 151 — Porosity and general fitness for concrete, 152.

CHAPTER VII.

THE SILICATING PROCESS.

Desire for indurating agents, 154 — Analysis of compound silicates, 154 — Soluble glass, its properties and methods of manufacture, 155 — Papin's digester, 156 — Gossage's process, 156 — Dr. Graham's experiments, and the dialysing apparatus, 157 — The character of silicates and their action, 158 — Houses of Parliament experiments, 159 — Silication of mortar, Mr. Spiller's experiments, 159 — Aluminate of potash, 160 — Condition of concrete for favourable influences from silicates, 161.

CHAPTER VIII.

ESTABLISHED PROCESSES OF CONCRETE MANUFACTURE.

Coignet's béton aggloméré, 163 — Machinery of mixture; "The Malaxator," 165 — Works done in Coignet béton, 168 — Proportions of béton mixture, 171 — General Gillmore's estimate of the cost of a Portland cement manufactory, 172 — American cement manufactory, 173 — Tables of tests of Portland cement mortar at the Philadelphia Exhibition, 174 — Description of the mode of testing, 175 — Non-absorptive properties of béton aggloméré tests by Dr. Isidor Walz, 176 — Greyveldinger mortar mill, 177 — Frear artificial stone, 179 — Foster's stone process, 180 — Van Derburgh system of block-making, 181 — Tests of Van Derburgh stone, 181 — Ranger's system of artificial stone making, 182 — Sorel stone, 183 — Anglesey magnesian limestone, 183 — Description of Sorel process, 185 — Dr. Jackson's tests, 185 — Reports on Sorel stone by Middlesex (U.S.A.) Agricultural Society, 186 — Proportions of various kinds, 187 — Frear stone tests, 190 — Foster and Van Derburgh artificial block tests, 191 — Tests of Ranger's artificial stone, 192 — General remarks on American artificial stone industries, 193 — Ransome's siliceous stone, 195 — Description of the process, 195 — Dr. Frankland's experiments on Ransome's stone, 197 — Buckwell's "granitic breccia stone," 199 — Tests and experiments on it by Mr. Andrews, 200 — Experiments with Ransome's artificial grindstones by Messrs. Bryan, Donkin, and Co., 201 — Magnesite or union wheels, 202.

CHAPTER IX.

ENGLISH CONCRETE INDUSTRIES.

Ranger and Buckwell's pioneer operations, 203 — The Buckwell "granitic breccia," 204 — Its use for mooring blocks of large size, 204 — Impact process and its advantages, 205 — Imperfect appreciation of the impact process, 205 — Dean Buckland and Professor Faraday's connection with the "granitic breccia" stone, 206 — Faraday's lecture, 207 — Cessation of "granitic breccia" stone operations, 207 — Victoria patent stone process, 209 — Its origin and history, 210 — Mr. Highton's efforts and progress, 211 — Analysis of Leicester (Groby) granite, 213 — Mr. Kirkaldy's tests, 213 — Size of aggregates used, 214 — Analysis of Portland cement used in the Victoria stone industry, 215 — Careful preliminary examination of cements, 216 — Proportions used, 217 — Description of process of manufacture, 217 — Description of articles made, 219 — Silicating process, 220 — Method of making the silicate, 221 — Analysis of Victoria stone, 223 — Rock-concrete pipe making, 223 — History of its introduction in England, 226 — Messrs. Henry Sharp, Jones, and Co.'s manufactory at Poole, 227 — Description of the process employed, 228 — Analysis of broken pottery, 228 — Mixing machine, 229 — Description of the moulding process, 230 — Analysis of rock-concrete, 232 — Size and weight of pipes, 232 — Experiments on fractured pipes, 233 — Description of the briquettes and cubes experimented on, 235 — Bournemouth experiments on rock-concrete and glazed stone-ware pipes, 237 — Mr. Ellice-Clark's experiments, 239 — Pipe-testing machine, 240 — Objects of sewering and responsibilities of engineers, 241 — Mr. Baldwin Latham's suggestion, 242 — Silicated stone manufacture, 243 — System of manufacture, 243 — Lincolnshire slag, 243 — Kentish ragstone, 243 — Value of slag as an aggregate, 244 — Silicated stone from Belgian granite, analysis, 245 — Messrs. Hodges and Butler's machinery, 245 — Pipe-moulding machinery, 248 — Silicated stone from Thames gravel, analysis, 250 — The "Trembler" machine for slab-making, 251 — Egg-shaped sewer, 255.

CHAPTER X.

CONSTRUCTIVE CONCRETE APPLICATIONS.

Character of artificial concrete, 257 — Lascelles system of construction, 258 — Description of slabs, 260 — Advantages of system, 261 — Greenhithe examples of cottage buildings, 262 — House at the Paris Exhibition, 1878, 263 — Mode of construction, 264 — Frieze panel, 265 — French rewards to Mr. Lascelles, 266 — Studio window, 267 — Country house, 268 — Character of labour employed, 269 — Architect's estimate of the system, 270 — Portland cement concrete tiles, 271 — Review of encaustic tile processes, 272 — Wall embellishments, 273 — Italian mosaic, 273 — Messrs. Patteson's tiles, 274 — Advantages of concrete tiles, 276 — Hornblower's fireproof floors, 277 — Illustrations and character, 279 — Description of the system, 286 — Works executed, 287 — Birmingham Free Library, 287 — Fire at Clumber, 288 — Capacity of fireproof construction in aiding decoration, 288 — Lish's patent Z blocks, 289 — The advantages which they possess for easy transit, 291 — Description of mode adopted in building with them, 292 — Wide field for extension in block direction, 293.

CHAPTER XI.
IMPORTANT ENGINEERING CONCRETE WORKS.

The Tay Bridge, 294 — Mr. Grothe's description of it, 295 — Floating the piers, 296 — River Liffey improvements, 297 — Difference between the procedure at the Tay Bridge and North Wall works, 298 — Messrs. Milroy and Butler's concrete cylinder building, 299 — Birkenhead Docks, 300 — Concrete mixing machine, 301.

CHAPTER XII.
GERMAN PORTLAND CEMENT.

"Stern" Works, Stettin, 303 — Reputation of Messrs. Toepffer, Grawitz, and Co., 304 — Guaranteed cement, 304 — Extent and character of "Stern" cement works, 305 — Analyses of chalk and clay used, 306 — "Stern" cement analysis, 306 — Philadelphia Exhibition award, 307 — Method of manufacture, 308 — Tests of the Society of German Engineers, 309 — Dr. Michaelis' tests, 310 — Messrs. Toepffer, Grawitz, and Co.'s tests, 311 — Guaranteed cement, 312 — Dr. Michaelis' table of relative values of cement, 313 — Character of guarantee, 314 — Technical director of cement works, 315 — Author's first experiments, 316 — Uniform quality of "Stern" cement, 317 — Author's second experiments, 318 — Table of German and English measures and weights, 319.

CHAPTER XIII.
BUILDING FRAMES.

Use of building frames, 320 — Henley concrete building apparatus, 321 — Its description, 323.

CHAPTER XIV.
CONSIDERATIONS AS TO TESTING GENERALLY.

Causes leading to carelessness of testing, 325 — Character of works executed without proper supervision, 326 — Clerks of works and their qualifications 327 — Disasters caused by use of improper materials, 328 — Duties of district and other official surveyors, 329 — Character of mortar used in London, 330 — Sir J. Bazalgette and the Metropolitan Board of Works, 331 — Machinery of testing, 332 — Laboratory experiments, 333.

b

CHAPTER XV.

CHARACTER OF BUILDING MATERIALS.

Selection of materials with reference to their capacity of heat influence, 334 — Dangerous absorptive properties of walls, 335 — Permeability of building materials, 336 — Specific gravities, &c., of various materials, 336 — Conducting and cooling power of building materials, 337 — Professor Pettenkofer's experiments, 338 — Professor Doremus's experiments, 339 — Devey's 'Hygiène de Famille,' 340 — Mr. Teale's 'Dangers to Health,' 341 — Dr. W. Chambers' observations, 342.

CHAPTER XVI.

SPURIOUS OR BASTARD MORTARS.

Smeaton's Eddystone experiments, 344 — Aspdin's patent, 344 — Character of adulterations and the objects for which made, 346 — General Gillmore's experiments, 347 — Mr. F. Kuhn's experiments, 348.

CHAPTER XVII.

ORIGINAL EXPERIMENTS ON CONCRETES, Etc.

Character of the concretes, 350 — Victoria stone compressive experiments, 351 — Victoria stone tensile tests, 352 — Leicestershire granite tests, 356 — Guernsey granite, 356 — Rock-concrete experiments, 357 — Potteryware tests, 358 — Rock-concrete compressive tests, 359 — Silicated stone tests, 360 — Magnesite tests, 362 — Ransome's stone tests, 363 — Granitic breccia tests, 364 — Spinkwell stone tests, 366 — Loch Etive granite tests, 366 — Mount Sorrel granite tests, 366 — Penmaen-Mawr basalt tests, 367 — Carrara marble, 368 — Slag brick tests, 368 — Ransome's stone and apauite tests, 368 — Mr. Haslinger's experiments, 369 — Ditto tables, 372.

LIST OF PLATES.

PAGE

FRONTISPIECE—
    FIG. 1.—SECTION OF VICTORIA STONE, SHOWING SIZE OF AGGREGATES USED NINE YEARS AGO.
    " 2.—DITTO, DITTO, USED SIX YEARS AGO.
    " 3.—SECTION OF ROCK CONCRETE, WITH BROKEN POTTERY AS AN AGGREGATE.
    " 4.—DITTO, DITTO.
    " 5.—SECTION OF SILICATED STONE, WITH THAMES BALLAST AS AN AGGREGATE.
    " SECTION OF DITTO, WITH BELGIAN GRANITE AS AN AGGREGATE.
        ALL THE ABOVE SECTIONS NATURAL SIZE AND COLOUR.

PLATE 2.—SECTION OF VICTORIA STONE AS NOW MADE 214

    " 3.—PATTESON'S CEMENT TILES 274

    " 4.—VIEW OF "STERN" CEMENT WORKS 308

b 2

LIST OF ILLUSTRATIONS.

		PAGE
Fig. 1.—	Templar Rock, Lundy Island	16
,, 2.—	Deacon's Extemporized Testing Apparatus	54
,, 3.—	Bailey's Testing Machine	56
,, 4.—	Pallant's Testing Machine	57
,, 5.—	Briquette Press	58
,, 6.—	German Testing Machine	59
,, 7.—	Hall's Multiple-action Stone Breaker	128
,, 8.—	,, ,, ,, ,, and Combined Crushing Roller	129
,, 9.—	Hall's Multiple-action Stone Breaker and Portable Engine	130
,, 10.—	Broadbent's Drawback-motion Stone Breaker	131
,, 11.—	Gimson's Duplex Stone Breaker	131
,, 12.—	Goodman's Single-action Stone Crusher	133
,, 13.—	,, Double-action ,,	134
,, 14.—	Patterson's Elephant Stamp	136
,, 15.—	Sholl's Pneumatic Stamp	141
,, 16.—	,, ,,	142
,, 17.—	Motte's Patent Crushing Machine	144
,, 18.—	Hall's Disintegrator	147
,, 19.—	Horizontal Grinding Mill	149
,, 20.—	Papin's Digester	156
,, 21.—	The Malaxator—Elevation	165
,, 22.—	,, —Plan	166
,, 23.—	The Greyveldinger Mortar Mill	178
,, 24.—	Victoria Stone Moulding	218
,, 25.—	,, Productions	219
,, 26.—	,, Silicating Tanks	221
,, 27.—	,, Post Office Moulding	222
,, 28.—	,, Stone Vase	223
,, 29.—	,, ,, ,,	224
,, 30.—	Concrete Mixer	229
,, 31.—	Pipe-moulding Machine	231
,, 32.—	Mr. Ellice-Clarke's Tests	239
,, 33.—	,, ,, ,,	239
,, 34.—	,, ,, ,,	239

ILLUSTRATIONS.   xxi

		PAGE
Fig. 35.—Pipe-testing Machine	....	240
„ 36.—Messrs. Hodges and Butler's Pipe-making Machine	..	246
„ 37.— „ „ „ Paving-slab Making Machine		251
„ 38.— „ „ „ „ „ „		251
„ 39.— „ „ „ „ „ „		252
„ 40.— „ „ „ Tamping Machine for Paving-slab Making	....	253
„ 41.—Egg-shaped Sewer Elevation ..	....	255
„ 42.— „ „ „ Isometrical View..	.. ..	256
„ 43.—Queen Anne House, Paris Exhibition	....	263
„ 44.—Panel of Frieze of Ditto	....	265
„ 45.—Studio Window ..	....	267
„ 46.—Country House ..	....	268
„ 47.—Hornblower's Fireproof Floors	....	279
„ 48.— „ „ „	....	281
„ 49.— „ „ „	....	282
„ 50.— „ „ „	....	283
„ 51.— „ „ „	....	285
„ 52.—Lish's Patent Blocks	....	291
„ 53.— „ „	....	292
„ 54.—Floating Tay Bridge Piers ..	....	296
„ 55.—Le Mesurier's Concrete Mixer	....	301
„ 56.—Henley's Concrete Apparatus	....	322
„ 57.— „ „ „	....	323
„ 58.—Briquette Section: Granitic Breccia..	......	366
„ 59.—Enlarged Section: Spinkwell Stone ..	......	367

LIST OF TABLES.

	PAGE
Tests of Portland Cement submitted to various Periods of Exposure	66
Similar Mortar Tests	67
Tests of different Finenesses	72
„ with different Quantities of Water	72
„ of different Grained Sands and in Mortar	76
„ of Portland Cement used in last Tests	76
Breakings of exceptionally Heavy Portland Cement	77
„ of the same Cement and Thames Sand	77
American Tests of English Cement at different Temperatures	82, 83
Roman Cement Tests	94
Pasley's Experiments of Ditto	96
„ „ and Sand	96
Compressive Values of different Stones: Portland, Cornish Granite, Kentish Ragstone, Yorkshire Landing, Craigleith Stone, and Bath Stone	96
Weights of same per Cube Foot	96
Their Resistance to Rupture	97
Keene's Cement Tests	100
Parian „	100
Tests of Wearing Value of different Granites	107
„ „ „	108
Values and Weights of various Building Stones	111
Permeability of Slag Bricks and other Materials	123
Table of Experiments with Patterson's Elephant Stamper	146
Tables for Proportions of Coignet Béton Aggloméré	170–71
Table of General Gillmore's Estimate for a Cement Works	172
Philadelphia Exhibition Tests of Portland Cement	174
Tests of Van Derburgh Stone	181
Proportions for Sorel Stone	187
Tests of Sorel Stone	188
„ Coignet Béton	189
„ Frear Stone	190
„ Foster and Van Derburgh Stones	191
„ Ransome Stone	197
„ „ and other Stones	198
„ Buckwell's Granitic Breccia Stone	200
Thicknesses and Weights of Rock Concrete Pipes	232
Experiments on Bournemouth Pipes	233, 234
„ by Hydraulic Pressure on ditto	237
„ „ „ on Glazed Stoneware Pipes	237
„ by Mr. Ellice-Clark on Bournemouth Pipes	239

LIST OF ANALYSES.

	PAGE
Tests of Stettin Cement by German Engineers	309
,, ,, Dr. Michaelis	310
,, ,, Messrs. Toepffer, Grawitz, and Co.	311
Guaranteed Tests of Ditto (Old)	312
Dr. Michaelis' Table of Relative Values of Cement	313
Guaranteed Tests of Stern Cement (New)	314
Author's ,, ,,	316–318
Table for the Conversion of German Weights and Measures into English	319
Tables of Specific Gravities, &c., of Building Materials	336
,, Conducting Power ,, ,,	337
,, Cooling ,, ,, ,,	337
,, Absorptive Values ,, ,,	339
,, Experiments with Bastard Mortars	348
Compressive and Tensile Tests of Victoria Stone	351
Rock-concrete Tests	357
Silicated Stone ,,	361
Ransome Stone ,,	362–368
Mr. Haslinger's Experiments (Tables 1 and 2)	370

LIST OF ANALYSES.

Egyptian Granite	12
Portland Stone	17
Houses of Parliament Stone	20
Average Portland Cement	50
Impure ,,	62
Carbonic Acid (Test)	62, 63
Imperfect Portland Cement	64
London ,,	65
Wallsend ,,	65
Boulogne ,, (Natural)	65
Stettin ,,	65
Bonne ,,	65
Average Analysis of Ditto	65
Lyme Regis Lias Limestone	87
Aberthaw ,,	87
Holywell ,,	87
Rugby ,,	87
Larne (Ireland) ,,	87
Thiol (France) Limestone	89
,, Lime	89
Roman Cement Stones	93
Harwich	93
Sheppey	93

LIST OF ANALYSES.

	PAGE
Calderwood	93
Dovoholes	94
Felspar	104
Mica	104
Kentish Ragstone	110
Portland Stone	112
Buxton Limestone	114
Mansfield Sandstone	114
Craigleith Stone	116
Soluble Silica	154
Leicestershire Granite	213
Medway Portland Cement	215
Thames "	215
Victoria Stone	223
Broken Pottery	228
Rock Concrete	232
Lincolnshire Slag	243
Kentish Ragstone	243
Silicated Stone (Slag)	245
Silicated Stone (Thames Ballast)	250
Chalk used at "Stern" Works	306
Clay " "	306
Stern Portland Cement	306

A

PRACTICAL TREATISE ON CONCRETE.

CHAPTER I.

INTRODUCTORY.

THE word Concrete has, like many other of our scientific names, a Latin origin, the meaning of which is *together grown*, implying thereby a natural rather than an artificial process. It is a term, however, which is now almost universally adopted, and in its application to our subject is specially significant, conveying as it does the idea of a mass indissolubly bound together, such as the nodules of chert and ironstone, the grains and spherules of oolite, and the grape-like clusters of magnesian limestone. These various conditions of naturally compacted mineral masses are the result of atomical aggregation and distinct in character from those produced by crystallization, the operation in the one case being mechanical and in the other chemical.

Concrete therefore signifies a mass held or bound together by an agent of cementation of varied character and quality. Natural concretes are of almost unlimited extent, and they are found in every geological formation, from the finest sandstone to the coarse millstone grit, or surface puddingstone; whether the binding agent be siliceous, ferruginous, argillaceous, or calcareous, and compacted by superincumbent pressure as in the sedimentary deposits, or by heat as in those of plutonic origin, the result obtained is commensurate with

B

the natural force that was primarily exerted. In the case of gradual and protracted aqueous deposition, the minute particles have been placed in the closest mechanical contact in combination with the cementing agent itself, or of the materials from which it was finally produced, and sometimes also by succeeding infiltration from other sources; the heat agent again exerting its influence by previously melting the mass, so as to prepare it for the arrangement of its crystals in the concluding and final process of cooling.

These natural concreting forces have exerted their respective influences under the most changeable and conflicting circumstances, producing results of sometimes very uncertain value, hence the variable quality of those stones usually employed for building purposes and the difficulties which stand in the way of guarding against the dangers surrounding their use. The long-buried treasures of the earth's crust from which are obtained all our building materials undergo great change, more especially on exposure in a vitiated atmosphere, and when subjected to the influence of changeful and varying climates.

Much danger therefore attends the indiscriminate or careless use of new and untried materials, more especially in works of an extensive and elaborate character. In past times the builder's attention was more particularly directed to the facility with which a stone could be converted rather than to the value of its durability when placed in the building. Modern builders are still to a certain extent influenced by the same considerations, although they have at their command information unknown to the ancients, based on practical and scientific knowledge, by which they may be governed and safely guided. Under the best conditions, however, the dangers surrounding the use of natural stones for any constructive purpose are numerous and in some cases almost unavoidable. The climate and the quality or condition of the atmosphere have much to do with the wear of stones and other building materials.

The present almost unimpaired condition of the Egyptian Pyramids is due to the equability of the Eastern climate in which they have for so many ages stood, and also to some extent to the excellent quality of the rock by which they are externally clothed. The granite used in the embellishment of the Great Pyramid was obtained from Syene in its neighbourhood, and is erroneously called by some authorities Syenite. That rock, however, is now more accurately defined by modern mineralogists as one of the hardest in nature, and differs in some essential particulars from that with which these apparently everlasting monuments were built. The numerous relics of Egyptian architecture maintain their pristine freshness of lines, and still in many instances exhibit the marks of the tools by which they were originally fashioned.

Many of those monuments of a past *civilization* have during the several military subjections and dominations under which Egypt has passed been transferred to the capitals of the conquerors, and the Romans, in their ambitious desire to decorate their public buildings and luxurious homes, incurred vast expense in transporting them to Rome and other cities. In this country we have not had the same opportunity of testing the capacity of the materials from which these ancient monuments were made, but the arrival in London of the so-called "Cleopatra's Needle" promises us a chance for that examination. One of the Egyptian sphinxes (removed to Paris by Napoleon the Great) sculptured from porphyry, a rock of compact texture, was inadvertently placed under one of the rain gutters of the Louvre and soon thereafter showed indications of decay. This piece of ancient sculpture continued to maintain its normal form comparatively unimpaired in the climate of its origin, but on being placed in one more humid and in the position described, immediately succumbed to the chemical change of its surroundings. The celebrated statuary marble of Carrara, first used a hundred years before the Christian era, and placed in the atmosphere of Rome, continues to the present time unchanged in form and exactness of outline. In some of the oldest examples of statuary, discoloration has arisen, in all probability owing to the slight and almost imperceptible amount of oxides of iron in the analysis of the marble. Works in this fine marble executed in the reign of the Emperor Augustus are now but little degraded or altered. In London, unfortunately, we are unable to record any approach to such a satisfactory condition, for some of the finest creations of Chantrey's genius exhibit, after only fifty years' erection, indications of decay and premature disaggregation.

Chemistry can readily estimate the causes from which such changeable results arise, and the geologist's and mineralogist's knowledge of the source and origin of these rocks determines their character, whether igneous, metamorphic, or sedimentary. But the combined guidance of the most advanced specialists in these several scientific departments have hitherto failed to secure for us enduring structures in a London atmosphere; stone of the best apparent quality, capable of being easily tooled, and superficially perfect in character, speedily deteriorates, and results in disfigurement of the building in which it has been placed. Even metals, similarly exposed, are liable to like degradation, unless protected by specially prepared paints which prevent their oxidation.

The artificial stone or concrete maker having thus placed familiarly before him these evidences of failure in natural concretes, readily avoids similar dangers by using only such materials as will when combined with cement resist deteriorating influences.

The two distinctive terms Béton and Concrete need no longer be used, because the one may conveniently be merged in the other, for practically there is little, if any, difference between them. A concrete mass in the constructive sense means the binding together of variously selected aggregates by a cementatious matrix, but there is not now usually any definition of the particular process by which the effect is realized; whether the materials are first in their desired proportions mixed in a wet or dry state, or whether they are put together by impact with the smallest possible amount of moisture, or simply allowed to settle and harden in a natural and unaided manner, is of comparatively slight consequence. Modern preparations of Portland cement concrete are intended to supersede, as they doubtlessly excel in quality, the old-fashioned *concrete* prepared with lime and any description of aggregates "tipped" from great heights and buried in positions where its demerits and imperfections were not likely to be seen or challenged.

Increasing and reliable knowledge of the valuable constructive properties of Portland cements and their action in concrete preparations, has naturally led to a much more sensible recognition of the qualities of the aggregates. It is now well understood that gravels, sands, and other varieties of minerals vary greatly in their mechanical character and require careful attention in their use; any thoughtless selection accompanied by a hap-hazard combination of one or other of the required ingredients will not result in a satisfactory concrete mass. In the case of the best and most successfully prepared artificial concretes, much regard to the scientific or technical principles has been displayed, otherwise the satisfactory results with which we are familiar could not have been obtained.

"Granitic breccia stone" is a favourable example of a concreted mass produced by a careful selection of the aggregates, and cemented by a previously prepared artificial matrix; adhesion in this case being assisted through the influence of a mechanical blow or impact, favouring its induration by placing the various ingredients in the closest possible contact with each other.

"Béton Aggloméré" is another product obtained by an accurate combination of suitable cementing agents and approved aggregates; any careless amalgamation of the materials would prevent the most accurate impinging process from realizing the required effect; simple ramming is not in itself sufficient to produce a good "béton aggloméré."

The above reference to two well-known and valuable artificial stone preparations is to show that their success is due to other than the usual clumsy and irrational system of concrete-making. Neither of these processes resemble the ordinary practice, for primarily great attention is bestowed on the character of the materials of combination, and their subsequent accurate amalgamation is carefully performed.

These two artificial products afford the best illustration of what a concrete for constructive purposes should be, and the nearer we approach the practice of these operations, the more satisfactory will be the results obtained. When a better knowledge of these indispensable and accurate principles of concrete-making is acquired, the reprehensible practice of "pouring in" and "tipping" concrete will no longer prevail, but become, as it ought to have been long ago, "a thing of the past." It cannot be too forcibly impressed on concrete-makers that the materials should not only be accurately mixed, but that such primary combination should be permanently secured by the adoption of the most careful precautions. The required moisture should only be so much as shall by the best treatment secure perfect crystallization, anything beyond that being merely superfluous and wasteful.

There is much confusion about ancient concretes and the

methods by which they were prepared. The conflicting opinions as to the quality of Roman concretes in this country, still leave the question in a somewhat doubtful position. Before the "constructive" Roman period we find no concrete remains, owing doubtless to the fact that all previous civilizations were ignorant of the properties of mortar obtained through the agency of carbonate of lime. The Egyptians, Babylonians, and Jews used large stone blocks in their more important structures, and the small amount of mortar (of various kinds) used may be regarded simply as a bedding agent to secure the accurate position of the stones when placed in the walls.

Many cracks in large blocks of stone, but more especially those made from granite or other crystalline materials, are due to the pressure of the hard substances against each other in the absence of some accommodating intermediary to act as a cushion. The old builders were well acquainted with this fact, and prevented such fractures by a joint of mortar, clay, or asphalte. No advantage beyond securing this necessary protection from fracture was expected, for indeed the character of the masonry and its great solidity ensured permanency without the aid of mortar or other joints. This view is somewhat confirmed by recent and varied observations amongst what are usually considered the "Ruins" of the "Old World." It is not quite clear, however, that these civilizations whose gigantic constructive remains attract our attention and command our admiration, are entitled to the priority usually accorded to them. The "New World," which the genius of Columbus brought to the knowledge of modern times, contains ruins from civilizations whose antiquity cannot be defined with any degree of accuracy. The ancient Peruvians were, however, ignorant of mortar and its properties, as the following account of the Ruins of Cuzco clearly shows.

Mr. Squier in his work, recently published, at page 435 says:—

"In the buildings I am describing (in Cuzeo, the ancient capital of the Incas) there is absolutely no cement of any kind, nor the remotest evidence of any having ever been used. The buildings in which tenacious clay, mixed perhaps with other adhesive materials, was used to bind together rough stones into one enduring mass of wall, are of a character quite different from the edifices of Cuzeo. In dismissing thus peremptorily the stories and speculations about some wonderfully binding, almost impalpable, cement, which is said to have been used by the Incas, and the secret of whose composition has been lost, I am quite aware of the responsibility I assume. No man has ever investigated, or can more thoroughly investigate, this mooted question than myself; and I give it as the result of an inquiry carried on over nearly all the centres of Peruvian civilization, that in their structures of cut stone the Incas consistently depended, with rare exceptions, on the accuracy of their stone-fitting without cement for the stability of their works,—works which, unless disturbed by systematic violence, will endure until the Capitol at Washington has sunk into decay, and Macaulay's New Zealander contemplates the ruins of St. Paul's from the crumbling arches of London Bridge. The exceptions to which I have referred are those where, as in Tiahuamico, the chulpas of Sillustani, and in the Fortress of Ollantaylambo, the stones were fastened together by bronze clamps, interfitting grooves and projections, and by other purely mechanical devices, bearing in no way on the question of the use of mortar. It is only right that I should say that Humboldt says distinctly that he found a true mortar in the ruins of Pullal and Cunmas in northern Peru."

"The statement of the old writers that the accuracy with

which the stones of some structures were fitted together was such that it was impossible to introduce the thinnest knife-blade or finest needle between them, may be taken as strictly true."

The existence of bronze in these buildings, and its application for the purpose of clamping together carefully tooled stones, indicate a knowledge of the comparatively indestructible nature of the metallic alloy, leading to the natural inference that the pottery art preceded its use, because it must have been melted in vessels made from a fire-resisting clay.

The recent discovery of a buried city in the Grecian Archipelago is particularly interesting from the fact of its proving that, during some remote volcanic eruptions, a whole population was suddenly and as completely annihilated as were the inhabitants of Herculaneum and Pompeii during the Roman period.

At Santorin and Therasia pumice-stone has been quarried for building purposes from time immemorial, and the gradual clearing of the surface rocks which the increasing demand for this material involved, has disclosed the walls of an entombed city belonging in all probability to the "stone period" of geology.

The buildings were formed of irregularly-sized lava blocks put together without mortar, the interstices being filled with a kind of red ash. Many interesting relics, such as pottery, lava vases, &c., were found, but no trace of metal having been used as a constructive agent was discernible. The roofs of the houses were timbered; and although the wood was found in a recognizable condition, no nail of either bronze or iron was forthcoming. Two small gold rings were found, but they are supposed to have been brought from some of the gold-bearing districts in Asia Minor. Obsidian or volcanic glass appears to have been used in the form of knives or scrapers, and it is indeed so employed at the present time by the women of Peru for scissor-blades.

This city now so curiously brought to light is supposed by geologists to have existed at the early period of Egyptian civilization, and if so, must be at least from four to five thousand years old.

We may therefore conclude from these curious discoveries that a knowledge of concrete, other than that prepared with clay, was unknown to the civilizations of either of the modern geographical divisions of the world.

The Inca architects clearly placed the best value and dependence on accuracy of stone-dressing, and in the absence of cementing agents, accomplished results which command our respect, and admiration of their ingenuity; the favourable character of the Peruvian climate, as in the case of Egypt, contributing largely to the preservation of the ancient and interesting ruins of both countries.

It is almost impossible, even if it were desirable, to trace the origin and development of concreted structures erected by man during his various stages of improvement and civilization. Even during the rudest times he must have been influenced by observing the instinctive actions of the lower animals, and profited from their example. The small and almost puny but active martin in his busy flights during the nest-building season would show how the little architect could render the most primitive materials coherent and weather-proof by the action of a bird's saliva.

The "cob" or clay dwellings still existing in many of our English counties are of the rudest and most objectionable character, injurious to those human beings who live within their shelter, and discreditable to the intelligence and humanity of the nineteenth century. If these primitive vermin-breeding and health-destroying buildings sheltered prize oxen, sheep, or swine, all legal difficulties surrounding their demolition would speedily vanish, and sumptuous edifices stand in their stead. But the labourer—the tiller of the soil, the worse than serf of modern times—has no chattel value to the lord on whose land he is hardly permitted to live, and he can with difficulty in some places maintain his "parish settlement,"—not his birthright, for he has none—but that of dependent poverty.

In modern times, with the difficulties surrounding the control and management of skilled labour, it is practically impossible to use large blocks of stone for other than special and exceptional purposes. Were we in these times so *favoured* as to possess the facilities commanded by ancient Egyptian tyranny, we could then comfortably, because no obstacles would exist, succeed in obtaining stone blocks of greater magnitude than we now employ. There is no probability, we fondly hope, of realizing any such advantage, for Egyptian labour was essentially ruled and controlled by the lash. Its cost was merely nominal, although it is recorded that the value of the garlic, onions, and other vegetables consumed by the labour employed in building the "Great Pyramid" was 200,000*l.* of our money.

The Egyptian granite, known generally by the name of syenite, has been used for nearly four thousand years, and the numerous examples of obelisks and other ornamental structures now in existence testify to its valuable and enduring properties. This granite possesses certain special characteristics leading in some measure to confusion as to its exact mineralogical position; the name "syenite" being due to the locality of its original obtainment, although mineralogists under the present arrangements do not regard it as a syenite proper. It consists of large crystals of red orthoclase, occasionally in a dual or twin form, with a porphyritic development, some yellow-tinged oligoclase, quartz, and dark mica, with sometimes a small quantity of hornblende. The general and characteristic colour is red, and although exceedingly coarse in grain is susceptible of a high polish. Its analysis is:—

Silica	70·25
Alumina	16·00
Oxide of iron and manganese	2·50
Lime	1·16
Magnesia ⎫ Potash ⎬ Soda ⎭	9·00
Water	0·63

The binding agent of this durable rock is mainly siliceous, with an unusually large amount of the alkaline ingredient, indicating that in a wet or moist situation it would speedily become degraded.

The world-famous Pyramids had, it is supposed, their origin in the desire of certain Egyptian kings or rulers to secure a resting-place for their bodies capable of withstanding all elemental or systematic violence. Like Egyptian history itself, however, some obscurity exists as to the exact object of their erection. A belief in the immortality of the soul and its future re-animation and rehabilitation of the body influenced them, doubtless, in resorting to the most effectual means of securing it from decay; hence the practice of embalming and secure entombment.

These monuments of a decayed and lost civilization continue to remain as evidence of the care bestowed in their erection, while more recent Roman edifices in the Nile delta have perished, leaving no traces of their site or character behind.

The granites of our own country are perhaps not so good as those of Egypt; at all events we do not experience equal advantages from their use. In London, the examples of the best-selected varieties used in the most important structures exhibit, in a greater or less degree, unmistakable signs of premature decay.

Waterloo Bridge, built of Cornish granite, and indeed London Bridge, where Devonshire granite was used, are well-known examples of both mineralogical defects and debasing atmospheric influences. In London we can find but few granite buildings entirely free from the blemishes which usually accompany the use of a material so susceptible to climatal influence.

The whole of Waterloo Bridge is built of Cornish granite, except the balustrades which are of Aberdeen granite, affording a favourable illustration of the difference in constructive value between these two granites. Some estimate the relative durability as 14 for Cornish and 22 for Scotch. At all events an examination of these two granites will prove the fact that, in point of durability, in a London atmosphere the Scotch exceeds in value the Cornish.

Engineers and architects, neither in this country nor France, had at the time of the erection of Waterloo Bridge considered the chemical constitution of this favourite building material. Indeed, shortly after the erection of this undoubtedly important structure, a celebrated French engineer, in a memoir addressed to the Institute of France, eulogized the building in the following terms:—

"If from the incalculable effect of the revolutions which empires undergo the natives of a future age should demand one day which was formerly the new Sidon, and what has become of the Tyre of the West, which covered with her vessels every sea? the most of the edifices devoured by a destructive climate will no longer exist to answer the curiosity of man by the voice of monuments; but the Waterloo Bridge, built in the centre of the commercial world, will exist to tell the most remote generations, here was a rich, industrious, and powerful city! The traveller, on beholding this superb monument, will suppose that some great prince wished, by many years of labour, to consecrate for ever the glory of his life by this imposing structure. But if tradition instructs the traveller that six years sufficed for the undertaking and finishing of this work, if he learns that an association of a number of private individuals was rich enough to defray the expense of this colossal monument, worthy of Sesostris and the Cæsars, he will admire still more the nation in which similar undertakings could be the fruit of the efforts of a few obscure individuals lost in the crowd of industrious citizens."

Long before any visible signs of our commercial decadence has arisen, the bridge, thus so favourably lauded, not only exhibits evidences of decay, but some *leading* engineers deem it unsafe for public use; we will say nothing of the "private individuals" whose patriotism has not certainly been recognized in the negotiations for the transfer of this "colossal structure" to a public representative body.

Indeed there are few if any examples of granite, whether used decoratively or constructively, free from the blemishes which its alkaline constituents occasion. One of the more unfortunate examples of granite degradation may still be seen in Essex Bridge, spanning the river Liffey, at Dublin. This bridge was built by Semple in 1776, from granite obtained from the Wicklow mountains, a few miles distant from Dublin. An examination of this structure exhibits in the most aggravated form the results of using a granite unsuited, from its chemical constitution, to withstand the dissolving influences of a wet climate like that of Ireland.

In these examples of what are by some authorities considered successful applications of granite in structures of great national importance, there is ample evidence of the uncertain and unreliable character of even the best-selected varieties. London and Waterloo bridges, the former of which receives an unparalleled amount of traffic, already show indications of decay on examination of their granite surfaces, now becoming rough and uneven. Waterloo Bridge is in much the worse condition, and indicates the difference of building value in a London atmosphere between the Devonshire and Cornwall granites.

In house building, because the surfaces are vertical and therefore less liable to the lodgment of moisture, the effects are not so strongly developed, and the decay and disaggregation make slower progress. The tooling and polishing of the stone in some degree predispose the surface to injurious chemical and weather action, although it is found that even some varieties in their natural site and bed of original deposition become decomposed.

The deposits of kaolin in Cornwall and elsewhere are the result of atmospheric action in reducing the rock, out of which are washed the potash, soda, &c., leaving only the felspar. Water is the active chemical and mechanical agent in effecting these changes, and many granite deposits exhibit superficial marks of its debasing influence. The "rock basins" in Cornwall show in a remarkable manner the eventual result of wind and water power in scooping out fantastic forms, which in past times were attributed to Druidical influences and agency. The Templar Rock in Lundy Island is a remarkable example of weather action producing such picturesque outlines as that shown in the following woodcut, Fig. 1.

There is some difficulty in deciding as to which of the mineral ingredients of granite are first liable to the solvent action of water. On a careful examination of an exposed piece of granite it is impossible to determine whether the quartz or felspar is first acted upon, for they are apparently affected in an equal degree. We thus find degraded granites in the pure atmosphere of their original formation, and far removed from the influences of a vitiated atmosphere.

Oxygen exerts an injurious action on rocks having oxides of iron in their composition, for it readily unites with them and soon destroys their original cohesion. Carbonic acid is, however, a more powerful solvent, more especially in the case of those rocks which have a calcareous base or matrix.

Dartmoor granite is irregular in quality, that which is quarried from near the surface being especially liable to

FIG. 1.

TEMPLAR ROCK, LUNDY ISLAND.

early disaggregation. The prison on the moor, built of this stone, exhibits both externally and internally ample proof that such a material is not adapted for building purposes. The crumbling walls endanger its structural stability, rendering the occupancy of the convict cells impossible, unless the use of cement plaster is resorted to, thus preventing danger from the dampness attracted by the pulverulent granite. As evidence of the irregularity in quality of granite quarried from the same locality, it may be stated that the stone used in building the Nelson Monument, Trafalgar Square, London, was also obtained from Dartmoor. Notwithstanding the atmosphere in which it is placed, there are no prominent indications of early disintegration.

D'Aubuisson gives us much information on the granites of Auvergne and the Eastern Pyrenees, which become so decomposed as to appear superficially vast tracts of gravel. In a hollow way or tunnel that had been made by the operations of blasting, in the course of six years after, a depth of three inches of decomposed granite was produced, principally attributed by him to the action of carbonic acid.

Dolomieu, another authority, describes these results as being due to *la maladie de granite*, or disease of the granite. In districts where this peculiar action prevails, masses of granite when touched by the hand or pressed upon by the foot crumble to powder.

The "Kettle and Pans" of the Isle of St. Mary, Scilly Islands, which De la Beche regarded as singular specimens of natural sculpture, have been, like the "Rock Basins" of Cornwall, attributed to Druidical handiwork. These singularly curious excavations are apparently hewn out of the solid rock, some of them measuring eighteen feet in diameter and six feet in depth.

When we examine the most successful building stone in London, illustrated by such examples as the noble pile of St. Paul's Cathedral, Greenwich Hospital, the Custom House, and other less important structures, we are not much more fortunate. The Portland stone of which these buildings are constructed is practically a pure limestone, from the Oolitic series of rocks, whose analysis is—

Carbonate of lime	95·16
Silica	1·20
Carbonate of magnesia	1·20
Iron and alumina	0·50
Water and loss	1·94
	100·00

C

In its massive condition the specific gravity is 2·145, and when reduced to grains or particles, 2·702. The texture is uniformly regular, its binding medium being calcareous. The best examples remaining to us, after the removal of so many of Wren's city churches, are those buildings of the Queen Anne period to be found in the now unfashionable quarters of the metropolis.

There are a variety of beds of this famous Portland stone, differing greatly in character and durability; and the architect or builder, who in former times bestowed the required care in its selection, was rewarded by the enduring character of his handiwork. In the present neglected condition of our public and private buildings, the insidious chemical destructants are afforded easy shelter in reserve for the wet weather to enable them to exert their most effective and damaging influence. The carbonic acid of the atmosphere acts prejudicially in this case and in all other carbonate of lime rocks, as it dissolves out the carbonate of lime. In old buildings, where this action has continued through many years, and especially in the London atmosphere, thin films of a stalactitic character are to be seen in those parts of the building subject to the drip from projecting ornamentations. In such cases it may be affirmed that there is no absolute loss from the building, as the disaggregation in one part has been balanced by the coating of stalactite in another. The result, however, is disfigurement, as is well shown on the western side of St. Paul's Cathedral, where many of the more highly-wrought sculptural embellishments are almost illegible. This building affords also good evidence of the partiality of the destructive influence, for it is more intense in the places which are dirtiest, owing to the conservation of the atmospheric destructants, which become at once developed in all their chemical activity on the occurrence of damp winds and rain.

In the manufacture of concrete, the inferior qualities of this stone may be used, giving the quarriers thereby an opportunity of selecting, with more chance of profit and success, the best for building purposes or masonry. An outlet for the present refuse, even at a small profit, would tend much to the advancement of the modern reputation of Portland stone, undoubtedly the best in this country of calcareous derivation.

The various beds of Portland stone are well defined, and their peculiarities clearly understood by practical men. The most marked failures in its use have arisen either from a desire to use the cheaper sorts to do the work of the best, or from negligence in their selection.

Wren and Smeaton, for their respective great buildings, took the necessary pains to secure the best the island of Portland could at that time produce, but the selection was not carelessly deputed to irresponsible or ignorant workmen. Modern builders have, however, long since used up the stones which these great constructors rejected.

Its comparative scarcity has doubtless diverted attention to other building stones, and we have now to deplore a great blunder in using an improper stone for the building of the Houses of Parliament.

The stone used in the construction of this important national building is a magnesian limestone, from Bolsover, in Derbyshire. The best varieties of this formation are those which, in the chemical analysis, represent nearly equal proportions of the carbonates of lime and magnesia. When in that state it is highly crystalline in character, and exhibits externally all the qualities of a durable stone. A commission, composed of what may be called scientific experts, was formed to examine and report on the various building stones of England and Scotland, and a most exhaustive report or blue-book was published describing their labours.

The stone selected, after much deliberate inquiry and examination, was of the following analysis:—

Carbonate of lime	51·1
Carbonate of magnesia	40·2
Silica	3·6
Iron and alumina	1·8
Water and loss	3·3
	100·00

In a short time, however, unmistakable proof was given that the stone so carefully selected was unsuitable for building in a London atmosphere, and that the generation who witnessed the erection of the Houses of Parliament would also have practical evidence of their premature decay. There were numerous specifics of a chemical character, and a great number, no doubt, of wonderful nostrums forthcoming to preserve, if not regenerate, the crumbling stone. All, however, proved futile, and the building in its most ornate parts still exhibits rapid and uncontrollable decay, rendering it dangerous for those requiring to pass to and fro along its external fronts or internal quadrangles.

Doubtless the error of such selection was due to a confiding reliance on the condition of those buildings standing in the locality of the quarry, and surrounded by a pure Derbyshire atmosphere. The cause of the deterioration is in a great measure due to the chemical character of a London atmosphere, which exerts a destructive influence, encouraging a separation of the tenderly cemented atoms or grains of the stone, and probably also in some degree to an antagonism between the unevenly balanced carbonates.

The holy city of Jerusalem, in its magnificent mountain site at an altitude of 2500 feet above the level of the Mediterranean Sea, must have been in the days of its glory a gorgeous spectacle. Built of the local tertiary limestone, of dazzling whiteness, and surrounded at the time of its final overthrow by natural and artificial defences, it was regarded at that period as impregnable. Famine, however, together with the dissensions among its defenders, hastened its downfall, and it fell a comparatively easy prey to Titus, who, when he obtained the much-coveted prize, exclaimed:— "It is manifest that the Almighty has fought for us, and has driven the Jews from their towns; since neither the utmost human force, nor that of all the engines in the world, could have effected it." A clearly expressed testimony by a heathen to the existence of a great God.

The remarkable quality and character of the masonry of this famous city show what reverent and heartfelt industry and care the builders bestowed on its construction. In the recent explorations of its ruins, much interesting information has been obtained, not only in regard to the extent of the city itself, and its Temple, but also regarding the sanitary arrangements for the welfare of its citizens. Aqueducts, fountains, drains, cisterns, and other structures calculated to improve their comfort or secure their health, were built on an extensive scale and in a most substantial manner.

The explorers in their researches found one stone nearly 40 feet long, weighing something like 30 tons; also courses of masonry 6 feet in depth, carefully worked, with joints so fine as scarcely to exhibit any appearance of mortar or concrete. All the masonry, including arches and other important details, was put together in the most careful and solid manner, indicating a firm belief in the minds of the builders that the city which they had gratefully dedicated to their God would endure for ever.

These ancient records of past greatness in other countries are not without somewhat analogous evidence in our own. Christianity, through its most influential time, the middle ages, has bequeathed—many of them, it is true, in a now ruinous condition—buildings of high architectural merit and importance. These remains, generally of a Gothic type or style of architecture, and usually built of local materials, indicate with much accuracy the durability of many sandstones which in those times were generally used.

In England and Scotland, in widely separated districts, and subject to variety of climates, is exhibited the instability of the most carefully selected stones. At Exeter in the west, Norwich and Ely in the east, Chester in the north-west, and York and Glasgow in the north, are to be found cathedral buildings, from a study of which it will be seen that even the most favoured district failed to supply what may be regarded as a moderately durable building stone.

The Norman strongholds built by William the Conqueror and his barons show with almost equal significance the uncertainty of the wear of the stones of which they were built, although the bolder character of their architecture prevented, in some measure, equally rapid decay.

The richly decorated Gothic cathedrals were, owing to the minuteness and variety of detail of their architecture, more liable to injurious action from the weather, which readily separated the laminæ of stones obtained from the sedimentary deposits.

While history records with some degree of accuracy the battles and strifes of these feudal times, but little is said of the details of building, or the labour by which these gorgeous edifices were erected. There is something truly mysterious in the ecclesiastical influence, which was all-powerful in such times of turmoil, to construct and hand down to our times, buildings of high architectural magnificence. During the turbulent revolutions and religious contentions of modern times, we have to deplore the destruction or mutilation of many buildings, which it is our bounden duty to rebuild and restore in a proper manner.

From these numerous examples of ancient buildings it is apparent that, more especially in the vitiated atmosphere of industrial and populous centres, no natural stone at our command is practically capable of withstanding its effects. Stones quarried from almost every suitable geological formation are more or less injuriously affected when exposed in their worked condition to weather influence. Indeed there is every reason to believe that other rocks as well as some of the granites (as we have already shown) become disintegrated in their natural bed beyond the influence of atmospheric action.

From the crystallized granite suddenly created (at least comparatively speaking), and without adventitious improvement through time, to the more recently formed sedimentary rocks, slowly deposited, having a more reliable cementing agent, and during long periods of repose improving in indurative value; all alike in this wide range seem incapable of permanent duration in any building, however placed.

The metamorphic rocks are apparently the best suited for wear, and some descriptions of lavas possess good durable properties. The Carrara marbles and the lavas from Vesuvius furnish us with, perhaps, the best examples of durability. The lava pavement discovered under the ruins of Pompeii, laid probably a thousand years before the Christian era, is apparently indestructible. This pavement was laid in large polygonal blocks or slabs, and their high state of preservation when exhumed, after being so long buried, was calculated to excite our surprise.

The above is a good example of the constructive value of an erupted rock. Volcanic products, and in that class we include every rock produced by internal eruptive action of the earth, receive their cementing agents at the instant of their expulsion, becoming in due time hardened or crystallized by the cooling process.

The limestones receive their cementing agent simultaneously with the deposit of carbonate of lime through the organic medium of their deposition. Stony corals, whose composition resembles that of shells, are undoubtedly composed of carbonate of lime and cemented by a gelatinous paste. Cowry shells are constructed from carbonate of lime and animal gluten, being analogous in character to the enamel of the teeth.

In whatever direction we turn, and after the most careful examination of all the various geological series of rocks, we are unable to find any, even the most favoured, having its cementing agent good enough to resist the disintegrating influence of abnormal atmospheres. Even granite paving in its natural state, and without other preparation than that of the required tool dressing, soon gives indications of rapid wear, although, from its slippery state, it is apparently as hard as adamant. When worn smooth, this class of pavement is roughed at great expense, and its new condition favours still more the incipient destructive water action, for the rough holes act as so many receptacles in which the water is lodged. The kerbing, laid originally with a roughened surface, is also in a greater degree liable to the same degrading operation.

Bricks and tiles of all kinds and descriptions may be regarded as concretes produced from the débris of rocks, and re-converted by means of heat into substances more or less resembling the character of the rock from which the clay was originally derived.

The innumerable varieties of clay abounding in nature, from the finest kaolin to the loamy brick earth, have afforded through all historic times great facilities for the production of the most beautiful porcelain and other equally useful wares. Long anterior to reliable chemical knowledge, a good appreciation of the properties of clays must have existed, otherwise the ancient remains of such excellent pottery, &c., could not have been forthcoming.

Our earliest and most reliable information as to the origin of burnt brick fabrication is obtained from sacred history. Thoroughly burned bricks were used in the building of Babel more than four thousand years ago, but there is some obscurity in the transmitted records, whether fire was at first used in the manufacture of pottery and bricks. Great natural facilities were taken advantage of by the Egyptians in the conversion of Nile mud into all kinds of pottery and bricks. On these sun-dried and fire-baked bricks and slabs we find indelible records of past histories, of which without such aid we should have remained ignorant.

The sedimentary deposits from which are obtained the paving flags with which our large towns are paved, when seen *in situ* and carefully examined, show the thin and clearly defined laminæ that prove so destructive to the stone when laid. Exposure to moisture and frictional wear from the passer-by at an early period destroys the cementing agent, permitting the splitting of the hitherto compressed *leaves* of the rock. Even the most careful laying on the "quarry bed" fails to secure that permanency so much desired. When the first thin skin is cut, those underneath are soon destroyed, for the uneven surface once begun is speedily reduced by the combined action of water and boot and shoe friction. The unsatisfactory condition of London pavements indicates that the expense of their maintenance and renewal is too great for even a rich and populous city to bear, otherwise they would not be allowed to continue in their present dangerous and disgraceful state.

In these necessarily hurried and somewhat imperfect references to an examination of building and other stones from divers geological formations, it has been shown that climate and locality are important factors in the question of wear and tear, and that bricks and pottery of ancient date alone seem competent to withstand such baneful influences. We have thus far dealt with the subject in its relation to natural building stones, but before closing this chapter we consider some reference is necessary to artificial products more or less intended to supersede those obtained from nature. Although there are many of these for building and other purposes, we shall limit our present observation to the four following well-known qualities, viz. :—

Ransome's siliceous stone.
Terra-cotta, in all its varieties.
Victoria stone, for pavements and buildings.
Silicated or rock concrete pipes, for sewers, &c.

First. Ransome's stone.—Produced by chemical ingenuity from the simplest of natural materials. This manufacture has not met with the commercial success its merits deserved, owing probably to the difficulties attending the command of the required chemical skill so necessary to its profitable prosecution. In the best examples of this stone are exhibited all the essential requisites for building purposes; colour and texture being excellent in those specimens where the exact chemical combinations had been secured. Otherwise, in an English climate at least, it fails to resist the action of frost, and becomes disintegrated and unsightly. In India and other warm climates it has been used in considerable quantities with much satisfaction. It may not be able to compete in price with the average natural building stones, but it affords great facilities for the preparation of ornamental blocks at a cost much less than hand work, and with great and unerring accuracy.

The manufacture of this artificial stone is based on chemical reaction, and flint first rendered soluble is the prime agent in the indurating process.

Second. Terra-cotta.—Another product obtained by the aid of fire or heat, cannot be regarded as a modern industry, for it was known to the ancient Greeks, who produced ornamental articles of a high class. There was no attempt at that time, nor for long after, to use it for constructive purposes. In more recent times and during what is known as the Renaissance period, as well as in the time of Queen Anne, many excellent examples were produced for architectural decoration. These examples not only indicate a high degree of artistic excellence, but show also that the potters of those days possessed an advanced knowledge of their art. The terra-cotta of that time was both artistic and durable. It is upon the lines of these old architects and potters that the comparatively modern industry of terra-cotta manufacture is based. Doulton, Minton, and other eminent potters have entered on a system of decorative manufacture which bids fair to equal, if not surpass, the best examples of the early masters of the art. Particular attention is now directed to produce a quality of ware competent to resist the most insidious action of vitiated atmospheres, and the no less damaging influence of frost. The great advance in chemistry, as applied to this important industry, enables the modern manufacturers to compound their clays and apply the exact degree of required heat, so that the baked ware shall, while retaining its desired accuracy of form, be of the greatest density.

There can be no question of the advantages secured to the architect through this improved and reliable terra-cotta industry, for the most elaborate ornamentation is attainable with a certainty that cannot be reached through any process of carving natural stone. The difficulties surrounding the use of ornamental details of carved stone, not only from the costly character of the required labour of conversion, but also from their susceptibility to atmospheric degradation, are now therefore in a great measure overcome by the substitution of terra-cotta. The numerous recent examples in London of extensive buildings in which terra-cotta plays an important structural and ornamental part, indicate that its extending use is consequent on a reliance on its quality and capacity for wear. The chemical forces brought to bear in fabricating a first-class terra-cotta, secure a dense and homogeneous substance, the coherence of which is beyond the degrading action of frost or gas.

An examination of buildings in London built during the Queen Anne period, will afford undoubted proof of the superiority of the terra-cotta of that time over the Portland stone. The artificial compound maintains its original lines and character, while the natural stone ornamentations are nearly obliterated, all their elaborate details being almost unrecognizable.

From these examples of the incompetency of stone to resist the wasting action of water, the architect should restrain his desire to seek his ornament through the agency of limestone of any geological formation. Siliceous stones are not so liable to degradation from water or humidity, and terra-cotta is also less prone to be influenced by its wasting action, although both are more subject to loss by frost action than the limestones.

Third. Artificial concretes, such as the Victoria and silicated stone used for pavements and other purposes.

This industry has been a long time in arriving at its present satisfactory position, and the many difficulties it has had to encounter since the early exertions of Mr. Buckwell in the manufacture of his "Granitic breccia," have been of a most formidable character. There was first the difficulty about quality, and unfortunately in addition to opposition which it encountered from this cause, the cementing agent was not at the time of the first experiments of a good or reliable character. With the progress of cement improvement, artificial pavement industry received a healthy impetus, and it may now be affirmed that these pavements excel, in every particular, those obtained from the most famous natural sources. Improved mechanical appliances and increased chemical application in some of the branches of artificial stone industry have resulted in the establishment of accurate rules, the following of which secures satisfactory results. The many miles of Victoria pavement afford to the most unconcerned practical evidence of the advantages derived from its use, nor is there any doubt regarding the durability of Victoria concrete under heavy and incessant traffic of the most trying kind.

Fourth. Silicated and rock concrete sewage and drain pipes are now extensively manufactured, and some important town drainage operations have been carried out with them. The mechanical means adopted in this manufacture secures the production of pipes of large diameter, of the most accurate form, and great density. The facility with which all kinds of aggregates can be utilized in this new industry, is not one of the least advantages realized.

In this hurried reference to the various formations, and the structural creations built from their wide-spread stores, we have said enough to show that a great deal depends on the selection of the rock, and the locality in which it is laid after its required conversion to the desired form. All of the natural cementing agents are more or less liable to degradation, and with their waste or decay, the aggregate with which they were incorporated must also suffer.

The process of the formation of some of the rocks is indeed truly wonderful, not only from the instinctive accuracy of the medium, but the marvellous results obtained. As an illustration of this kind of natural action and as evidence of operations from which the concrete-maker may be instructed, we will shortly refer to the coral-forming zoophytes.

Although familiar with the results realized in the deposition of great formations by the most insignificant vitalities, we have not seen these operations in progress, as we can in the case of the coral reef and island-forming constructions. These tiny engineers are the madrepora, or coral polypi, of lowly organism possessing high constructive instincts. They are busy building new islands and reefs, more especially in the Pacific Ocean, and their persistent and never-ceasing efforts must in some remote age change the face of continents. Their task seems an unselfish one, for in the construction of coral reefs, there is no apparent purpose of their own attained, except obedience to the instincts with which they are endowed by nature. All their industry is of a marine character, for out of the water they cease to exist. They seem the pioneers who lay the foundations of future islands, that, after their labours are finished, become clothed with a low type of vegetation, the growth and decay of which produce a better quality suited to the existence of higher and more advanced organisms.

Captain Beechey, a close observer of the works and results of these coral-forming polypi, found that the method pursued indicated a full appreciation of the difficulties surrounding their operations. The coral structure assumed the form of an inverted cone, such figure being best calculated to resist the mechanical thrust and agitation of the sea. He further observes:—" The north-eastern and south-western extremities are furnished with points, which project under water with less inclination than the sides of the island, and break the sea before it can reach the barrier to the little lagoon formed within it.

"It is singular that these two buttresses are opposed to the only two quarters whence their structure has to apprehend danger, that in the north-east from the constant action of the trade-wind, and that on the other extremity, from the long rolling swell so prevalent in these latitudes; and it is worthy of observation that this barrier, which has the most powerful enemy to oppose, is carried out much farther and with less abruptness than the other."

These tiny semi-vegetable engineers, without plan or rule, thus instinctively arrange the lines of these structures, so that the natural forces with which they have to contend strike harmless against them.

We are astonished at the beautiful hexagonal cell of the bee, and the ability of the little architect who constructs it, but even its marvels are excelled by the work of the lowly organized madrepora; these remarkable creatures living in the sea, out of the water of which they not only extract their food but obtain also the materials wherewith they construct the coral islands and reefs. Hydraulic engineers like these toil without drawing and receive no specification of the materials to be used; but their works may exist when the more elaborate structures of highest organized beings shall have crumbled to dust and be forgotten.

There is much in the work of these apparently insignificant builders which the concrete-maker can profitably imitate. For instance, there is one uniform quality of material, and its adaptation is controlled by a never-changing agency. The carbonate of lime, silica, and other chemical ingredients exist in accurate proportions, and the converter, satisfied with their fitness, varies not in the method of his task. The sea-water is his chalk and clay quarry, and his own imperfectly-defined body is kiln, grinder, and general converter by which he accomplishes results that even the most expert chemist could not improve.

Hitherto geologists and other observers have agreed that the operations of the coral-forming zoophyte are very slow, but such conclusions have been arrived at from an imperfect examination of them. The extent and character of these formations are influenced by the localities in which they are created, and so protracted is their growth in some places, that observant authorities compute it at something like two feet during a century. Recently, however, a gauge of measurement has been obtained through the agency of telegraph-cable laying. The facts are as follows:—

"A remarkable piece of coral has been taken from a submarine cable, near Port Darwin, Australia. It is of the ordinary species, about five inches in height, six inches in diameter at top, and about two inches at the base. It is perfectly formed, and the face bears the distinct impression of the cable, while a few fibres of the wire rope, used as a sheath for the telegraph wire, still adhere to it. As the cable has been laid only four years, it is evident that this specimen must have grown to its present height in that time; this seems to prove that the growth of coral has been much more rapid than scientific men have hitherto admitted."

In these introductory remarks we have endeavoured to show the difficulties surrounding the use of all the best building stones. The failures in our most important buildings clearly indicate that there is no existing remedy by which the engineer or architect is enabled to avoid their dangers. On all sides practical evidence abounds, by the unsightly and unsafe condition of works of high architectural excellence, that the most vigilant precautionary measures are futile in averting the consequences of using unsuitable building materials.

From what has been said, it must be apparent to all engaged in building operations that in the future more attention must be paid to the fabrication of artificial building stones on whose durability perfect confidence can be placed. In showing the weak points of the best natural stones, we intend that comparisons should be formed between what nature has done during protracted periods of time and what may now be accomplished with the help of well-established scientific rules.

Until but recently there were few good examples of concrete which could equal in value the commonest natural stones. It is now very different, for we will, before closing these pages, make it quite clear that first-rate concretes have been made, and a daily increasing knowledge of the conditions which should regulate their manufacture, bids fair to reach a point of excellence hitherto considered unattainable. All the old and heedless processes by which the preparation of concrete has been for many years conducted, have not only failed to attain satisfactory results, but they have tended much to retard the more sensible system which is based on the truest technical formulæ.

Having said thus much of an initiatory character, we will now proceed to enter on the practical consideration of concrete-making, beginning with the description and character of concrete and the matrices and aggregates used in its preparation.

CHAPTER II.

WHAT IS CONCRETE?

WHAT are the conditions which should be observed in the preparation of a good concrete? To answer this question satisfactorily, it is necessary to examine what constitutes a concrete compound in the sense in which we wish it to be understood. Primarily the object is to put together previously selected materials of the proper quality, and to do so in such a way that, when compacted, the mass will withstand all the ordinary injurious actions of wear and climate of whatever kind.

To accomplish what at first sight appears a simple operation, it is necessary to be clear as to the relative duties required of the different ingredients. The cement undoubtedly is the prime agent, and presuming all precautions have been adopted in securing its quality, care must be exercised in the manner of its application, for the aggregate is to be bound together through its agency. This operation is generally performed in a careless and irrational manner, and not likely by such means to result in a good and faultless concrete.

The Romans were undoubtedly the earliest concrete builders, and in some of the countries which were under their domination, the practice of using lime in cementing various kinds of stones together prevails at the present day. In Italy especially, the practical knowledge of concrete-making, in some of the most out-of-the-way places even, is worthy of our consideration. We do not of course recommend the adoption of the exact mode we are about to describe, but we refer to it for the purpose of illustrating the subject we are discussing.

In ordinary building operations in some of the Italian provinces, the use of concrete is very general. Much attention is given to the thorough amalgamation of the matrix with the aggregate. There are no rough heaps of gravel and lime or cement put together in any pernicious and hap-hazard manner, but the aggregate is placed upon a board whereon is a prepared paste of lime, which is carefully handled so as to produce a thin film or coating of lime on the gravel or stone. The action required to accomplish this, is similar in character to the process of coating confectionery. By this mode of manipulation, all the surfaces of the aggregate are, as it were, painted, but not too thickly, with a sufficiently adhesive covering to produce, when the pieces are brought in close enough contact, a compact concrete without any superfluity or deficiency of matrix. These observations are only applicable to the use of a rich lime paste, in which there is in reality no property of cohesion, a successful result being only possible when the exact quantity of lime is used. For instance, if too much paste or mortar is mixed in a careless manner, the aggregate would be improperly coated, and the excess of matrix tend to weaken the concrete by its interposition in thick joints, that would, from want of the requisite cohesion, be incapable of hardening. Cements are not open to the same objection, for they are cohesive as well as adhesive in their action.

This familiar illustration is given to show that, whether the means of mixture be manual or mechanical, the main object of the operation is to bring the materials into accurate combination. Failing the accomplishment of the required accuracy, a concrete will be produced imperfect in character, and deficient in compactness. The remains of old Roman concrete will, on careful examination, show that no superfluity of the matrix was permitted, and every precaution adopted to control and regulate its accurate proportion; otherwise the successful results obtained would have been impossible.

Cost must in all cases be a factor of more or less importance in the production of concrete, and although, in that made from Portland cement especially, no danger would arise from an excess of the matrix, still no advantage would accrue, but, on the contrary, as we hope to show, a waste of material and labour, with an attendant depreciation of quality in the mass itself.

There are now several modes of preparing concrete, and in the more improved systems of paving and pipe-making, both mechanical impact and hand labour are in fashion. The best systems, however, although much in advance of the ordinary careless method of hand treatment, fall short of the easily attainable perfect impact process. It will take time before this desirable end is accomplished, but it is nevertheless gratifying to see the advance made in concrete-making, and still more the improved intelligence on the subject generally.

There are a great many kinds and varieties of work now being executed in concrete under varied circumstances and for divers purposes.

We may with advantage classify them in the following order:—

1st. Ordinary concrete, as sub-structural and for foundation purposes in building.
2nd. Concrete, for streets and roadways, including cellars, basements, and courtyards.
3rd. Floor concrete, in cements and plaster of Paris, generally in combination with iron girders of various sections and designs.
4th. Concrete for building, either by the use of frames,

as monoliths, or in blocks of various shapes, both hollow and solid.

5th. Concrete used in sewer or drainage works, whether in combination with bricks and stones or by itself.

6th. Concrete in pavement slabs or for other similar purposes, being subsequently treated with solutions of silicate of soda or potash.

7th. Concrete in pipes, also submitted to a silicating or hardening process.

These several classes of concrete work may be capable of further division, but they sufficiently define the various processes for our present purpose of examination.

We will shortly describe the several kinds of concrete as above classified and the mode or methods by which they are produced.

No. 1. Concrete for the purpose of ordinary foundations is usually made with lias lime or other cheap matrix, and the quality of aggregate used is generally anything that comes handiest and, above all, cheap. There is in most cases an utter disregard of the conditions which should regulate the selection of the materials.

No. 2. The concrete used in foundations for streets is now more carefully prepared, and this improvement is in some measure due to the introduction of asphalte, for laying which a sound and solid base is indispensable.

In Liverpool, Manchester, and indeed in some of the metropolitan districts, Portland cement concrete is frequently used as a foundation, whether to receive asphalte, granite setts, or wood. In Liverpool especially the concrete preparation receives at the hands of the engineer to that corporation much sensible attention. The Portland cement is first submitted to the most searching tests, and the gravel or other aggregate carefully selected. Even the mixture of the ingredients is accomplished by the employés of the corporation, and the concrete thus so well looked after is carted direct from the corporation premises to the street where it is to be used. This concrete so carefully made is properly laid to the accurate contour of the intended surface of the street, and allowed to set sufficiently before the paviers are permitted to begin their operations.

As illustrative of some of the still existing and ordinary methods of producing Nos. 1 and 2 concretes—familiar to those taking the smallest amount of interest in the subject. An old building has been pulled down to make way for an imposing new edifice, and far under the lowest depth of the old walls the foundation for the modern structure, a bed of concrete, has to be laid, whereon is to stand the whole superstructure and its accumulation of dead weight. The "modus operandi" as recently witnessed by the author in the centre of the City of London is as follows.

A trench has been dug, the gravel (unwashed) wheeled on to the site at a high elevation, from which it is raked down to a rough platform made of planks. A quantity of gravel having been brought down by one operator, it is sprinkled with lime by another, and a third with a bucket of water washes the lime all but out again. This mass, or rather *mess*, water and all, is thrown down into the trench, and a labourer at one of its angles throws on the *concrete*, from a considerable height, old bricks with their surroundings of mortar and worse. This last and final operation is done probably in the belief that the concrete by this weighting will be kept on its good behaviour, in fact, "sat upon." In the specification directing this concrete-making operation there was doubtless much ingenuity displayed in the description of the materials, and the method by which they were to be mixed; the result being as described, and all done under the eye of a clerk of the works, looking on admiringly at his *indus-*

trious agents—manipulations which to the uninitiated looker-on must be regarded as a grand constructive performance of mysterious importance.

Another familiar illustration of concrete-making arises in relaying some of the streets with granite setts. Locality, the great city, and during the early morning.

A bed of lime concrete is carefully laid under the direction of the foreman of paviers; its breadth, depth, and accurate contour of surface properly defined. The paviers follow immediately after the concrete-makers, and when a thin grout of lime has been poured in between the joints, the *scientific* pavier proper or rammer, with his ponderous ram, follows, and takes care by his industrious and well-aimed blows, that the concrete shall not be allowed to set, or, at all events, be so disturbed in the indurating operation as to render it absolutely worthless as a concrete.

These two operations have been so long in fashion that those entrusted with their execution regard the whole business as one of prescriptive right, and beyond the influence of improvement. These things were done in "old time before us," and, like the City institutions generally, are incapable of modern or intelligent improvement; at least so they of the City stronghold consider.

No. 3. This particular application of concrete for the construction of floors is now becoming very general in new buildings, whether for warehouses or habitation. In the many plans of this comparatively modern adaptation of iron in combination with concrete several important advantages are secured, the greatest of which may be considered that of preventing, to a considerable extent, the liability to destruction by fire. It would be superfluous in these pages to particularize the many systems, patented and otherwise, of girders and arches of varied character used with concretes of all kinds. Some of these floors so constructed have stood the test of many years, and the advantages of such floor construction are now generally appreciated by the building profession at large. Some attempts have been made to cover considerable widths without the supporting aid of iron, but such daring use of concrete unassisted is attended with much risk. The numerous failures which have arisen through this indiscreet application of concrete, will tend to check its general use in that direction. Up to 10 or 12 feet span spaces may be successfully covered when the concrete is carefully made and of the proper thickness. It is desirable, however, to attend carefully to the support of these monolithic floors, and every precaution should be adopted to ensure that perfect setting has taken place before the withdrawal of the centres on which they are laid.

No. 4. In this direction there has been during the last few years a large increase of the use of concrete for all kinds of buildings. The first important impetus given in this adaptation of concrete was by the introduction of building frames, through whose assistance houses were promised to be built at a cost much under those produced by hand labour in the ordinary way. The inventors of these machines (whose names are now legion) promised such wonderful results that many were used, and more or less adopted, with an utter disregard in too many cases of the quality of the concrete itself, resulting in a considerable amount of disappointment and loss. The adoption of building frames in many instances has not reduced the cost of construction, but it provides against some of the difficulties surrounding the vexed question of skilled labour and its erratic and suicidal tendencies. There is no question about the desirability of erecting dwellings of ordinary extent by mechanical aids, but their careless use is surrounded with much danger. Primarily, the great object in any structure is stability and endurance, neither of which is safely reached by a too confiding dependence on the moulding or framing machine. Hence we have on all occasions, more especially in times past, when a disregard of the quality of the cements used was more general than it is at present, insisted on the lesson of Portland cement, and its product in the form of concrete, being more carefully studied. In the absence of this most important knowledge it would be much wiser, and we are sure much safer, to pursue the plan of building concrete houses by the block system, ensuring thereby the quality of the materials before any serious structural outlay has been incurred.

No. 5. This application of concrete is perhaps the most important of all the various systems, for by its means the vital question of perfect sewerage and drainage is eventually to be reached. It is only of late years that sanitary engineers have given their serious attention to the valuable properties of Portland cement concrete, as an important agent in the realization of a perfect air and watertight sewer or conduit. In the important works of the London main drainage, Portland cement concrete played but a subordinate part, and many advantages which it commanded and secured were thus thrown away. Since that time, however, engineers outside the charmed circle of the highly-privileged Metropolitan Board have more fully appreciated its valuable constructive properties, and in England and abroad considerable sewerage schemes have been successfully accomplished. Hitherto works of this class have been executed in a monolithic form, and not always with a due regard to the exact conditions which should be observed in the preparation of a sound and non-porous concrete; the cement, doubtless of the best procurable quality, being too frequently mixed with unwashed gravel or equally unsuitable aggregate. In some important arterial drainage works now in progress, the concrete is so porous that already some of the drains are partially filled with water obtained from the soil in which the sewer is laid. Such a condition of drain does not promise, when dedicated to its intended purpose, that security against leakage which above all things should be the desideratum in a sewage conduit. In addition to the use of concrete pure and simple, it is now in many cases applied as backing to brick or stoneware linings. A somewhat expensive and, from our point of view, irrational adaptation of this material. As foundations for large brick sewers it is of course most valuable, and in strengthening the sides of drains where stoneware pipes are used no objection can be made where the knowledge of how to produce a good concrete is wanting. It is doubtful, however, whether such a combination is advisable; for the numerous joints in the case of brick sewers, and the faulty or imperfect collar joints of the stoneware pipes, are unavoidable weaknesses, which the most careful treatment is incapable of overcoming.

No. 6. We now come to the production of concrete prepared under more scientific conditions, and which, from its superiority, has obtained a position of acknowledged usefulness. The use of concrete in a monolithic form for pavements has been a long time in vogue, although it is found that, owing to the impossibility of securing a true and sound foundation, and from the impracticability of using any impact force in its preparation, the results have been hitherto of a very unsatisfactory character. As such work requires to be done hurriedly a quick-setting cement must be used, which unavoidably increases the risk of mechanical disturbance. Not only are these pavements unsightly, but when the cracking, expansion, contracting, or whatever the disturbing force may have been, becomes finally developed, rapid and destructive disaggregation of the mass ensues. The improvements on this imperfect system of paving were first introduced by Mr. W. Buckwell, who produced slabs, prepared from various aggregates and Portland cement, compacted together between iron moulds, their perfect amalgamation being secured by the impinging blows of hammers or rammers. There were numerous examples of this kind of pavement laid in various parts of London about twenty-five years ago, and some of them still exist in an almost perfect state, to testify to the efficiency of the principle of manipulation. This manufacture, under the inventor's supervision, ceased about the year 1857, and since that time none of these slabs have been made. In 1868, however, Mr. Highton patented his process of Victoria stone making, by which the concreted mass was subsequently hardened by being placed in a bath of silicate of soda, obtained by treating with caustic soda the natural silica of Farnham. We will not further refer to this manufacture here, as its importance demands a more special reference, which will be found in Chapter IX.

No. 7. This may be regarded as the latest scientific application of concrete, and although now acclimatized, as it were, in England, was borrowed originally from the United States of America, where pipes made by machinery have been used on a large scale for years. We purpose also in this case to devote some space to a full and detailed description of the process of manufacture in Chapter IX.

In the several heads or divisions we have generally referred to the commonest and best known processes of concrete-making, but outside of these various adaptations there are many others, of a too subordinate character, however, to warrant our making any special allusion to them. Many ornamental as well as useful articles are produced in considerable quantities, such as architectural embellishments of various kinds, sills, sinks, and other equally useful domestic appliances. Indeed, a gradual desire to use well-prepared concrete is now becoming very general, and much of the ornamental work of some buildings, usually entrusted to the sculptor or mason, is being produced in silicated concrete in its best and most durable form.

Before closing this chapter we shall shortly refer to the effects of rain on aggregated road materials, and its influence in disturbing the surfaces and rendering them liable to unnecessary wear and tear. There is no outside beneficial action to assist the compacting of the gravel and other materials generally used for the maintenance of city and suburban roads. "Macadam" prepared from granite, basalt, ragstone, and other suitable rocks is, perhaps, the best method of obtaining a good surface, but in the following remarks we shall endeavour to show that the present mode of their application is not attended with such beneficial results as might otherwise be easily reached.

Water has a considerable influence in the consolidation of road-making materials, and a favourable illustration of its effects is to be witnessed in connection with the maintenance of country roads. Gravel or other similar materials laid on loosely have little, if any, inherent property of binding in themselves, and do not secure, without some natural or artificial assistance, good or comfortable roads. Moderate showers of rain produce a beneficial influence, and tend, by the water action, to consolidate the mass, by bringing into closer contact the several particles of which it is composed. Storm-rains and heavy showers, on the other hand, have an injurious tendency, and by the mechanical disturbance of the surface reduce the coherency of the mass by removing the finer particles, without which the road becomes disaggregated and incoherent. Its restoration to its normal condition is only possible by the application of fresh material, equivalent in character and value to that washed away.

From these two common and well-understood phenomena

of very frequent occurrence, the concrete-maker can derive some instruction, the practical lesson from which would be as follows:—

1st. That imperfectly cemented masses are readily disturbed or degraded by the action of water if the ingredients of which they are composed are inaccurately or ignorantly proportioned, and,

2ndly. That an excess of water used during the preparation of the mass would, by washing out the more valuable soluble ingredients of the mixture, render the ultimate induration a matter of uncertainty, if not impossibility.

The whole process of road-making on the ordinary macadam system is attended with much that is wasteful. The insistance upon uniform cubes of $2\frac{1}{2}$-inch or 2-inch size involves a subsequent operation of consolidation before beneficial results are attained. An accurate estimation of the mechanical wear and tear of vehicles, besides the waste of animal power, would be not only instructive, but would also show that a smooth road surface is only possible at a cost in comparison with which the materials themselves are but a bagatelle.

The use of macadam has done good service, and at the time of its first introduction proved most beneficial and useful, but its continuance in these times of advanced scientific knowledge is not at all creditable to our intelligence. The process of macadamizing is so familiar to everyone that it must almost appear an act of supererogation to describe it. The continuance of the system, however, from the writer's point of view, appears so antiquated in character as to induce him to say a few words by way of criticism.

Examine carefully the whole process, from the quarrying of the granite until the accruing dust is either blown away or carted off in the shape of mud. Powder and much steel and iron were consumed at the quarry in fitting the cubes for use, and much more steel and iron was wasted in reducing the obdurate granite to fine and almost impalpable dust. Look at the section of a macadamized road when a trench is opened by the gas or water companies' men, and you will find that those orthodox-sized cubes have all their angles rounded, as if they had been subjected to the triturating action of the sea on a sloping beach. The barbarous character of a newly-macadamized road is known only to the unfortunate charioteer or equestrian who is obliged to use it. How can it be otherwise than objectionable when the whole operation is carefully examined? It can be understood well enough why the cubes require to be large if you are dealing with a road passing over a boggy or spongy subsoil, but to insist on the use of such materials on hard, well-formed roads is, to say the least of it, wasteful and extravagant.

The cubes are carefully gauged, and none over or under the regulation standard are used. The interstitial space occurring in a 6 or 9 inch thickness of such "macadam" cannot be less than 30 per cent. This void space has to be reduced, and indeed should disappear if the road is to be made profitably solid for heavy traffic. Ordinarily this filling up of the interstices and consolidation is entirely dependent on the action of vehicular traffic, and in the course of its performance a large amount of the angular portions of the cubes are rubbed off, which fill up the superficial voids and prepare the surface of the road for the succeeding operation of horse-foot and carriage-wheel degradation, or blown away in dust.

An *improvement* has taken place in the above method, and in specially favoured districts sand is now used in addition to the cubes, and the ordinary compacting anticipated by the use of a steam roller. The result, however, does not, to the mind of the writer, warrant the increased cost, for leading thoroughfares so treated do not exhibit any better results than those obtained by the original plan. If a solution of this problem be desired, the authorities should turn their attention to the use of small chippings instead of sand. On roads where this plan is adopted much less expensive and more durable and comfortable roadways are secured. The stones also would be more likely to get compacted together if instead of 2 inches or $2\frac{1}{2}$ inches they were reduced to $1\frac{1}{2}$ inch. The object in making a good macadam road should be to brecciate the stones together, and so preserve their angularity, which would secure eventually an interlocked mass, free from all excess of interstitial vacuities.

CHAPTER III.

MATRICES.

IN discussing the properties of the various cementing agents, we will continue, as in a former work, to designate them by the above heading, which is derived from the Latin, signifying "womb," regarding that term as sufficiently indicative of the influence exerted by the cement, and quite characteristic of the purpose it serves in any aggregated concrete mass.

Although having good reason to be satisfied with the favourable results obtained with the well-known "Portland cement," we cannot regard our task as complete if some of the other cements are disregarded or their merits and peculiarities left unnoticed. It is not our intention, however, to enter into a discussion of all cements, but to limit our remarks to those only which are familiar to the architect and builder, and which have an established or well-known reputation. We shall as nearly as possible notice them according to their order of merit from the concrete point of view, and as Portland cement is undoubtedly the most valuable of them all, we shall give it precedence in these pages. Following it in constructive value, although not strictly speaking cements, are the blue lias and other hydraulic limes, which, unlike the Portland cement, are of direct natural derivation, being obtained in several districts of England, but of comparatively limited extent in Scotland and Ireland. The Roman cement, also derived from natural sources, is well known, and for some special purposes may still be considered a good binding agent when used in its proper and suitable place.

We shall for the present confine our observations to these three leading cementing agents, for their properties are now pretty well understood, although we may in another division of our subject comment on some of the "new-fashioned" cements, whose merits are unreasonably extolled by their *inventors* and producers. The natural thirst for novelty has, we are afraid, obtained for these mixtures a position which they do not deserve.

First in order as in value is Portland cement. This cement, when produced on true and accurate lines, is beyond question the most valuable cement at present within our profitable reach.

Of an artificial character, and its quality capable of accurate controlment during the process of manufacture, it offers to the concrete-maker the safest, under reasonable and judicious treatment, and best cementing agent for the preparation of mortars and concretes.

It may be regarded from a chemical point as a double silicate of lime and alumina, possessing the valuable property of maintaining its normal value, when protected from damp or moisture, for a lengthened period of time. Although of comparatively modern origin, it has during its history of upwards of fifty years established for itself a deservedly high reputation as a hydraulic cement of maximum indurative capacity.

The sources of the raw or mineral ingredients from which it is prepared are of almost universal extent, and practically speaking we may consider them as beyond the possibility of exhaustion.

When it can be obtained without the heavy charges attending its transit to distant points from the source of its manufacture, it may be regarded, taking into account the amount of work it can perform, as being cheaper than the commonest hydraulic limes or natural cements.

The leading characteristics and properties of a first-class Portland cement are as follows:—

1st. Weight. Should not be less than 112 lb. per straked Imperial or Winchester bushel.

2nd. Strength. Measured by its tensile capacity of resistance to accurate vertical tension, should not be less than 300 lb. per square inch.

3rd. Texture or fineness of powder should be as nearly as possible equivalent to 2500 atoms to a square inch, or, in other words, a sieve of that gauge (50 square) should not reject more than 10 per cent. of the cement when gently shaken by the hand. Should feel roughish to the touch.

4th. Hydraulicity, or capacity of resisting the solvent action of water during at least six days' immersion. When taken out of the water it should be perfect in character, and exhibit neither crack nor fracture of any kind.

5th. Colour, which is best tested by an air sample, as the water obliterates any tendency to an abnormal shade. The true colour resembles the best quality of the well-known Portland building-stone, namely, a greyish blue or steel-grey.

6th. Setting. When mixed neat with a proper quantity of water, it will set in from one to six hours, according to its weight or density. Setting means when the moistened mass has lost its plasticity, and can be handled without injury.

7th. Chemical analysis. A good average Portland cement, capable of passing the six preceding tests, would produce within a moderate divergence of range the following analysis of the main ingredients:—

Lime	60·05
Magnesia	1·17
Alumina	10·84
Silica	24·31
Alkalies	1·54

Portland cement, from its prominent position as a constructive material, has received considerable attention from engineers, and in this, as well as other countries, extensive and elaborate experiments have been performed, more especially on works of great magnitude. These tests were at first instituted to guard against the risks attending the use of inferior products, and to ensure accurately-guided dependence on its best quality only. Indeed, in well-conducted engineering works none other than the best could be used, and that which failed to pass the challenges of any member of the triple or original test was rejected and sent off the works.

The original, and not unduly stringent test, enabled the reflective engineer, while securing the safety of the works under his charge, at the same time to extend his inquiries beyond the limits by which they were at first controlled. These extended inquiries have produced results of an unusually interesting character, with the particulars of which the concrete-maker should be acquainted.

Portland cement in its best condition appears so simple in character, that many heedlessly disregard the dangers surrounding the use of those qualities of an abnormal or inferior kind. Even outside the safeguard of intelligent testing, simple and well-defined indications are apparent of the faultiness of improperly prepared sorts, but the two most dangerous qualities may be described as follows:—

In any case the dangers of a faulty cement are primarily due to a defect in the controlling chemistry of its manufacture. That we may the more clearly illustrate this important point, we shall shortly describe the process of manufacture, and thus more readily be able to indicate the causes which lead to the production of an improper quality.

The mineral ingredients from which Portland cement is produced are carbonate of lime, and clay or other silica and alumina-bearing deposits, with variable quantities of oxides of iron, magnesia, &c. These, however, play so insignificant a part, that they can be disregarded in our examination. Carbonate of lime, silica, and alumina may therefore be considered as the essential ingredients, and their accurate and proportionate blending produces this now well-known cement. Various methods are adopted, according to the circumstances of each particular locality, but they all aim at the production of a hard, lava-like "clinker" (the technical name of the burnt mixture). The result obtained depends very much on the quality of the raw materials and the intelligence by which they are converted. Imperfection in the proportions or insufficient calcination in the kiln are frequently the cause of inferior products, which, from being uncertain in their effects on the quality of the cement, we shall shortly particularize.

Carbonate of lime in excessive proportion, and therefore chemically uncombined, produces a hard and dark clinker, difficult to reduce to a powder, and excessively rough to the touch. The lime in excess becomes, either by exposure to air or from moisture in the mortar or concrete, hydrated, and in swelling produces a mechanical disturbance in any mass with which it may be incorporated.

Excess of silica or alumina produces a brown dusting clinker, easily ground into a soft smooth powder, which, however, does not endanger the mass in which it is used, although it is weak in character, seldom attaining a sufficiently high value of induration.

From neither of these qualities of cement can a concrete-maker hope to realize satisfactory or beneficial results. The first quality will at an early period give proof of its inferiority by fracturing the work in which it may be used; while the second quality will prove passive in character, failing to produce a compactly indurated mortar or concrete. The qualities of the setting energy are distinct in each case.

The over-limed quality sets most feebly and cannot stand the action of water, but the over-clayed cement sets quickly and is not injuriously affected by immersion in water.

A good and faultless Portland cement may be described as being of intermediate character between the one and the other of these two faulty kinds, and of good colour, high indurating capacity, and will set in water, maintaining under all circumstances its true normal form unchanged.

From the experiments which have been recorded during the last ten years, unerring rules for guidance in the selection and use of Portland cement are all but established. Originally instituted by the consumer as safeguards against the carelessness of the producer, they became eventually beacons encouraging the manufacturer to improve the quality of his products, thereby promoting confidence in their use. Modern English testing is based upon the practice of French engineers, and the form of briquette first used was that which had for a long time ruled in France for testing English Portland cements. This system of testing enabled the consumer to ascertain the quality of the cement he was about to use, and rendered him independent and free from the risk attending the misrepresentations as well as the ignorance of some of the less careful manufacturers. The English system of testing was an improvement on the French by adding an element of capacity, making it therefore a simple test of three members, viz.:—1st, Measure of Capacity; 2nd, Hydraulicity; 3rd, Tensile Strength. If unable to pass the challenge of any of these tests, the cement was rejected.

The above test, first instituted by the Metropolitan Board of Works under the author's advice, has done much good service to engineers, as well as improved the quality of the cement and compelled the manufacturers to give more heed to the chemistry of their art. Owing to the comparative costliness of the testing machines first introduced, and the waste of cement and labour consequent on their employment, a desire has arisen to substitute less costly instruments of technical precision, but having sufficient regard to an accurate result.

There is a lurking wish among engineers and architects to adopt an approximate test that will in a speedy manner prove the fitness of the cement for ordinary purposes. In this direction Mr. Deacon, borough engineer of Liverpool, has suggested a testing arrangement as shown by Fig. 2.

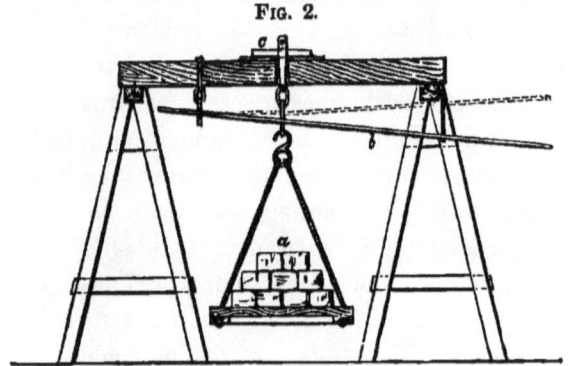

DEACON'S EXTEMPORIZED TESTING APPARATUS.

The specification for this mode of testing is as follows:— Samples of pure cement are to be gauged with water, and pressed into a mould 10 in. × 1½ in. × 1½ in. The block of cement thus formed shall be placed in water, and after seven days tested by putting it on supports 9½ in. apart, and gradually applying a load of 150 lb. on its centre. If more than one out of three blocks are broken within one minute of the application of the load, the cement from which the sample has been taken is to be at once rejected. The 150 lb. may be considered as equivalent to a tensile strain of 800 lb. on the same breaking surface of 2¼ sq. in., or practically 355 lb. per sq. in.

The mode of using this simple machine is thus described :—
" The block of cement C is to be tested by transverse

breaking, and the specification requires that it shall not break with a less weight than 150 lb. applied at the centre. One hundred and fifty pounds of bricks or materials of any portable kind are placed on the board (*a*); it deflects the iron or wooden bar (*b*), but is prevented by that bar from bearing on the shackles hanging from the test block until the end (*b*) is lowered. This is done gradually until it takes the load. If the whole weight or load of 150 lb. is carried for a few minutes, the required strength has been reached. The dotted lines indicate the position of the bar at rest. The block or briquette of cement is made 10 in. × 1½ in. × 1½ in., and may be moulded in a wooden mould, which can readily be made by any joiner."

This mode of testing and the machinery by which it is performed should only be regarded as approximations, and ought not to supersede the other more precise and accurately constructed testing machines now commonly used.

In the above specification it will be observed that the duty required of the briquette is defined in a clear and exact manner. If the block resists the weight prescribed, the conditions of the test have in reality been performed, and there is no object in prosecuting the test to the point of rupture. Briquettes or blocks thus reasonably treated may be preserved for future experiment, and such an arrangement would result in a knowledge of the progressive value of cement and its mixtures. Engineers and architects should be satisfied with obtaining the desired strength which they specify, for there is not much value or use in experiments which travel beyond the point originally contemplated. Indeed, this desire for unreasonable results has led to the assertion of some makers that their cements have attained certain imaginary breakings which, instead of having an alluring influence on the minds of consumers, have a tendency to create caution in adopting cements of such superexcellent quality.

56 A PRACTICAL TREATISE ON CONCRETE.

Mr. Deacon's test may be termed an easily extemporized one, and will no doubt prove acceptable to many small consumers who have neither time nor opportunity for adopting the more complicated and expensive machines.

In large works it is, however, more advisable to use the higher-class testing apparatus, so that an accurate reliable record shall exist of the quality of cement used and its behaviour when submitted to examination by a good and properly constructed machine.

FIG. 3.

BAILEY'S TESTING MACHINES.

Fig. 3 represents a Bailey's single lever machine, the testing weight of which is applied by a stream of water passing from the fixed cistern. When fracture arises, the value is indicated on the graduated glass or tin cylinder.

The plan recommended for moulding the briquette used in testing by this machine is simple. There is no pressure applied, the mould being hinged, and when the cement has become firm enough the mould is opened and removed, leaving the briquette on the pallet or board, from which it is in due course taken and placed in the water cistern.

There is much advantage realized in obtaining briquettes

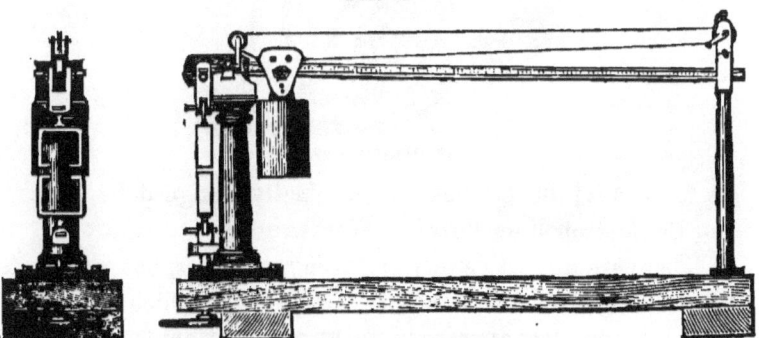

FIG. 4.

PALLANT'S TESTING MACHINE.

by this accurate method, as it has been found that those briquettes moulded by the spring press are frequently defective when any carelessness has arisen in the process of moulding. Unless the spring press is used with discretion, and the briquette released at the exact time, two kinds of danger arise: one being the result of irregular pressure, and the other occurring if the cement has not lost enough of its plasticity. In both cases almost imperceptible hair-like cracks are produced, which at the time of testing exert an injurious influence in the resulting breaking, and as they generally occur at the angles and weakest point, their prejudicial influence is the more dangerous.

Fig. 4 is Pallant's single lever machine, of simple design,

which has been much used on public works in this country and abroad. It is probably the earliest machine that was adopted by English engineers. The present form is that which was first made, the briquette press used with it being represented in Fig. 5. The makers continue to manufacture

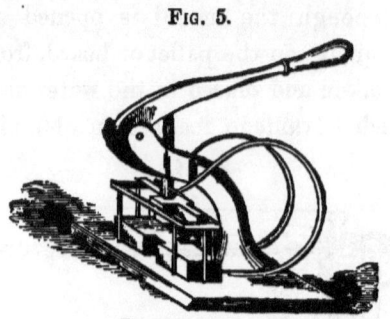

Fig. 5.

Briquette Press.

it for the 2¼ in. briquette as originally designed by them for the Metropolitan Board of Works.

There are a great variety of testing machines, but we will confine ourselves to the notice of the above, which are the simplest and most approved kind at present manufactured in England. It is unnecessary to describe that made by Adie, for it is well and widely known to all engineers, and has been used in almost every country where engineering works of any magnitude have been constructed.

The testing machine now very much used in Germany, was adopted at a general conference of engineers, architects, cement-makers, and others interested in the subject; it is represented by Fig. 6. Its use is regulated by the following conditions, which may be regarded as the specification controlling its employment for testing, but these instructions differ from the English system of testing, inasmuch as the cement is incorporated with certain proportions of sand. In fact, it is a mortar test, and one of a very satisfactory and practical character.

The tests are to be made with briquettes of 5 square centimetre section. The cement to be examined must, when mixed with three parts of clean sharp sand (proportions being weighed), be competent to stand a maximum tensile strain of 8 kilos. per square centimetre. The briquettes are allowed to remain in the air for twenty-four hours, and then put in water, where they remain for twenty-seven days, so that the briquette is in reality twenty-eight days old when submitted to the test.

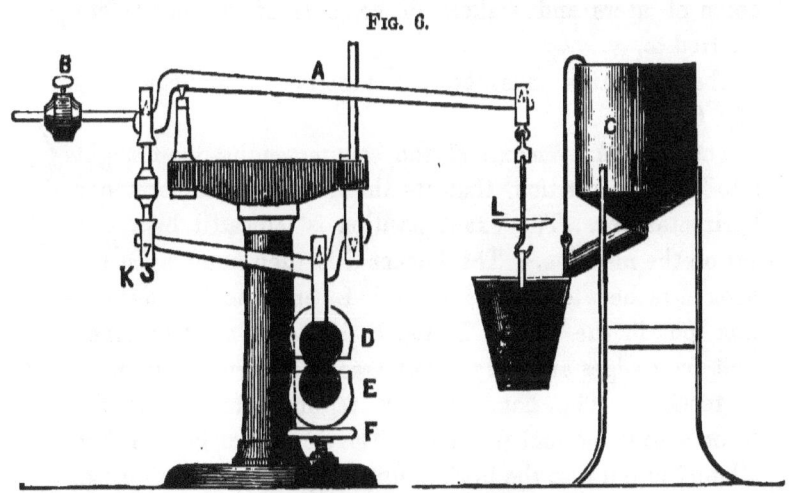

Fig. 6.

German Testing Machine.

The required fineness of the cement in powder is equivalent to a gauge of 900 meshes to the square centimetre, and should pass through a sieve of that fineness, leaving a residuum of not more than 25 per cent.

The preliminary treatment of the sand is performed by sifting it in a natural state through a 60-mesh to the square centimetre sieve. The coarser particles being thus excluded, the remainder is again passed through a sieve of 120 meshes, by which operation it is freed from the finest particles.

Quick-setting cement under such treatment cannot be

expected to reach the prescribed strength in twenty-eight days, nor indeed with some qualities of inferior cement would such a breaking ever be realized.

The primary object of this excellent system of testing is to encourage the manufacture of high-class heavy cements, to the exclusion of inferior quick-setting qualities.

This machine being compound in character, requires some little guidance as to the mode of using it. The following description is given in the instructions published by the union of users and makers of cements at Berlin, before referred to.

The apparatus must be put together as indicated by Fig. 6.

The lever A by means of the counterweight B is brought into such a position, that its three knife-edges form one horizontal line. The exact position is indicated by a line cut on the machine. The bucket C is then hung upon the hook attached to the scale-board L, and the briquette is then put in the clamps D and E, but care must be taken that their edges are truly parallel, so as to secure uniformity of tension. The hand-wheel F regulates or adjusts the lever A to its normal position. The small shot in G is then allowed to fall into the bucket, until rupture of the briquette takes place, which is caused by the combined weight of the bucket and its contents.

To ascertain the breaking weight, place the bucket and the shot it contains on the hook K, and weigh it by weights placed on L. The machine being a decimal one, if the number of grammes upon L be a, the weight of the bucket and shot will be $10 \times a$. This latter weight multiplied by 50 gives the weight which caused the fracture, the proportion of the upper lever being 1 : 10, and of the lower 1 : 5. The weight which broke the briquette is therefore $50 \times 10\, a$: the section of the briquette being 5 square centimetres at the point of fracture, the breaking weight per square centimetre is $\dfrac{50 \times 10 \times a}{5} = 100 \times a$ in grammes, or $\dfrac{1}{10} \times a$ per square centimetre in kilogrammes.

If, for instance, 105 grammes were upon the scale-board, the breaking weight would be 10·5 kilogrammes per square centimetre. The same calculation is applicable to any weights or measures, but we here give those used in Germany by the introducers of the machine.

The moulds used for this machine are hinged, and when closed ready for filling, are held together by a clip or screw, so that they are readily withdrawn from the briquette after it has sufficiently hardened to be removed. It is necessary to keep the moulds clean, and a little oil rubbed on their inner surfaces, not only secures a true briquette, but facilitates its accurate separation from the mould.

On some engineering works it is not convenient to have a testing machine (we do not regard any excuse for such supineness however), and in such cases it is desirable to resort to the following most primitive and simple method.

Make up a small circular cake or disc, about 4 inches in diameter and half an inch thick, slightly thinned at the edges. After it is set, and when all signs of plasticity have disappeared, place it in water, where it should remain for three or four days. If on careful examination after that time there are no indications of cracks or fissures, it may be regarded as safe to use. This is merely a test of the cement's hydraulic value, but affords no evidence of its strength or other qualities. It is advisable to make duplicate pats, so that one may be in the air, while the other is in the water. The air sample affords a proof of the colour of the cement, which if of a brown or "foxey" character should be carefully watched, as such indication is proof of weakness. This approximate test is of doubtful character, but even its application will avoid the risk of using bad or over-limed cement in the absence of proper apparatus.

In the analysis at page 50 of a good and unexceptionable Portland cement, carbonic acid has no place, but in some inferior and defective qualities, its presence is attended with much danger. The following analysis of faulty cement used in building a group of concrete cottages, indicates its objectionable character, which caused their eventual downfall, luckily before they were finished. The disaster led to litigation, the author being professionally engaged in the matter, and in his required examination made the analysis as under:—

Lime	34·56
Magnesia	1·44
Alumina and oxide of iron	11·26
Alkalies	1·93
Silica	20·63
Carbonic acid	11·19
Coke	3·21
Water	15·05

In the first place, not only was the cement of inferior quality, but it was aggravated by adulterants, such as coke and burnt shale. It is to be hoped that before long such an occurrence will be impossible, for increased and more accurate knowledge of the properties of cement will enable consumers to protect themselves against such accidents by accurate testing.

The presence of carbonic acid in Portland cement has an injurious tendency, and the following experiments were made with the view of ascertaining its extent and character.

The cement used when first obtained from the manufactory was analysed, and the following results obtained:—

Lime	65·0
Silica	23·7
Alumina and oxide of iron	9·0

For the purposes of experiment the cement was spread, to a thickness of 3 centimetres, on a floor in an enclosed warehouse, for eight months.

At the end of five months it was again analysed, with the following result:—

Lime	63·8
Silica	23·3
Alumina and oxide of iron	8·8
Carbonic acid	1·8

And at eight months it was:—

Lime	63·5
Silica	23·2
Alumina and oxide of iron	8·8
Carbonic acid	2·2

Portland cement in absorbing carbonic acid changes in character, and an alteration takes place in its specific gravity as follows, viz.:—

When fresh the specific gravity was	3·20
At five months	2·96
At eight months	2·85

The cement used in these experiments was of German manufacture and of good quality.

An English Portland cement, of high repute, after being stored in Germany for twelve months, absorbed 2·1 per cent. of carbonic acid, and its specific gravity fell from 3·09 to 2·85 in consequence.

Experiments of the above kind are necessarily conducted by skilled chemists, untrained hands being, perhaps, incapable of making them with the needful refinement; they are referred to here for the purpose of showing there are lurking dangers beyond ordinary control or challenge, and we desire that their character should be known and understood.

An English Portland cement, kept for some months in Dublin in a cask, when opened and analysed gave the following analysis:—

Carbonate of lime	8·66
Oxide of iron	2·59
Alumina	8·22
Sulphate of lime	3·03
Lime	50·62
Silica	22·50
Alkalies	1·62
Combined water	2·13

There must have been some carelessness in the fabrication of such a cement as the above, and the presence of sulphate of lime indicates that the fuel employed in burning it was impure in quality.

Some makers of cement have an idea that a proportion of carbonate of lime acts beneficially in Portland cement, and in such cases deliberately use it as an adulterant.

A few years ago the author, in making an examination of an English cement works in the north of England, found that about 25 per cent. of blue limestone was ground up with the "clinker" for the purpose, as the manager gravely asserted, of improving the strength of the cement.

The limestone cost 8s. per ton and the clinker was worth about 25s. per ton at the place in question. The cement so produced, and notwithstanding its supposed improvement by the addition of the spurious ingredient of excellence, obtained such a bad reputation, as ultimately to necessitate the breaking up of the establishment: a suitable fate for all such manufactories whose conductors aim at obtaining profit regardless of the quality of their products.

Uniformity of results in testing Portland cements is good evidence of quality, and it will be found that first-class cements closely approach each other in chemical value. In illustration of this point we give below analyses of six cements of good reputation:—

       Nos. 1 and 2 .. .. .. London.
       No. 3 .. .. Wallsend.
       „ 4 .. .. Boulogne (natural).
       „ 5 .. .. Stettin.
       „ 6 .. .. Bonne.

The analyses were made by different chemists and for manufacturing or constructive purposes.

	1.	2.	3.	4.	5.	6.
Lime	59·06	60·40	60·41	65·13	62·81	57·13
Magnesia	0·82	1·17	1·19	0·58	1·14	1·35
Alumina..	6·92	9·14	10·30	13·87	5·27	9·38
Silica	24·07	21·84	20·10	20·42	24·63	23·81
Iron oxides	3·41	3·21	3·47	trace	2·00	5·22
Sand	1·47	0·36	0·57	..	2·54	..
Potash	0·73	0·59	0·58	..	1·27	0·59
Soda	0·87	0·53	0·42	..	..	0·71
Carbonic acid	..	1·40	0·82	..	..	..
Sulphate of lime	2·85	..	..	trace	1·30	1·11
Sulphuric acid	..	1·43	0·82	..	..	..

The average analysis from the above gives:—

       Lime .. .. 60·82
       Magnesia .. .. 1·04
       Alumina .. .. 9·10
       Silica .. .. 22·48
       Iron oxides .. .. 2·88
       Sand .. .. 0·82
       Potash .. .. 0·62
       Soda .. .. 0·42
       Carbonic acid .. .. 0·37
       Sulphate of lime .. .. 0·87
       Sulphuric acid .. .. 0·37

This average approaches pretty nearly the analysis of an undoubtedly first-class Portland cement given at page 50.

Even the handling of the briquettes intended for experiment should be carefully attended to. Some experiments in relation to the difference in value arrived at from irregular exposure of the briquettes to air after being taken out of the water produced the following remarkable results. The briquettes under examination had been in water for three months and were made with neat cement.

The breakings were as under:—

		Broke at		
After 1 day's exposure	..	30·0 kilos. per square centimetre.		
,, 2 days' ,,	..	17·1	,,	,,
,, 4 ,,	..	33·0	,,	,,
,, 5 ,,	..	40:2	,,	,,
,, 6 ,,	..	40·8	,,	,,
,, 7 ,,	..	44·0	,,	,,

Showing that after a week's interval of exposure they again resumed, or rather recovered, their normal value.

These experiments carefully performed, if hereafter confirmed, indicate a source of error which would help to explain away some of the discrepancies exhibited in past tests and their records. It appears, however, from these experiments that the greatest loss in tensile value occurs on the second day of air exposure, and a gradual recovery takes place by the end of the seventh day; the reading of the machine on immediate removal from the water having been 43·2 kilos. per sq. centimetre, or 0·8 less than the strength after seven days' exposure.

We find some difficulty in accounting for the eccentricities here recorded, but presume the hurried evaporation of the moisture of crystallization had much to do with it. It is satisfactory, however, to find that the danger was only of a temporary character. Had these briquettes been weighed at the several stages of their examination, we might have been better able to reconcile the differences.

Other experiments instituted with a similar object, in which the cement was mixed with various portions of sand, did not develop similar results, and such differences were only observed when the briquettes were of neat cement. The mortar briquettes were uniformly and under all circumstances of a continuously improving value.

Another important question in regard to Portland cement

is its quality of fineness. In some experiments made to elucidate this point, the following interesting and valuable information was obtained.

A parcel of cement was selected and divided into the four following qualities of fineness:—

1st. All of which passed through a sieve, each of whose perforations measured 0·023 of an inch.

2nd. Ninety per cent. only passed through.

3rd. Eighty-five per cent.

4th. Eighty per cent.

Representing No. 1, fine; No. 2, 10 per cent. of coarseness; No. 3, 15 per cent.; and No. 4, 20 per cent.

These several qualities of cement thus treated were each mixed with equal proportions of good sand of uniform quality made into briquettes (2¼ in. breaking surface) immersed in water, and after seven days broken with the following results:—

	No. 1.	No. 2.	No. 3.	No. 4.
Per square inch ..	207½ lb.	215 lb.	212 lb.	220 lb.

These experiments were made with great care.

Another series of experiments with a similar object gave equally conclusive results.

No. 1. Cement as it came from the manufactory weighed 116 lb. per imperial bushel.

No. 2. Sifted through a sieve of 1296 meshes to the square inch, which rejected 20 per cent. When thus treated it weighed only 102 lb. per imperial bushel.

No. 3. Sifted through a sieve of 2500 meshes, rejected 30 per cent., and then weighed 99 lb. per imperial bushel.

The breaking results were after three months as follows:—

	No. 1.	No. 2.	No. 3.
Per square inch	504 lb.	457 lb.	449 lb.

In these conditions of fineness, the roughest gave the highest results, for in that state the coarser particles acted as

F 2

sand or aggregates. Similar results were obtained in the first experiment.

This is the place to examine from a consumer's point of view, the different value of coarse and fine cements, and the advantages arising from high grinding; and for the purpose of rendering this important question more clear, we will follow the experimenter in his examination with the same kinds of cements mixed with sand.

It should be stated that all the proportions were measured and in a dry state.

No. 1. Mixed with one of sand, at three months broke at 286 lb. per square inch.

No. 2. Mixed with one of sand, at three months broke at 324 lb. per square inch.

No. 3. Mixed with one of sand, at three months broke at 377 lb. per square inch.

The finest cement set quickest, required most water in gauging, and gave the most coherent mortar, although of less bulk, from the same quantity of materials.

What are the conclusions arising from these experiments, and how do they affect the concrete-maker?

The cement received from the manufactory (No. 1) was composed of the clinker as drawn from the kiln, ground imperfectly, and the powder of various degrees of fineness, as well as of varied specific gravity.

The finest powder of the experiment was competent, from its atomical regularity and evenness, to exert in a mass of sand the greatest superficial influence, and each particle was in a position to become accurately hydrated when the water of plasticity was applied. Yet this finest cement was not the *best* of the manufacturer's bulk as delivered, for it was only the lightest and most easily ground, and therefore of least value, although performing the most valuable work. Nos. 2 and 3 had more valuable ingredients in their constitution, but the mechanical agency necessary for their perfect utilization was not forthcoming. It may be said by the manufacturers that it is not possible to reach the limit of this required fineness. Under the present system of cement-making, not only is the desired end unattainable, but such excellence is very far from being aimed at by any of our English manufacturers. It is only a question of expense, however, and if the consumers insist upon it, the extra cost must be borne by those who exercise the necessary "pluck" to obtain it. There can be no wish to deprive the manufacturer of his fair profits, but unfortunately hitherto the standard of the producer has been forced on the consumer, and he, regardless of his own interests, rests satisfied with the existing state of things.

The author is much amused, when brought in contact with certain cement-makers, at their egotism and conceit, presuming on the antiquity of their firms, or the success and opulence obtained by their industries. The poor consumer, or the credulous engineer or architect, abashed by the display of worldly prosperity of some manufacturers, and impressed by their conceitedness, fears to stand up and manfully assert his position. When he has sufficient courage, he will simply say to the cement-maker, "I require a certain quality of cement, the test of which shall be its capacity to pass a reasonable ordeal of testing so as to fit it for my purposes. They are not such tests as to impose upon you undue obligations, but I prescribe them from a knowledge of their fitness for my work, and they are based on an accurate practical acquaintance with the properties of Portland cement. I do not so persistently press the 'weight condition,' but I insist that you must not expect in future to be paid for your coarse unground clinker useless to me, because I can buy coarse sand at a tenth part of its cost. In short, my intelligence informs me that your coarse particles of cement are worthless. Grind them to the fineness I require, and if the cost of such reduction is increased, I am quite willing to pay a reasonable advance on the price. I refuse to be under the manufacturer's thraldom any longer, as I am the best judge of what is required for my purposes, and you must be guided by me. The time has passed when you were perhaps unduly hampered by ignorant and ambitious engineers' specifications. I have now read the lesson of Portland cement, and I cannot longer subscribe to your supineness. Your manufacturing process is little better than that practised by Aspdin fifty years ago, and it is now time you should be awakened to a knowledge of your position. There are mechanical as well as chemical improvements, not only capable of cheapening the cost, but improving the quality, and I insist upon your adopting them. In the demand for a heavy cement it was never expected that 30 or 40 per cent. of the bushel would be made up of grains so coarse as to practically make them useful only as so much sand, but the proper cement power was wanted, and its not being forthcoming makes all the difference to me as a consumer."

Not only does this fineness secure the best cementing power, but it also beneficially guards against the risks attending the use of an imperfectly combined cement.

In the author's varied experience in the manufacture as well as use of Portland cement, the fineness of powder has been forcibly brought home to him, more especially in the fabrication of experimental compounds in new districts, and where, as a matter of course, the necessary machinery is not obtainable.

A recent instance will show the great value of fine reduction.

In the first experiments made on the ground and far distant from any machinery or even suitable kiln aid—

indeed a hand experiment almost—it was necessary to reduce the clinker to powder by imperfect means. It was passed through a 2500-mesh sieve, and the pats of neat cement failed both in the water and air; but when sifted through a 5000-mesh sieve all danger was removed, and it continued to harden under every examination.

We have shown elsewhere that all cementing agents in nature have been applied in the finest and almost impalpable condition, otherwise the results we are familiar with would have been impossible.

Engineers and architects disregard their best interests when they lose sight of this important question; for if they would direct their attention to the points now so apparent and indisputable, instead of first rushing at high breaking strains, they would be more satisfied with the results, and more likely to obtain them without difficulty.

Consumers cannot afford to pay the high price of cement for unground stuff that is really only so much aggregate, and it will be found hereafter that cement which cannot pass through a 5000-mesh sieve is not economical. We state this most unreservedly, for there is not so much difficulty in doing this fine grinding if the desire existed. Better have lighter cement well ground than a heavy powder something like sand in its texture. Some intelligent manufacturers have for years insisted upon this, but the object in the writer's belief was more for the purpose of reducing the cost of manufacture than improving the quality. Both requirements are, with our increasing knowledge and intelligence of the cement question, not only desirable but *easily possible*.

In a series of exhaustive and interesting experiments made by Lieut. Innes, R.E., it is shown that the quality of the powder has much influence on the strength of neat cement, and is confirmatory of the above-quoted experiments.

A 2500-mesh sieve was used, the cement being uniform in quality, and weighed 117 lb. per imperial bushel. All the samples were made of neat cement and immersed in water as soon as possible after being moulded.

The qualities were as follows:—

    No. 1. Was all passed through sieve.
    „ 2. Ten per cent. was rejected.
    „ 3. Twenty „ „
    „ 4. Thirty „ „
    „ 5. Forty „ „
    „ 6. Fifty „ „
    „ 7. Sixty „ „
    „ 8. Seventy „ „

The breakings were per square inch in pounds:—

	1.	2.	3.	4.	5.	6.	7.	8.
Three months	464	402	499	507	500	490	512	489
Six months	492	508	510	515	520	525	527	535

The quantity of water used in mixing up cements, mortars, or concretes has also a marked influence on the breaking or tensile results, as will be seen by the following experiments:—

Proportion.	7 days.	3 months.	6 months.
1 water to 3 cement	1036	1508	1544
1 „ 2½ „	804	1314	1278
1 „ 2 „	637	1161	1094

These breakings were obtained from the 2¼-inch section.

A careful attention to the quantity of water used is desirable if mortar or concrete mixtures are wanted to be of the best character.

It should be remarked that these water experiments were made with chalk cements, and go to prove of course that the least quantity of water used produces the best results. Lias and like cements made by other than the wet process are better, or at least produce more satisfactory results when an excess of water is used. There are no reliable experiments, however, obtainable in this direction, and the author only refers to it for the purpose of preventing disappointment to those who use the cements referred to.

In these experiments we have pointed out four leading dangers in the use of Portland cement, and they may be classed thus :—

First. Evidence of the damaging effects produced by the presence of carbonic acid, whether produced in the process of manufacture or absorbed by subsequent careless air exposure.

Second. The necessity of careful treatment of the briquettes between their withdrawal from the water and the period of testing, showing that no interval of time should elapse between these operations.

Third. Indicating the great importance of fine powder, and warning us against the use of heavy cement where the condition of fineness is absent.

Fourth. That the careless use of water is attended with loss, and showing that the minimum quantity is the most advantageous.

From these four sets of experiments much information may be gained, and the peculiarities and dangers they describe and indicate should be regarded by practical concrete-makers as lessons for their consideration and guidance.

Some valuable experiments have been made by more or less competent authorities, and we give farther on a few of the results obtained. It is to be remarked, however, that the fact of our recording such tests is not from a desire to insist on a blind confidence in their teachings, but to encourage direct personal testing by the consumer. Everyone using Portland cement for concrete or other purposes should regard testing as a most indispensable condition of success. The methods of single and compound testing which we have described and the instruments by which they are performed are so simple and inexpensive, that no excuse can be made for neglect of this very important safeguard.

Another source of danger is liable to be incurred under the following circumstances.

Dr. Heintzel thinks that the influence of light on cement has not hitherto been sufficiently considered. He instituted some experiments upon a quantity of cement, which he divided into three parcels, exposing parcel A to the air and full light, B to the air and diffused light, and excluding C in darkness from the air. After six months he found that A made a weak mortar by absorbing 38 per cent. of its weight in water, and it had become crumbly; B with $33\frac{1}{2}$ per cent. of water made a mortar which was too adhesive to a trowel, and it yielded up none of its water; C with $33\frac{1}{2}$ per cent. of water made an excellent mortar, easily stirred and flowing, and it relinquished some of its water. After setting for twenty-eight days, the relative strength was A 3 : B $37 \cdot 9$: C $44 \cdot 6$.

The object of testing cement is to prove its capacity to hold in energetic embrace certain sands or other aggregates. Of course the quality and character of the materials with which it may be eventually combined increase or diminish (according to their fitness) the value of the mortar or concrete.

Much depends on the kind of aggregate used, and in elucidation of this subject we give some experiments made by Lieut. Innes, R.E., who has treated this department of our subject with much attention and carefulness.

There were selected for the purposes of experiment seven different qualities of natural sands or other artificially prepared aggregates, and the cement used was of the best quality and weighed 120 lb. per imperial bushel. The following is the description of these aggregates and their peculiarities:—

No. 1. A sea-sand, with roughish and uneven grains, chiefly siliceous and quite clean. Mortar produced, very short and incoherent; strength of different samples, somewhat variable.

No. 2. A siliceous pit-sand, quite clean, with smooth, uniform, semi-transparent grains. Mortar somewhat "short," strength very uniform.

No. 3. A siliceous pit-sand, containing a number of minute shells and a small quantity of some orange-colouring matter; grains semi-transparent, brownish yellow, and of unequal size. Mortar rather "short," strength tolerably uniform.

No. 4. Portland stone-dust, a mixture of roach and whitbed; grains rough and irregular, quite clean. Mortar tolerably coherent; strength very uniform.

No. 5. Drifted sea-sand, with pure, siliceous, semi-transparent, and almost colourless grains, and quite clean. Mortar rather short; strength tolerably uniform.

No. 6. Smithy ashes, containing a good deal of unburnt coal-dust; grain rough and irregular. Mortar moderately coherent, set slowly, and apt to shrink and crack; strength very uniform.

No. 7. Clay ballast, burnt and ground, of a pale brick-red colour, with a rough uneven grain, and containing a good deal of dust. Mortar very coherent, apt to shrink and crack; strength very uniform.

The above very carefully noted particulars may be regarded as the mineralogical characteristics, and we will now examine them in combination with cement before and after the water of plasticity had been added.

All grains or particles exceeding one-twelfth of an inch in size were excluded from the sands and other aggregates, with the object of obtaining more even and reliable results in the necessarily limited breaking area of 2¼ square inches. A preliminary examination of an important character was made to determine the amount of interstitial space or voids, represented by each sample of the sands, &c.

The result being as follows:—

No. 1.—38 per cent. dry, 34 per cent. wet, shrinkage 6 per cent.
„ 2.—41   „   34   „   „   11 „
„ 3.—32   „   19   „   „   16 „
„ 4.—46   „   34   „   „   18 „
„ 5.—43   „   36   „   „   11 „
„ 6.—64   „   52   „   „   25 „
„ 7.—50   „   40   „   „   17 „

The percentage of grains exceeding $\frac{1}{50}$th of an inch was

No. 1 94   No. 5 8
„ 2 76   „ 6 56
„ 3 15   „ 7 40
„ 4 56

The tensile breakings obtained after the briquettes had been in water for three weeks were as under:—

	1.	2.	3.	4.	5.	6.	7.
Per square inch ..	140	94	108	165	60	38	260 lb.

After three months' immersion:—

Per square inch ..	249	175	248	254	193	91	385 lb.

The Portland cement used in these experiments was of good quality, and when mixed neat, gave the following results:—

At 3 weeks .. .. 450 lb. per square inch.
At 3 months 529   „   „

The fineness of powder was equivalent to 31 per cent. of its particles only exceeding $\frac{1}{50}$th of an inch in size; or, in other words, a 2500-mesh sieve rejected 31 per cent. of roughness.

The variableness of the readings of these experiments was doubtlessly caused by the presence of fine or almost impalpable dust in some of the aggregates. The importance of accurate attention to this dangerous element in an aggregate will be more particularly referred to when the subject of concrete mixture comes to be discussed.

These experiments are worthy of our best consideration, for they appear to have been conducted under intelligent guidance and painstaking care.

In some experiments made with cement weighing 123 lb. per imperial bushel, the following breakings were recorded:—

1 week	363 lb. per square inch.
1 month	416 ,, ,,
3 months	469 ,, ,,
6 ,,	523 ,, ,,
9 ,,	542 ,, ,,
1 year	546 ,, ,,
2 years	589 ,, ,,

The briquettes were in water during the periods named, as also were those of the following experiments.

Equal quantities of the same quality of cement and Thames sand.

The results were

At 1 week	160 lb. per square inch.
,, 1 month	201 ,, ,,
,, 3 months	244 ,, ,,
,, 6 ,,	284 ,, ,,
,, 9 ,,	307 ,, ,,
,, 1 year	318 ,, ,,
,, 2 years	351 ,, ,,

Those desirous of further study of Portland cement, will find more detailed particulars in the author's Portland Cement book. Enough has been said in these pages, however, in our opinion, to guide the concrete-maker in the selection of the right quality of cement. Numerous tests are accessible, but we recommend all users of cement to test for themselves; by so doing they will obtain information which is only forthcoming through practical observation.

It is advisable to remember that according to the density of Portland cement, is its property of setting reduced or increased. A cement weighing 100 lb. per bushel may be regarded as quick setting, and its plasticity will have disappeared in about half an hour. Cement of 120 lb. per bushel has its set prolonged, and probably six or eight hours will elapse before the indurating process begins.

By the sifting treatment (starting with a cement as supplied from the manufactory, whose smoothness is equal to a rejection of 25 per cent. through a 2500-mesh sieve), it will be found as you eliminate the coarser particles through a finer sieve that the product thus obtained increases in initial setting energy. The reason is quite obvious, for at each successive sifting the coarser particles are excluded, and as these are in reality the heaviest and best portion of the cement, the lightest and finest powder is probably 30 per cent. less in weight than the heaviest. This has been well illustrated by Lieut. Innes's experiments, described at page 67, where it is shown that cement as received from the factory weighed 116 lb. per bushel, when sifted through a 2500-mesh by which 30 per cent. was rejected, weighed only 99 lb. per bushel, or a difference of 17 lb. per bushel. If the process of reduction was further pushed, say by its passage through a 5000-mesh sieve, the difference would be much more striking.

In the United States of America civil as well as military engineers have for some time given the subject of Portland cement great consideration. Until comparatively recent times the natural cements so abundant in that country prevented the merits of foreign Portland cements receiving much attention. Mercantile enterprise, however, may be said to have forced upon American engineers the use of English Portland cement in particular, and at the present time New York alone receives 200,000 barrels, or 40,000 tons, from abroad every year. That quantity does not, however, include the considerable shipments to other

North and South American ports. While readily accepting these foreign supplies, American engineers have not lost sight of the importance of a cheaper and more reliable source from which to obtain a material possessing such valuable constructive properties. They have accordingly turned their serious attention to a native supply of first-class Portland cement from the bountiful stores of raw materials existing in almost every district of the States. In our remarks in Chapter VIII. it will be seen with what success so far this movement or desire has been realized.

General Gillmore, one of the earliest American writers on limes, mortars, &c., has put before his countrymen during some years most reliable records of the progress of scientific cement and mortar making. He has been followed in his course of instruction by other writers also entitled to our consideration; the facilities and assistance afforded by the United States Government in the prosecution of all scientific researches contributing to the profitable discussion of the Portland cement question. A recent paper by Mr. W. W. Maclay, of the Docks Department of New York, places in our hands numerous and interesting examinations, which indicate that the American Portland cement inquiry has already travelled beyond the farthest point of English investigation. This paper was read before the American Society of Civil Engineers, who awarded its author the Norman medal in testimony of its great value, thus recognizing with honours the merits and labours of Mr. Maclay in his exhaustive and instructive inquiry. The tables appended to the above paper contain experiments which, in the aggregate, embrace no less than 7000 tests of American, English, French, and German Portland cements made during a period of several years.

In the course of his inquiry Mr. Maclay proposed certain questions to English and French cement-makers, in the hope, no doubt, that their answers would indicate, if not uniformity, at least an approach to some well-defined system of testing. The answers, however, generally indicated that no very defined or exact method existed, even in the establishments of the largest and best-known cement manufacturers in England. The reason for this information appeared very desirable from the conflicting and unaccountable difference between the tensile values of the cements received from the foreign makers and their recorded breakings at the point of consumption. The following extract illustrates this difficulty in a very singular manner. Mr. Maclay says:—

"Contracts have been made for Portland cement of a certain weight, fineness, and tensile strength. After the cement has been manufactured in England, and tested there, it has generally been found fully up to the requirements. Delivered here and tested, the cement has frequently been found deficient in tensile strength, and sometimes in weight. On one occasion, upon a large shipment of cement to New York, the results of the tests made in this city gave a tensile strength per square inch of less than one-half of that obtained and recorded by the English manufacturer."

It was with the view of reconciling these differences that Mr. Maclay set about his experiments. In the course of his inquiry many new and interesting facts have been developed regarding the behaviour of Portland cement, and mortar, made from it, under varying circumstances of temperatures, showing conclusively that some of the eccentricities in recorded tests are due to the condition of the air and water in which the briquettes are placed. All Mr. Maclay's experiments were surrounded with the most accurate conditions to secure reliable results, and they certainly place before us most valuable information in a direction hitherto unnoticed by previous experimenters. It is impossible for us here to enter into any lengthened detail of Mr. Maclay's labours, but we cannot resist the desire to notice a few of the more striking peculiarities he records. Before entering into these details, however, it will be well to examine the peculiar position of the American engineer using imported Portland cement, and to show under what disadvantages he labours in comparison with his more favoured English, French, or German brother practitioners. Generally speaking, the importation of Portland cement into the United States is undertaken by merchants, and not only has the consumer to run the risk and cost of transit, but in addition the mercantile profits of the original importers and their distributors. Under such circumstances it is the more essential that the American engineer should exercise his controlling influence to secure the best of an article so loaded with extraneous and, under the existing circumstances, unavoidable charges.

The main feature, and most striking to our mind, in Mr. Maclay's examination is the question of varied values from different temperatures, and the following extract exhibits in a remarkable manner its influence on the tensile value of neat Portland cement:—

"In February and March of 1876 deliveries were made in this city of 5000 barrels of Portland cement, guaranteed by the manufacturer to exceed in tensile strength 250 lb. per square inch. When the cement was tested, the tensile strength was less than 200 lb. per square inch. To reconcile this inconsistency the above experiments were undertaken, and in their progress very clearly indicated where the fault in testing lay. The room where the testing was carried on was heated by a stove, and the temperature during the day ranged between 60° and 70°, but at night, after the fire was extinguished, it sometimes fell below 40°. The gauged cement, after immersion in water, was thus often subjected to a very low temperature. As soon as it was ascertained that a variation of 30° in temperature of the water could affect the tensile strength at the end of seven days to the amount of 168 lb. per square inch, or 127 per cent., all the cement received was re-sampled and re-tested, the water in which the briquettes were kept being then maintained at a uniform temperature of 60° day and night. The following were the results :—

	lb.			lb.
1370 barrels first tested	185 per sq. in.	re-tested	317 per sq. in.
1373 ,,	,, 192	,,	,,	285 per sq. in.
1373 ,,	,, 236	,,	,,	289 per sq. in."

From these experiments, therefore, it is proved conclusively that temperature has an important influence on the results of tensile strength.

In another series of experiments it was found that cement paste moulded into briquettes in a room having a temperature of 32°, and afterwards placed in a temperature of 40° for twenty-four hours, gave the following breakings in water of variable temperatures. The mean tensile strength of five briquettes gave in

	40°	50°	60°	70°
Per sq. inch ..	156 lb.	186	259	299

The first briquettes, made of paste at 32°, were moulded when the air was 30°. All the others were made when the air was of 60° temperature.

The whole of these briquettes were immersed in water during a period of seven days. A continuation or extension of the tests in the direction of increased temperature of paste and the air in which the briquettes remained before being immersed in water shows in a similar manner the improved value obtained when the temperature is raised, but at a uniform temperature of 40° of paste and air in moulding and in the probationary room a much enhanced value was reached, for an average of ten breakings was thus recorded :—

Water	..	..	..	40°	50°	60°	70°
Per sq. inch	..	..		147 lb.	206	275	314

Corresponding improvement was realized in briquettes twenty-one days old, and others made of cement and sand. From these experiments it is shown that there is an undoubted increase of value when the temperature of the water of immersion is raised. There is very probably some improved activity of the soluble silicates in the cement from the higher temperature, and consequently a tendency to an increase of their fluidity, thereby rendering them more competent to permeate the mortar mass. This view is somewhat confirmed by Mr. Maclay's observations, who declares, in reference to the improvements from increased temperature:—

"The effect of changing the temperature of the cement paste or mortar to be much less upon the tensile strength than the effect of changing the temperature of the water in which the samples are kept immersed prior to breaking."

This subject is more generally discussed in the "Silicating process," Chapter VII.

Mr. Maclay, in his lengthened and exhaustive inquiry on the subject of Portland cement, had access to a mercantile correspondence spreading over many years between a large importer of English Portland cement and a manufacturing firm. A considerable amount of recrimination occurs in the various letters, and the manufacturer often threatens, and the author of the pamphlet says: "That sooner than be held accountable for the unsatisfactory testing of their cements in this country, they would much prefer to abandon the American trade entirely, and confine their shipments to the colonies, where the principal works being under the supervision of English engineers, the cement testing would be conducted in the same manner as at home. In the work with which the writer has been connected this same trouble has been experienced year after year."

A bit of English manufacturer's "bounce," having many a parallel in the writer's home experience. For instance, in page 327 of the Portland cement book of 1877, the writer relates the following occurrence, which was more of a threat than *bounce*, but it fell harmless, and luckily for the reputation of Portland cement, the author, from being practically acquainted with its manufacturing details, furnished the confidence which checkmated the presumption of the "great" cement-maker. The quotation is:—

"When, under the author's advice, 400 lb. per $2\frac{1}{4}$ square inches was adopted, and when the first specification for the main drainage was in proof, a leading maker waited upon Mr. Bazalgette and informed him that the test was too onerous, and incapable of fulfilment. An appeal to the writer reassured the engineer, and the test was launched with the result, after nineteen years' experience, of being nearly doubled."

These valuable experiments unfortunately lose much of their intended value in the absence of analogous data from the manufacturers, for without it no reliable deductions can be hazarded. The author's experience of the English cement-producing interest does not encourage him to hope that the desired information which Mr. Maclay took such trouble to obtain will be, for the present at least, forthcoming. There is too much reason to believe that the bulk of English cement-makers still tenaciously adhere to their antiquated policy of controlling according to their own intelligence the make, testing, and general treatment of the Portland cement made by them. From their point of view there can be no improvement in the manufacture.

Mr. Maclay refers to "laying cement down," meaning thereby the necessity of cooling or seasoning it after leaving the horizontal or grinding millstones, and attributes much of the damage which the cement sustains during its Transatlantic voyage to the "hurry to supply large orders." There is too much truth in this observation, and much of the evil reputation which surrounded Portland cement in past times (and indeed in some cases now) was due to the helter-skelter system of hurried shipment. Even if the clinker from the kiln is faultless, there is the liability of frictional warmth from the millstones by their being too dull or the grinding carelessly performed. Cement in such a state is liable to cake if put direct into sacks or casks.

Freshly-ground Portland cement does not of necessity mean that in that condition it is unavoidably faulty, but, on the contrary, indicates that if the output from the millstones requires keeping, or "seasoning," before it can be used, that the controlling chemistry of the works where it was made is faulty. "Keeping" in reality means storing the *unsound* and *carelessly* made cement, which can only be purged of its excess of carbonate of lime or other impurities by an irregular and uncertain period of air exposure, so as to hydrate the particles of uncombined or free lime which it contains.

"Clinker" sent to New York fresh from the kilns and ground there to a fineness equal to a 13 per cent. rejection from a 2500-mesh sieve, broke at 416 lb. per square inch under the usual seven days' test. "Clinker" so supplied would necessarily be of a picked character, otherwise some of it would powder during transit, for any quality of "yellow" could not withstand the degrading influence of exposure. The weight of this cement is not recorded, neither does it appear whether the clinker was in bulk or packed in casks.

It was found by these American experiments that concretes and mortars made from Portland cement in cold weather suffered but little depreciation if they had not been previously heated "to keep out the cold." In some of the tables of experiments recorded it appears that concrete has been used in hydraulic work under conditions of temperature which in this country would alarm engineers. It is considered in England imprudent to carry on outside concrete operations if the thermometer ranges at or under the freezing point of Fahrenheit. Careful engineers generally suspend concrete works from November till April. The reason is obvious. Concrete or mortar when made with a heavy, slow-setting Portland cement, has not sufficient energy to eject its water of plasticity while exposed in a freezing atmosphere, and consequently the water which remains in the mass will, on frost expansion, disturb its uncompleted concretion. The damaging influence of frost on rocks is well known to be in proportion to their density or capacity to absorb water; mortars and concretes, whether from lime or cement, are subject to like influences; and those made from the latter, owing to their greater setting energy, are sooner beyond the influence of such risks.

Limes.

These are obtained from a variety of geological formations and various in quality, ranging from the fat limes of cretaceous origin to the poor and feeble siliceous limes of the oolitic deposits.

For concrete-making purposes, where lime matrix is intended to be used, all those limes having less silica and alumina in their analysis than 10 per cent. should be rejected.

Smeaton was the first engineer who intelligently investigated the hydraulic properties of English limestones, when searching for the best lime to use in building the Eddystone lighthouse, and his elaborate experiments indicated the source of hydraulicity in lias and other limes.

The Aberthaw and Barrow limes, both obtained from the

lias formation, were regarded by him as essentially hydraulic in character, and modern experience has confirmed the conclusions arrived at by the great engineer.

The lias formation is of great extent in England, ranging from Whitby in the north to Aberthaw in the south, with limited deposits in isolated districts of Scotland and Ireland.

The best known limes from this formation are obtained in Somersetshire, Leicestershire, Warwickshire, Dorsetshire, and Yorkshire. In North and South Wales and Ireland there are some excellent deposits, more especially at Aberthaw in South Wales, and Holywell in North Wales.

We give a few analyses of the more generally known limestones :—

        No. 1.—Lyme Regis, Dorsetshire.
        „ 2.—Aberthaw, Monmouthshire.
        „ 3.—Holywell, Flintshire.
        „ 4.—Rugby, Warwickshire.
        „ 5.—Larne, Ireland.

	1.	2.	3.	4.	5.
Carbonate of lime	79·20	86·0	71·55	68·25	71·66
Silica and alumina, and oxide of iron	17·30	11·0	26·00	23·01	24·03
Magnesia	..	2·0	1·35	1·44	2·67
Water and loss	3·50	1·0	0·50	6·06	1·10

The first four of these limestones produce limes of high value, having widely-known reputations, and which have been used in some of the most important engineering works in this and other countries.

The works at which these limes are produced, being either connected with railways, or commanding facilities of transit by sea, afford the cheapest and most convenient means for delivery to any point on our coast or in the interior.

All hydraulic limes, and indeed limes of all kinds, should be reduced to a fine powder by mechanical means only.

Their reduction by the ordinary slaking processes is not recommended, unless the powder is afterwards carefully sifted through a fine sieve. Grinding is most certainly the safest and cheapest means of converting all stone or hydraulic limes for concrete purposes, and when they are obtained from the " lias," much advantage is derived by this process, owing to the absorption in the powder of the shale adhering to the stone when dug in the quarry. Otherwise, and by the ordinary slaking process, the shale has to be rejected as useless, because it cannot be slaked.

Although Portland and other cements do not exhibit when of the proper quality dangerous expansive tendencies, limes continue to expand for a lengthened period, unless they have been reduced to a powder fine enough to extinguish all possibility of their undue expansion when hydrated.

From an economical point of view, therefore, it is equally desirable to reduce limes for the same reason as is given for reducing Portland cement to a fine powder. It is perhaps preferable in the case of limes, as they are more likely when ground coarse to exert an injurious influence in the mass in which they may be employed, and they certainly are when fine better able to enter into the most profitable contact with the aggregates.

All hydraulic limes develop their best properties when used in wet or damp situations, and many important engineering works have been executed by their aid. In England, these limes have been used in the preparation of concretes for foundations of some of our greatest architectural buildings, but no very important attempt has yet been made to utilize them in superstructures. French engineers and architects have, however, for many years built largely with the celebrated Thiel lime, and in all the works at Port Said, in Egypt, it was freely and almost solely used.

The analysis of Thiel limestone is as follows:—

Carbonate of lime	81·36
,, magnesia	1·00
Clay	14·90
Oxide of iron	1·70
Water and bitumen	1·10

The limestone when burnt produces lime of the following analysis:—

Carbonic acid	8·00
Silica	18·20
Alumina	1·20
Lime	60·00
Oxide of iron, &c.	0·80
Magnesia	1·32

The lime produced is light in colour, not being ground but water-slaked, and the almost impalpable powder thus obtained is carefully packed in air-tight casks. When so protected it keeps good for a lengthened period. The powder weighs about 72 lb. per bushel when thus slaked.

Smeaton had the lime he used in the construction of the Eddystone lighthouse water-slaked, and the powder packed in casks. So effectual was this mode of preservation, that some of this lime was kept and used seven years afterwards with much satisfaction in works far distant from Plymouth.

All slaked limes (whether produced by air or water treatment) have a greater affinity for water than the mechanically ground limes, which, for ordinary purposes, keep well in sacks.

The cost of grinding lime is not so much as slaking and sifting, and there is less waste; besides, some of our best water-limes are difficult to slake as before stated.

In using limes it is advisable to adopt weights instead of measures for determining the proportions, because it is almost impossible to ensure accuracy by the latter method, owing to the flitty character of the finely reduced dry lime. A bushel of Thiel lime weighs when poured into the measure loosely, 71¼ lb., and when shaken weighs 84¼ lb.

A mixture of Portland cement and lime is sometimes used, and in the United States of America it is frequently resorted to in concrete and mortar compounds. We do not, however, recommend such combinations, for in our own experience the results have not been at all assuring, and we are of opinion the application of such preparations for ordinary concrete purposes will be found unsatisfactory. In fact, such treatment results in a total loss of the distinctive features and best properties of each of the matrices so compounded.

Lias or other hydraulic limes must eventually fail to command much attention from builders, for in comparison with Portland cement their value is very low indeed, and even at one-half the price of cement they are dear cementing agents. The following experiment is corroborative of this opinion.

Two blocks were made of lias lime and Portland cement, each being mixed with six parts of good Thames ballast; the size of the blocks was 12 inches cube (12" × 12" × 12"), and they were kept in water. At the end of nine months they were submitted to a compressive test with the following results:—

The Portland cement block stood a pressure of 963 lb. to the square inch, while that made from the lias lime was crushed with a force of 93 lb. per square inch.

We have only thus cursorily referred to the best-known hydraulic limes, but there are great varieties of them in almost every locality worthy of examination.

In Lancashire there is a good and well-known hydraulic lime of high reputation, obtained from the coal measures in the immediate neighbourhood of Manchester. The stone is obtained by sinking shafts, and the cost of getting it with the limited quantity now left reduces the trade to a narrow compass. Before railways and canals afforded the means of cheap transit this lime was extensively used for engineering works, and many examples of buildings in which it was used testify to its value. Lias and other far distant sources have been now opened up by the means referred to, thus affording suitable facilities for the delivery of unlimited supplies at a comparatively low cost.

At Halkin mountain, near Holywell, as well as various points near Colebrooke Dale, in Shropshire, good hydraulic limes are of easy obtainment, as well as several other districts in Staffordshire and Northumberland.

In Ireland the "Calp" limestone produces a good hydraulic lime, but difficult to slake. Its conversion, under the author's and Mr. W. Smith's patent, into good Portland cement will open a new Irish industry, and render Ireland eventually independent of its supply of this now indispensable material from England at a much lower cost.

In Scotland there are to be obtained, in several localities, good hydraulic limes. The celebrated Arden lime, from the neighbourhood of Glasgow, is well and favourably known, and good hydraulic limes are also to be found in Fifeshire and elsewhere in the north.

We have, we trust, referred at sufficient length to the lime division of our argument, and will now proceed to discuss the properties of the well-known

Roman Cement.

This cement, taking its name from its inventor and patentee, Parker, was first produced in 1796. The patent was for the conversion of the nodules or "septaria" found in the London and Kimmeridge clay formations. These flattened nodules of indurated calcareous clay furnish, when calcined at a moderate temperature and finely ground, cement of a dark-brown colour which, when freshly made, possesses remarkable setting energy, although never eventually attaining any high indurating value or hardness.

The name "Roman" was given to this cement by its inventor, probably under the impression that during the Roman occupation of England it had been used by these enterprising and successful builders. Some of the old Roman mortars in various parts of the country are of dark colour, which we now know is due to another and quite different cause.

The supply of stone from which this cement was and is still made is obtained by dredging on the shores of Kent and Essex, in the districts where the clays in which it was embedded exist. On the Yorkshire coast, as well as in other localities, these "septaria" are abundant, and in the several districts distinctive names arose to distinguish the cements from each other. The name "Roman," however, may be regarded as a generic one, for although "Mulgrave," "Sheppey," "Harwich," and other local names appear, they all mean Roman cement, from the "septaria" of the London and Kimmeridge clay formations.

In London, the original and most important centre of the Roman cement industry, the stone is obtained by dredging, and its cost, delivered alongside works in the river Thames, is about 8s. per ton. The trade, before the establishment of Portland cement, was an important and lucrative one, and many of our early successful engineering works, the Thames tunnel among them, could only have been accomplished by its help.

There are other sources from which argillaceous limestones are obtained, and from which are manufactured other varieties of Roman cement. The more important deposits are those of Calderwood, in Scotland, and Lyme Regis, in

Dorsetshire, which latter locality furnishes the greater part of the stone for making "Medina cement." In addition to those sources there are, in various parts of the coal districts in England, Roman cements of limited local reputation.

There is much difference in the value of the "septaria" dredged from the sea, and those which have been longest subjected to the storm or water action are the best, owing to the abrasion of the less crystallized portions. The Sheppey stone was especially noted for making the finest cement, from having been subjected to the wave action and thereby purified from the excess of clayey matter surrounding the more highly charged carbonate of lime forming the original nucleus of the "septaria."

The following analysis indicates the relative values of Harwich and Sheppey stones:—

HARWICH STONE.

Carbonic acid	22·75
Silica	9·37
Oxides of iron	17·75
Lime	29·25
Sulphate of soda	7·50
Water	3·88

SHEPPEY STONE.

Carbonic acid	31·00
Silica	18·00
Oxides of iron	5·25
Lime	30·20
Magnesia	0·20
Oxides of manganese	6·75

The Calderwood Roman cement is made from stone obtained in a stratified condition, and which exists in considerable regularity. The analysis of the stone is as under:—

Carbonic acid	34·30
Silica	8·90
Alumina	3·40
Protoxide of iron	10·20
Lime	30·24
Magnesia	6·76
Phosphates	2·64
Water and loss	3·64

In the millstone grit formation of Derbyshire there are several deposits of argillaceous limestone capable of conversion into a good Roman cement. One of these strata, of the following analysis, was used by the author in making Roman cement, which proved of excellent quality.

The analysis was :—

Lime	25·49
Carbonic acid	33·34
Magnesia	10·85
Alumina	8·81
Sulphur	1·30
Silica	11·62
Manganese	0·52
Water and organic substance	1·68

The operation of burning the above stone requires to be conducted with much nicety, for when overburnt it becomes vitrified, and in such condition the initial setting energy of the cement is greatly impaired.

Before the discovery of Portland cement, and indeed long after that time, Roman cement held a position in the market which for years could not be affected by its new and, now proved, more valuable competitor. The special properties of Roman cement commend it to some builders, who can only believe that a cement is best when it sets fastest. This is undoubtedly one of the peculiarities of Roman cement, but beyond that there are no advantages. It fails to attain any valuable degree of induration, and cannot be used advantageously with aggregates of any kind.

The following experiments afford some information as to its constructive value.

The cement used was freshly ground, and weighed 80 lb. per imperial bushel.

Four breakings gave the following results :—

	Age 7 days.	1 month.	6 months.	1 year.
Per square inch	73 lb.	120 lb.	170 lb.	191 lb.

Such results as the above are only attainable when the cement is freshly burnt and finely ground. The same cement kept for six months would not reach anything like the above figures.

During the time of Pasley's experiments Roman cement was in general demand for important engineering works. Its merits and advantages were fully recognized by the elder Brunel and the younger Stephenson, and by both of these celebrated engineers it was used, by the one in the Thames Tünnel, and by the other on the works of the London and Birmingham Railway. The cement was used neat in the arch of the Thames Tunnel, and the mortar consisted of one of cement to one of clean sharp river sand for the other parts of the work. In the Kilsby tunnel and the arches of the bridges of the Birmingham Railway the cement was used neat, for it was found that much better results were obtained by this method.

But for Roman cement it would have been difficult to accomplish these important engineering works. Portland cement had not, at the time of the execution of the Thames Tunnel, obtained any position, owing to the imperfect character of the quality then produced.

Pasley, in his search after an artificial hydraulic lime, made many interesting experiments with Roman cement to prove its capacity of receiving various mixtures of sands for mortar. The result of these experiments was unfavourable however, for they proved that it was dangerous to use more than one of the best quality of sand to one of cement.

Pasley's mode of experimenting with his artificial limes and cements was to join together bricks or stones with the mortar under examination. These bricks, &c., were 10 inches long, 4 inches wide, and 4 inches deep, and were submitted to tensile strain until ruptured or separated.

The following are some of the results thus obtained:—

PURE CEMENT,
At 11 days old, broke at 1241 lb.
„ 17 „ „ 1003 „
„ 17 „ „ 1031 „

1 CEMENT AND 1 SAND,
At 11 days old, broke at 205 lb.
„ 17 „ „ 257 „
„ 17 „ „ 313 „

The experiments clearly point out the marked decrease in value by even an admixture of one of sand.

Pasley also instituted experiments to compare the relative values of cements and their compounds with natural stones, with the following results (these blocks were of the same dimensions as the above):—

	lb.	
Portland stone	4004	average breaking.
Cornish granite	3841	„
Kentish ragstone	3773	„
Yorkshire landing	3642	„
Craigleith stone	2439	„
Bath stone	1408	„

Indicating that the least valuable of these stones exceeded in value that obtained from neat Roman cement.

In other experiments made by the same authority to prove the resistive capacity of these stones, he prepared prisms 4 inches long, 2 inches deep, and 2 inches wide. These were placed in iron stirrups attached to a beam, and on the centre of them a blunted knife edge was placed, from which a scale-board depended. The weights were added gradually and carefully, and the breakings may be regarded as satisfactory when we consider the oscillating and vibratory character of the scale-board arrangement.

The weights of the stones per cubic foot were as follows:—

No. 1. Kentish rag .. .. 165·69 lb.
„ 2. Yorkshire landing .. .. 147·67 „
„ 3. Cornish granite .. .. .. 172·24 „
„ 4. Portland .. .. 148·08 „
„ 5. Craigleith .. .. 144·77 „
„ 6. Bath .. .. 122·58 „

The following were the results of the experiments:—

No. 1. An average resistance of 4581 lb.
" 2. " " 2887 "
" 3. " " 2808 "
" 4. " " 2682 "
" 5. " " 1896 "
" 6. " " 666 "

In recording the results of these experiments Pasley did not think it necessary to remark on the various peculiarities of the stones, except in the case of the Craigleith, in reference to which he says:—

"The Craigleith stone was torn asunder horizontally, according to the grain."

The other stones exhibited various kinds of fracture at the points of rupture, but not in the peculiar manner of the Craigleith stone. This stone is obtained near Edinburgh, and has a high local reputation from its quality of durability and fineness of grain and colour. Its densest qualities withstand a crushing force of nearly 5000 lb. per square inch.

These interesting experiments were made with the view of comparing the value of artificial cements, and also to examine the constructive qualities of the artificial stone of Mr. Ranger, and other well-known building stones.

Pasley, in summing up the results, added the following quaint foot-note:—

"Mr. Ranger having at all times, with great civility, afforded me every information I requested relative to his proceedings, and he being the only person in England who makes artificial stone, from these circumstances combined, I really feel considerable reluctance in giving my opinion of it, which is much more unfavourable than when the first sheets of this treatise were sent to press. But as he is not merely a maker of artificial stone, but a builder of skill and reputation, I sincerely hope that if the former branch of his business should decline, the latter may prosper, so as to make him ample compensation."

In the above quotation there is indicated the kind and amiable gentleman sorry that his duty requires him to decide against the artificial stone on its merits, but anxious that its fabricator should not suffer. Mr. Ranger did not, unfortunately, meet with the reward wished for him by Pasley, for after embarking in large railway contracts he was obliged to succumb to the difficulties which their non-success occasioned.

We have thus endeavoured to describe what may be regarded as the leading matrices of this country, and we trust that the information in these pages will assist intending builders to measure at their real worth all other concoctions of cements, promising unheard-of and indescribable advantages. If the builder will always bear in mind that no mixtures of various cements are safe, he will simplify his business and be proof against the allurements of marvellous compounds. With a good and faultless Portland cement almost everything in the constructive direction is possible, provided the safeguards of accurate testing are strictly attended to, the character of which we have endeavoured to explain and illustrate.

Much chemical ingenuity has been exercised in pointing out a ready means of correcting the blunders of the manufacturers of cements by extinguishing through acid and other treatments their inherent dangers. To adopt such remedies is simply an encouragement of a false system of manufacture, and as we now well know that a good and indisputable quality of cement is of the easiest possible accomplishment, it is simply unwise to accept any but the best.

These panaceas are of two kinds, one for neutralizing any excess of lime in the cement; and the other, to correct the errors caused from its being incorporated in the mortar of the building.

We need not at any length refer to the cements from

various preparations of gypsum or sulphate of lime, as they are only used for indoor work, and therefore unsuitable for the concrete of which we are treating. Parian, Keene's, and Martin's cements and ordinary plaster of Paris are so well known that we will content ourselves by short references to the more important.

Gypsum or sulphate of lime differs essentially in character from carbonate of lime, the base of those cements we have just described. In the case of Portland cement and other carbonate of lime cements, the hardening is due to the action of the silicate of lime during the process of setting, while the cements prepared from gypsum derive their indurating value from the absorption of water.

Gypsum, if exposed to the air for however long a period when mixed with water, will not harden. If, however, it is exposed to a sufficiently high temperature so as to eliminate the water of crystallization, and afterwards ground to a fine powder, it will, when the required amount of water is applied, speedily and with considerable energy harden. This is due entirely to chemical reaction.

Keene's cement, a well-known and admirable material for internal decoration, is made by mixing the powdered gypsum with a solution of alum, and afterwards baking such compound to a temperature sufficiently high to eliminate the water of combination.

Parian cement is prepared by using a ley of borax, and in the same way drying the compound by heat as in preparing "Keene's."

Martin's cement differs in some degree from the treatment applied to Keene's and Parian. Carbonate of soda or carbonate of potassa as well as alum is used, and a higher degree of heat is applied for the elimination of the moisture.

Chemists have for many years directed their attention to the conversion of gypsum by the addition of chemical ingredients, and the above-named cements may be regarded as the commercial result of these investigations.

There can be no doubt of the constructive (decorative) value of these cements, and it has been proved by experiment that they possess some degree of value besides their special internal and decorative usefulness.

The following experiments were made with the view of comparing them with Portland and other cements, without reference to their constructive value, as their incapacity to resist the action of frost or external exposure prevented the chance of their being used for outside building purposes.

Age.	In water.	In air.
	lb. per sq. in.	lb. per sq. in.
KEENE'S CEMENT.		
7 days	242	243
14 „	216	261
21 „	224	257
1 month	217	261
2 months	201	288
3 „	226	320
PARIAN CEMENT.		
7 days	253	274
14 „	267	298
21 „	241	312
1 month	242	331
2 months	222	322
3 „	232	380

It will be observed that the chemical treatment to which these two cements were subjected results in their being able to resist the action of water, which plaster of Paris is not competent to do in the ordinary way.

CHAPTER IV.

AGGREGATES.

THE term aggregate is usually applied to those materials which are cemented together by any artificial binding agent, its literal meaning being a collection of particles or atoms into a mass, and is applicable to any kind or denomination of minerals. Its general scope therefore embraces a range wide enough to include all natural or artificial substances which may be used in the preparation of mortars or concretes.

In discussing the merits and peculiarities of those minerals which we consider it necessary to refer to, we hope that what is said will be of use also in guiding the concrete-maker in the selection of any local aggregate which the circumstances of his particular case may compel him to use.

The abundance of minerals and the wide field from which may be selected the required aggregates is almost unlimited in extent, and any desired quality is generally easily obtained. Every locality almost exhibits a variety of the most favourable materials, the physical and chemical characteristics of which it is our desire to describe in a general way.

The examination of the properties of an aggregate is as necessary as testing the quality of the matrix. There may not be the same dangers attending the use of an unsuitable aggregate, but its quality has a considerable influence on the resulting concrete. It would be unwise under any circumstances to sacrifice or nullify the value of a good cement by combining it with a worthless aggregate. No

influence imparted by the cement can make it better, while certain unsuitable qualities in an aggregate can injuriously affect the cement.

There are three leading and almost indispensable qualities which should guide us in the selection of an aggregate.

First. Density or hardness.
Second. Moderate porosity.
Third. Angularity of fracture.

Granite is entitled to our first consideration, not only from its value as an aid in concrete-making, but also from its importance in a geological sense, as the original rock in nature and by the débris of which succeeding geological formations have been more or less influenced. Its original distribution is confined to mountain ranges, though more recent deposits are to be found protruding through comparatively recent formations, but wherever it occurs the physical character and aspect of the country is bold and precipitous.

In all civilized ages man has freely availed himself of granite for building purposes, and also of its disaggregated deposits for pottery and similar wares. In its natural state, as we have already in our introductory remarks stated, it develops certain defects, generally due in some measure to the climate in which it is situated. Difficult, from its crystalline hardness, of economical conversion, its use has been in consequence limited to structures of great importance where cost was a secondary consideration. The dangers surrounding its careless or unguarded use are capable of ready avoidance when it is used as an aggregate in well-proportioned and carefully prepared concretes.

Rocks are either agglomerated or conglomerated, the former being those of the primary or igneous formation, while the latter are of diluvial or sedimentary origin. The component parts of the one are angular and of the other rounded; the agglomerates being produced by volcanic action, by which they were originally expelled from internal sources, and the conglomerates in the condition as deposited after being more or less rounded by the action of water or ice.

We will now proceed to describe the character of

Granite.

The most primitive rock, its name being derived from the Latin signifying a grain; of igneous origin and highly crystalline in texture; and it differs essentially from all other eruptive rocks in being comparatively free from vesicular or porous characteristics.

Granite is practically a silicated rock, the silica ranging between 65 and 80 per cent. The minerals of which it is composed are easily distinguished, and according to their composition develop more or less beautiful effects. The peculiar position of the felspar and mica is evidently due to their having been imbedded in the quartz while the erupted mass was in a viscous state. Sorby and other microscopists have demonstrated that the quartz of granite contains minute cells, having within their cavities more or less water, inferring therefrom that the rock was deposited under great pressure and that steam exerted a powerful influence in the operation.

According to Ansted, granite contains in a state of nature $0 \cdot 8$ per cent. of water, and it is still capable of further absorption to the extent of $0 \cdot 2$ per cent. more. This basis of calculation, made from accurate observation, indicates a total absorptive capacity of $3\frac{1}{2}$ gallons to the cubic yard, the weight of which is about 2 tons. This calculation cannot, however, be regarded as a fixed standard, for granites vary greatly in character, but we may assume that these figures represent the minimum capacity of absorption.

None of the ingredients of granite exhibit the slightest trace of being rounded or abraded, but are invariably angular, and partake of the crystalline character in every sense of that term. Although, generally speaking, this rock occurs in mountain ranges, it is sometimes found imbedded in sedimentary deposits, and in this condition its character is changed by the action of the heat by which it was extruded. In these deposits its texture is irregular and indefinite in form, resulting in a variety of classifications, which from a concrete point of view may be regarded as of little value.

The leading and more distinct crystals of granite are quartz, felspar, and mica, and according as they predominate is its quality influenced. Quartz is practically pure silica, unless where its colour is changed by the presence of some metallic oxide. Felspar is made up of silica and alumina, with variable proportions of alkaline ingredients, those having a more or less injurious action on the mass according to their extent and value. Mica is a silicate of alumina, being composed of finely laminated plates or scales, having a metallic lustre or sheen.

The analysis of felspar and mica are as under :—

	Felspar.	Mica.
Silica	66·75	43·00
Alumina	17·50	34·25
Lime	1·25	..
Potash	12·00	8·75
Oxide of iron	0·75	4·50
Oxide of manganese	..	0·50

Granite and its various compounds and classifications have been a fertile source of discussion amongst geologists and mineralogists. In a rock bearing indubitably the clearest indication of an original molten state, it is almost superfluous to speculate on its exact mineral constitution or to dogmatize with much nicety of description on its varied and changing character. The violence of the initial force by which it was originally expelled so far exceeds anything within the scope of our intelligence, that we may judiciously decline any attempt at its estimation.

In a chemico-mineralogical sense there is much fascination surrounding the subject of granite compounds and their blending with basalt and other volcanic rocks, but the niceties by which they require to be distinguished are not of much value in the discussion of artificial concrete mixtures. We will confine our observations to the distinctive qualities of the native granites of England, Ireland, and Scotland, for within the range in which they occur the concrete-maker will find ample stores for his purpose.

The extent of granite deposit in England is comparatively limited, although ranging from Cornwall to Westmoreland. The Dartmoor granites have been largely employed in constructing works in London, and those from Devonshire as well. In the Channel Islands and the Isle of Man extensive quarries have existed for a long time. The more central deposits are not very extensive, but those of Leicestershire, from their nearness to London, are much used for road and street making purposes. Jersey and Guernsey also largely contribute their hard and durable granites for the same object.

Scotland has considerable granite stores of almost every variety, and the high reputation they have obtained for both building and paving purposes is the best evidence of their suitability. Aberdeenshire, Kirkcudbrightshire, Argyleshire, and the Islands of Arran and Mull may be regarded as the more important points from which, for a lengthened period, granites have been quarried and used for a variety of constructive purposes at home and abroad.

Ireland is bountifully supplied with granite of varied

quality found in nearly every county. From Donegal in the north to Wexford in the south it prevails within moderate distances of the sea shore, and few localities are beyond the convenient range of its supply. Donegal, Down, Armagh, in the north; Wicklow and Waterford, in the south; and Mayo and Galway, in the west, are the more important districts in which it is found. The granite of Fermanagh, from which the famous Beleek porcelain is produced, is purged, by an ingenious preliminary treatment, of the metallic oxides with which it is associated.

Granite has been the favourite aggregate used for nearly ten years in the manufacture of the now well-established Victoria stone paving, many miles of which have been laid in the footways of London and elsewhere. In the earlier examples of that pavement Jersey and Guernsey granite chippings were used, and with much success, but owing, however, to the too coarse character of the aggregate, used probably from a desire to secure as much of the granite surface as possible, the pavement was not so good, and less comfortable to the walker than the much improved recent manufacture, which is made of Leicestershire granite chippings of small size.

The variable character of granite is good evidence of the difference in its chemical value which influences, in a high degree, its capacity of wear under heavy traffic, either of carriage or foot.

Experiments were made on the wearable properties of different granites under perhaps the most severe test it was possible to institute. On the tramway in the Commercial Road, London, over which heavy traffic to the Docks passes, various blocks of granite were laid, and after seventeen months' wear, the following results were obtained.

The blocks were 18 inches wide and 1 foot deep, being laid a sufficient distance apart to take the wheels of the heavy wagons and carts. Between the blocks the street was paved with the ordinary granite sets.

Name of Granite.	Super. area in Feet.	Original Weight.	Loss of Weight by Wear.	Loss per Sup. Foot.	Relative Losses.
		cwt. qr. lb.	lb.	lb.	
Guernsey	4·734	7 1 13	4·50	0·951	1·000
Herm	5·250	7 3 24	5·50	1·048	1·102
Budle	6·336	9 0 16	7·75	1·223	1·286
Peterhead (blue)	3·484	4 1 7	6·25	1·795	1·887
Heyton	4·313	6 0 15	8·25	1·915	2·014
Aberdeen (red)	5·375	7 2 11	11·50	2·139	2·249
Dartmoor	4·500	6 2 25	12·50	2·778	2·921
Aberdeen (blue)	4·823	6 2 16	14·75	3·058	3·216

A similar set of experiments on the same tramway, spread over a period of four months, produced somewhat similar results; the relative losses under that examination being as follows:—

Guernsey, 1000; Budle, 1040; Herm, 1156; Peterhead, (blue) 1715; Aberdeen (red), 2413; Aberdeen (blue), 2821.

The Herm granite (sienite) was a highly crystallized intermixture of felspar, quartz, and hornblende, with a small quantity of black mica. The Guernsey granite, of the dark and hardest sort, is generally free from mica, which is replaced by hornblende, and in that condition is of great durability. The Budle stone is a species of whin or basalt from Northumberland.

These experiments, conducted under the most trying conditions, indicate the relative values of the selected specimens and their capacity of resisting the wearing action of traversing loaded wheels. The friction under such circumstances must have been considerable, while the exposed surfaces were prejudicially acted on by the water which in the line of blocks would readily collect and dissolve the alkalies of the cementing agent. The above experiments were supplemented by others, to ascertain the value of some of these granites' resistance to compression. The cubes were of irregular sizes, and the superficial area exposed to pressure was as under:—

	Superficial Inches.	Weight.	Per Square Inch.	
			Fractured.	Crushed.
		lb. oz.	tons.	tons.
Herm	16	6 6	4·77	6·64
Aberdeen (blue)	17½	5 0¾	4·13	4·64
Heyton	16	4 7½	3·94	6·19
Dartmoor	16	4 9	3·52	5·84
Peterhead (blue)	18	5 3¾	2·86	4·36
„ (red)..	18	5 0½	2·88	4·88

On comparing these two tables, it will be seen that Herm is the densest and stood the greatest pressure. It will be well to examine at this point the value of some of the natural and artificial cements, both in a neat and mortar condition, so as to compare them with the above results.

In some experiments made during 1866, with a variety of building stones and cements, among the results obtained were the following:—

A block of neat Portland cement, 3″ × 3″ × 1½″, three years old (the first seven days of that period it had been immersed in water), was fractured with a pressure of 8·4 tons, and crushed at 15 tons. The surface pressed upon was 4·5 square inches, giving an average pressure of 5287 lb. per square inch. In the course of the same experiments a block of granite (called granite stone without any description of its origin or peculiarities) was fractured under a pressure of 11·4 tons per square inch, and crushed at 19·4 tons or on the square inch 7865 lb.

The last experiments were made for the Metropolitan Board of Works, and the first by Bramah and Sons.

Granite in some of its deposits admits of being quarried in large blocks, the obelisks of ancient Egypt affording ample evidence of the facility with which great stones were obtained. Wonderful as were the results in the excavation of ancient monoliths, they are much eclipsed by modern ingenuity. A recent example of granite quarrying in the United States of America is worthy of notice. The account of this specimen of granite is as follows:—

"There was recently quarried without the use of powder, at the Barre granite quarries, for the use of the Olton granite works, of Rutland (Vermont, U.S.), a block weighing about 618 tons, being 40 feet long, 17 feet high, and 10 feet thick. This immense stone is said to be perfect in every respect, and is believed to be the largest block of granite ever quarried in the State."

The cubic contents of this block is according to the above dimensions 6800 feet, which if divided by the weight gives $203\frac{1}{2}$ lb. to the cubic foot.

In some careful experiments made with reference to the absorptive capacity of Aberdeen granite, having a specific gravity of 2·708, it was found that a specimen weighing 500 grains, reduced to coarse fragments, absorbed by weight 2·00 and by bulk 5·416 of water. Granite may be regarded as the least absorptive of all building stones.

Kentish Ragstone.

A well-known and highly prized geological deposit, principally obtained from the neighbourhood of Maidstone, near London. It is also quarried in other localities in Kent, near Hythe and Folkstone, as well as at various points on the coast. This stone has been for some centuries extensively used in church building, and from its great durability and pleasing appearance, has always been a favourite with ecclesiastical architects. The best beds of ragstone should only be used for building purposes, but the others of a less reliable quality, are well suited for a concrete aggregate. Many of the roads in Kent are macadamized with the rubble from the ragstone quarries, but it is only suited for roads on which the traffic is light, and cannot be compared to the more durable Guernsey granites used in the streets and roads of London and its neighbourhood.

Its analysis is :—

Carbonate of lime, with a little magnesia	92·6
Earthy matter	6·5
Oxide of iron	0·5
Carbonaceous matter	0·4
	100·0

The ragstone is a calcareous sandstone, varying in colour from a bluish grey to a brown or light yellow, the former being considered the hardest and best suited for building purposes. The stone should not be used for internal work, as it develops with atmospheric changes globular moisture of an unsightly character. Mediæval builders, who used the ragstone extensively, avoided the somewhat objectionable method of *overdressing* the external surfaces, owing in some measure to the cost of labour, and the tendency to early decay when highly polished. Hence in nearly all the ancient structures where ragstone was used, as well as in modern buildings, the stones are laid in random courses and roughly dressed. The mortar best suited for bedding ragstone is a mixture of *ragstone* lime with sharp sand. Lime from common chalk, or even grey chalk, should be carefully avoided.

Ragstone was found by Pasley, in his celebrated experiments, to be equal in cohesive strength to some of the best building stones. Various cubes were made of each of the materials under examination, $9'' \times 4'' \times 4''$, and four holes bored (two on each side) by which the clamps could be fixed. The weight was suspended by a hook, on which the scale-board and weights were placed. The results were as follows:—

	Name of Stone.	Weight per Cube Foot.	Weight in lb. that tore them asunder.		Average Fracturing Weight in lb.
			No. 1.	No. 2.	
		lb.	lb.	lb.	lb.
1	Portland	148·08	2964	5045	4004
2	Cornish granite	172·24	3741	3941	3841
3	Kentish rag	165·69	3549	3997	3773
4	York landing	147·67	3597	3688	3642
5	Craigleith	144·77	2103	2775	2439
6	Bath	122·58	1549	1268	1408
7	Run chalk	94·99	323	623	473

These experiments were not conducted under the most favourable circumstances, and at a time when less attention was given than at present to the important question of quality of building materials, nevertheless they enable us to form a good idea of the relative value of the well-known stones dealt with.

For use in the neighbourhood of London, Kentish ragstone will be found an economical and desirable aggregate for first-class concrete.

The name, Kentish ragstone, is purely provincial, and prevails in the localities (almost limited to Kent and Sussex) where the lower greensand occurs, with which it forms also a member of the lower cretaceous formation. The depth of this deposit averages from 60 to 80 feet and the separate beds of which it is composed are variable in character and quality.

There are other ragstones in different districts of England, such as the coral rag of the middle oolitic formation, and the crags of Norfolk and Suffolk. Good aggregates can be selected from these deposits, and also from the siliceous stone found in the Bagshot sands.

The characteristic qualities of the ragstones are the roughness of their texture and valuable capacity of adhering readily and with great tenacity to the cement.

Portland Stone.

A well-known and valuable deposit of carbonate of lime, which has been used in London for some of the most important buildings. St. Paul's is built of Portland stone, and its tolerably good condition proves the value of this material for important works. Many failures have arisen in using this stone owing to the carelessness of the builders in neglecting the proper selection in the quarry, and unfortunately also from a desire to use the softest qualities, from their being more easily tooled.

Like almost every geological formation each bed varies in chemical and mechanical value, and unless some care is bestowed in selection, improper qualities must be used. It is not so important (the question of selection), from the concrete-maker's point of view, as the imperfections which would prove fatal to the mason are of comparative insignificance in an aggregate, the cementing agent exercising its beneficial influence in the correction of moderate faults. The best qualities of Portland stone are those having a greyish or bluish-white tinge, and which are in their chemical composition free from oxide of iron or other metallic impurities. It was the colour of the first-class Portland cement, resembling the best sorts of Portland stone, that led to the adoption of the name of that well-known and valuable cement by its inventor, Aspdin.

The analysis of a good Portland stone is as under:—

Carbonate of lime	95·16
„ of magnesia	1·20
Iron and alumina	0·50
Silica	1·20
Water and loss	1·94
	100·00

When in a dry state, Portland stone absorbs 8·86 per cent. of its own weight of water.

Portland stone is the hardest of the oolitic limestones, being better able to resist the injurious action of impure or chemically impregnated atmospheres than the softer sorts.

A careful examination of the Portland stone used in some of the old buildings indicates an amount of durability which even the dissolving action of rain fails to degrade to any serious extent. The oolitic limestone covers a large area of the south and middle of England, from which may be obtained unlimited supplies of a good average quality of aggregate.

Carboniferous or Mountain Limestone.

The hard and sometimes crystalline character of this abundant geological formation, secures in many districts the obtainment of a sound and very durable aggregate. It is now, and has been for a number of years, used as a mixture with coal-tar for footpaths, and many of the roads and railway-station platforms in Lancashire and Yorkshire have been formed from such a combination. The best examples of this kind of cheap and comfortable footway are produced from a carefully washed aggregate so as to eliminate the dust of fracture from the surfaces of the broken stone. It must be observed that the carboniferous limestone produces a large percentage of dust when crushed by a machine such as Blake's stone breaker, and the concrete-maker should be careful that, before mixing such aggregates, the dust is washed off, otherwise the cement will be unable to exert its most profitable influence in the concrete preparation.

There are many varieties of the mountain limestone, and the purest carbonates are generally the whitest in colour; indeed in the Buxton (Derbyshire) district an almost pure carbonate of lime is obtained, such as the following analysis shows:—

Carbonate of lime	97·13
„ of magnesia	0·18
Silica	0·69
Alumina and iron	0·40
Alkalies	0·16
Silicic acid	0·10
Sulphuric acid	0·24

In the variety of deposits in and around Buxton, all shades, from white to an almost black quality, are found, and many good marbles are quarried in the neighbourhood of Bakewell and Ashbourne.

The Magnesian Limestone.

This deposit occurs between the carboniferous and triassic beds, occupying a wide surface in the Midland and Northern counties of England. The variable character of this stone renders its careful selection absolutely necessary when required for external building purposes. The chemical composition of magnesian limestone is uncertain, for from the pure carbonate (magnesite) to the red Mansfield calcareous sandstone, a very considerable difference in its constituents occurs. The various tints of the Mansfield sandstones do not, however, show a very marked divergence in their analyses except in the quantity of iron and alumina, from which ingredient the colouring matter doubtlessly proceeds.

The following analyses of three specimens of this calcareous sandstone give:—

	White.	Roseate.	Brown.
Silica	51·40	49·40	49·40
Carbonate of lime	26·50	26·50	26·50
„ of magnesia	17·98	16·10	16·10
Iron, alumina	1·32	3·20	3·20
Water and loss	2·80	4·80	4·80
	100·00	100·00	100·00

Some of the magnesian limestones are highly crystalline in texture, approaching such quality when the proportions of the carbonates of lime and magnesia are nearly equal. There are no deposits of limestone where so much diversity in value takes place, and in consequence many errors have arisen in the selection of this stone for building purposes. In the comparatively pure atmosphere where it is met with, many durable examples of ancient buildings exist, but when used in large cities surrounded with vitiated atmospheres it does not always maintain its character for durability. Many failures through this cause have arisen in large buildings of an important character, notably that of the Houses of Parliament.

From this source may be obtained good aggregates; and when proper care is bestowed in their selection, using only those of a crystalline character, satisfactory concrete results may be attained.

Geologists and mineralogists have sometimes endeavoured to establish a line of division between the many varieties of these rocks, and it has been suggested that all those containing in their analyses more than 23 per cent. of carbonate of magnesia should be classed as dolomites, the pure or properly balanced magnesite having $45 \cdot 7$ per cent. of carbonate of magnesia, and $54 \cdot 3$ per cent. carbonate of lime.

Sandstones.

These deposits are numerous and, especially in the north of England and Scotland, abound in almost unlimited quantities. The term Freestone is sometimes locally applied to these rocks, probably from their being capable of easy and cheap conversion into the most elaborate forms. Unlike the cementing paste of the limestones, their cohesion is secured by a siliceous cement of more or less tenacity. It is, however, susceptible of comparatively easy degradation in some of the more porous qualities, such as the red sandstones of Cheshire and Lancashire. The quality of these deposits depends pretty much on the source from which they were derived. There are micaceous, calcareous, and felspathic sandstones, but nearly all are composed of rounded particles abraded from pre-existing geological formations. Some of them are agglomerate and others conglomerate, according to the size and quality of their particles; the former being represented by angular fragments, and the latter by the water-worn and rounded pebbles of diluvial débris.

Many of the sandstones are not suitable for concrete-making, because of their friable character; and the range of their hardness is a wide one, from the soft surface stone of Tunbridge Wells to the Craigleith (Edinburgh) highly indurated building stone.

The analysis of Craigleith stone is as follows :—

Silica	98·3
Carbonate of lime	1·1
Iron, alumina	0·6
	100·00

Some of the coarse sandstones of the millstone grit series have large crystals of quartz in their composition, and offer many advantages for concrete purposes.

Basaltic Rocks

Are readily accessible—especially in the northern counties, and in Scotland, and Ireland—at a great many points, and where the limestone or other suitable rocks are absent may be used as an excellent substitute. The extreme hardness of these igneous rocks renders their reduction by machinery imperative, but they will well repay any extra trouble and expense in their conversion for aggregates in a concrete mixture. There is great difference in the density of these rocks, some being porous and others extremely hard, so much so, indeed, as to preclude an examination of their composition by the naked eye. They are, however, generally speaking, susceptible of profitable employment by the concrete-maker. The fracture of basalt is highly angular, and therefore will secure the best quality of brecciated concrete, while the dust eliminated during the process of crushing is less than in the limestones.

In the above general remarks on the more prominent rocks we have endeavoured to show with what facility the most favourable aggregates may be obtained in almost every district of Great Britain and Ireland. We would, however, recommend a careful rejection of rocks which have been quarried any length of time, for the hardest in a very few years become dirty, being covered with lichen growth which it is almost impossible to thoroughly eliminate by ordinary means.

We will now proceed to consider other sources from which aggregates are derived, the best known being the gravels and sands obtained from river beds or inland pits.

Gravels.

Derived from the worn débris of almost every rock, but more especially by the deposit of river-borne flints. The rounded character of these deposits indicates the source and manner of their formation, but unfortunately they are generally intermixed with a large percentage of clay, loam, and carbonaceous matter, which for concrete purposes is very objectionable. In addition to these disadvantages, the form of the pebbles and smaller particles is always spherical or rounded.

The first process which should be performed is thorough cleansing the mass from all impurities; and the second, the careful fracturing of the whole, so as to reduce the rounded forms to angular ones. It is not always possible to accomplish this thoroughly, but any reduction of the circular form of the gravel not only increases the surfaces, but also reduces the sizes of the aggregates, adding thereby to the value and quality of the concrete.

Until comparatively recent times it was supposed that gravel, dug from the pit or dredged from the river, was the only available aggregate for concrete purposes. This practice was doubtless adopted from the difficulty of reducing other materials, which could only be accomplished by hand labour. Modern machinery, of special character devised for the purposes of stone crushing or breaking, has quite revolutionized the whole system of concrete-making, and, while reducing the cost, adds to the value of the results obtained.

Sands.

In the increasing desire to reach a perfect concrete, finer aggregates are gradually being adopted; and when a more perfect system of mixing the aggregates and matrices prevails, coarser qualities of siliceous sands will receive more attention. There are few deposits of sand so pure, however, as to render preliminary cleansing or washing unnecessary. Some of our best white sands, such as those used for glass-making, are quite clean enough for concrete work, and any soft ingredients with which they may be mixed, so long as they are not of a loamy character, will not prejudicially interfere with the best action of the cementing agent.

When an intelligent desire arises to obtain in the locality of operation the best available aggregates, many deposits hitherto regarded as useless will be examined and tested by the light of scientific examination, saving thereby a large amount of unnecessary expense.

Flints.

In some districts, more especially in southern England, these hard and durable minerals may be used with advantage. Weather-worn flints are, however, to be carefully avoided, for even their hard surfaces are liable to vegetable coatings. On a freshly fractured flint there is not much chance of vegetable growth of even the lowest type, but when it is exposed for a length of time to the atmosphere it becomes abraded and affords a foothold for algæ. These, however minute their organisms, further favour by the moisture which they absorb a more advanced lichen growth of a low class. On the flints of the Sussex Downs there are found at least four lichens, but this vegetable growth is not so wonderful when we consider that an old and partially decomposed flint flake will absorb nearly one-twelfth of its own weight of water. These observations are made with the view of showing that even the hardest siliceous material is liable, like everything else in nature, to the solvent action of rainwater, and, being so, favours the growth of vegetable matter, which the concrete-maker must carefully guard against.

Having thus called attention generally to the natural sources of aggregates, we will now consider those artificial preparations more or less used for concrete-making.

Burnt Shales.

In the coal-mining districts the abundance of suitable shales secures an aggregate of valuable character at a low cost. The numerous calcined shale heaps so prevalent in the northern coal-fields indicate that the quantity ready at hand is almost unlimited. For concrete-making many of these older heaps are not, perhaps, good enough, owing to the insufficient quantity of fuel used in their calcination, and therefore a tendency of the exposed portions to disintegration. When freshly burnt, some of the purest shales make the best aggregate, not even excepting the granites and basalts; but in those localities, where no attention except that of complete burning is exercised, aluminous shales are indiscriminately mixed, resulting in a flaky product, which must be carefully washed out before being used.

In Lancashire much use has been made of these calcined shale heaps, and more especially in the neighbourhood of Manchester good concrete floors have been made through their agency as well as ordinary footpaths.

Burnt Ballast.

On the clay districts in and around the northern suburbs of London great quantities of ballast are used for road and railway purposes. The preparation of this material is of the simplest kind, and the expense of labour and fuel is comparatively trifling, the cost per cubic yard being in most places under 2s. The red colour is objectionable for some purposes, but the bright red products indicate an insufficient calcination, which is best when the ballast is of a dark purple. The departure from the pale to the dark colour truly indicates the quality of the material, and that of a light red should be carefully rejected.

Broken Pottery,

Especially from vitrified ware, is perhaps the best aggregate yet employed. In some experiments made to test and prove the best kind of materials to employ for making concrete, broken pottery was found to be equal if not superior to the best gravels and other compounds. The source from which it must be procured is necessarily limited, and concrete-making from such a material must be confined to the localities where stoneware pipes and other inexpensive kiln products are produced. In the Poole district, in Dorsetshire, where the best qualities of potter's clay abound, and where, owing to the concentration of the pottery trade, a large quantity of broken ware is readily obtained, the concrete pipe industry has been introduced with much advantage. As we shall refer more particularly in Chapter IX. to this new and interesting outcome of concrete-making, we will not now further allude to it.

Iron and other Slags.

The numerous attempts to utilize this abundant waste have been attended with varying success, and although some satisfactory results have been reached through the direct treatment of the incandescent mass as it leaves the furnace, no very marked progress has been made in what may be called the "concrete proper" direction. At Middlesborough-on-Tees, extensive operations have been for some years carried on in the manufacture of slag bricks, but owing probably to the imperfect cementing agent used, the quality of those bricks which have come under our observation has not been very good in character.

In Germany the manufacture of bricks from slag has been going on for upwards of twelve years, and at Osnabruck more particularly the industry has been of a progressive character. In 1866, about half a million of bricks were made, and in 1875 upwards of six millions and a half.

The cost of producing these bricks is 16s. per 1000, the lime mixed with the granulated slag costing 3s. 4d. of that amount. In the early experiments, the machinery for compressing the bricks was found to suffer much damage from the hard and gritty nature of the slag; this difficulty has been completely overcome by the machines invented by Mr. Lurman.

Many structures have been built in Germany of these slag bricks, and the testimonials received from architects and others interested testify to the advantages derived from their use.

In two sets of experiments made to test the compressive strength of slag bricks made in two different districts, the following results were obtained.

Osnabruck Bricks.

Average of ten bricks, 1871 make, were crushed with a pressure of 1487·72 lb. per square inch.

Average of ten bricks, 1873 make, five months old, were crushed with a pressure of 1571·63 lb. per square inch.

Siegen Bricks.

Bricks made from granulated slag mixed with naturally disintegrated dust broke under a pressure of 2648·30 lb. per square inch.

Bricks of granulated slag only were crushed with a pressure of 2109·26 lb. per square inch.

These experiments were made for the purpose of comparing their strength with local red and pale hard and soft burnt clay bricks, and in every case slag bricks exceeded them in compressive value.

Experiments were made in 1874 to test the capacity of these slag bricks to resist the action of heat, and it was found that they remained perfect when red hot, thus proving their suitability for chimneys and furnace work.

In the investigations connected with this industry, a series of experiments were made under the direction of Professor Pettenkofer, of Munich, and the results are so generally interesting, that we quote them below.

The primary object of the experiments was to test the permeability of the slag bricks to air and water. The pressure of air applied was 0·0108 kilo. per square centimetre, and it was found that through every square metre of the following materials the quantity of air in litres passed through per minute was :—

	Dry.	Soaked.
Concrete	15·5	·0
Green sandstone (Upper Bavaria)	7·8	1·4
,, Switzerland	7·1	2·1
Tufaceous limestone	478·8	233·2
Hair mortar	54·4	3·9
Portland cement	8·2	·0
Slag brick (Osnabruck, 1871)	93·0	53·8
,, ,,	105·0	15·4
,, ,, 1873	110·4	10·2
,, (Haardt, 1873)	455·8	41·0
,, (English, 1876)	158·0	1·1
Clay bricks (pale) Osnabruck	23·3	5·1
,, slightly burnt, hand made, Munich	19·3	7·8
,, hard burnt, hand made, Munich	9·6	1·5
,, machine made, Munich	7·9	1·7

By the same experiments it was found that clay bricks were completely soaked with water in from 9 to 12 hours, whereas it took from 55 to 190 hours to saturate a slag brick.

An hospital at George-Marien-Hutte was built during 1870–71 of these slag bricks. It was opened in 1872, and in June, 1876, the medical superintendent testified to its salubrity. There is accommodation within its walls for thirty-two patients, twenty-eight being comfortably placed in four spacious wards, and four in separate small rooms. The walls are hollow, having an air space of $2\frac{1}{4}$ inches, and so successful is the result that no smell of hospital-air is observable even in the rooms where offensive suppurating wounds are treated. All these advantages are derived from "the extremely advantageous porosity of the slag bricks."

This new and valuable feature of porosity is entitled to some attention in this country, as we usually disregard the advantages to be obtained in this direction, and too frequently aim at an air-tight house.

We need not further refer to the various less important aggregates, except to caution intended concrete-makers against using broken bricks, old plaster, and any rubbish which some writers on concrete recommend for house-building purposes. There is just as much care required in the selection and treatment of the aggregate as the matrix, and unless this is carefully borne in mind there is not any chance of much success in the concrete mixture.

CHAPTER V.

MACHINERY FOR REDUCING THE AGGREGATES.

IN this chapter we shall give some particulars of the best-known machines used for breaking and crushing stones, as well as information regarding others of a modern and novel type.

The necessity of careful reduction of rocks for concrete purposes is now receiving much attention, and the concrete-maker should have the most reliable information on this all-important branch of his industry.

A knowledge of matrices and aggregates, their qualities and peculiarities, is indispensable to the intending concrete-maker, and correct information regarding the machines by which they can be tested and prepared is also of the utmost importance.

There are a great variety of reducing machines, but the leading peculiarities in their mode of dealing with materials such as are required in concrete-making may be thus classified:—

1st. Crushers (rollers).
2nd. Squeezers (Blake's stone breaker).
3rd. Impactors (Patterson's Elephant stamps), and Sholl's pneumatic stamps.
4th. Abraders or Triturators (millstones and cognate machines).

We will examine and describe them in the above order.

First Division.
Crushers, Rollers, &c.

The oldest perhaps of all the machines used in the reduction of obdurate materials. Their use is now very limited, owing to the great waste of power and the large percentage of wear and tear, as well as irregularity of results obtained through their agency. In those districts where the old arrangements of cast-iron rollers continue to prevail, their use may be regarded as a matter of convenience or necessity rather than evidence of their fitness for stone-breaking purposes; generally erected at a comparatively high cost, and their structural belongings also of an unavoidably substantial and expensive character, their abandonment would in most cases involve too serious a loss to warrant the adoption of less costly, although more efficient, machines. A very powerful set of rollers of this class is still at work near Buxton, in Derbyshire, where the "Buxton Lime Company" reduce the limestone of that district from the sizes supplied by the quarrymen to the smallest pieces required for the alkali works and the pavement layer. There are three sets of rollers placed vertically one above the other, and in the upper set, near the level of the quarries, the stones are tipped, which descend when crushed to the second and third rollers, and thus finishing by a continuous discharge of the broken material of the desired quality into the trucks placed on the rails underneath. An examination of these operations will fully illustrate the quantity of almost impalpable dust evolved in the operation, and which we have elsewhere referred to.

Small steel and chilled-iron rollers, in single pairs (both serrated and smooth), of various diameters, are still used for the reduction of minerals for a variety of purposes, but the high speed at which they must revolve to obtain anything like profitable results renders this class of crusher most objectionable and costly.

Edge runners have been for many years a favourite means of crushing stones, but their use in the present day can only be tolerated where a true knowledge of the principle of the reduction of materials is imperfectly understood. Fortunately they are now pretty generally superseded by more suitable machines, although this class of crusher still exists in a modified and perhaps less objectionable form in the now fashionable mortar mill and clay mixer.

SECOND DIVISION.

Squeezers, Blake Machine.

The Blake machine, the first in order of this class, has obtained a wide reputation for its simplicity of construction and effective operation. Originally introduced in the United States of America, and brought into England by Marsden, of Leeds, it became by his enterprise and perseverance an established and, at the present time, the best-known machine for the reduction of all kinds of minerals used in a variety of industries throughout the world. Many subordinate but useful improvements have from time to time been made in the original Blake's stone breaker, and among others the division of the movable jaw into two parts. Fig. 7 is a representation of Hall's multiple-action stone breaker, which is now much used for various purposes in cement works and for concrete-making. Those which have come under the writer's observation were performing the work allotted to them satisfactorily.

The manufacturers of this machine, in referring to the principle of the Hall's "Multiple Action," say:—

"To illustrate this principle in a familiar way, let us take the case of a man having a given weight to carry 100 yards, say 1 cwt., in a given time; it is clear he would be able to carry one-half of this weight (56 lb.) for 200 yards with much less strain *for each yard travelled* than the former weight for half the distance, although the total work done is the same in both cases. The operator in the case of our machine is the connecting rods, which collectively travel through twice the space of the single connecting rod of the

Fig. 7.

HALL'S MULTIPLE-ACTION STONE BREAKER.

Blake machine, in the same time for the same amount of work done. It follows therefore that we are enabled to make machines constructed on this improved principle, *lighter*, without in the least impairing their strength, but considerably adding to their efficiency over the Blake machine."

It is also claimed for this machine that it works with less vibration, and testimonials have been given that it performs a much larger amount of work than Marsden's Blake stone-breaker of equal size.

The special arrangements for the renewal and reparation of the moving jaws will tend to reduce the high cost of wear and tear usually attendant on the use of this class of stone breaker.

Fig. 8 is an illustration of a combined multiple Blake and steel rollers intended to perform in a direct manner the operation of reduction to its furthest limit. We do not altogether approve of machines so arranged, unless their

FIG. 8.

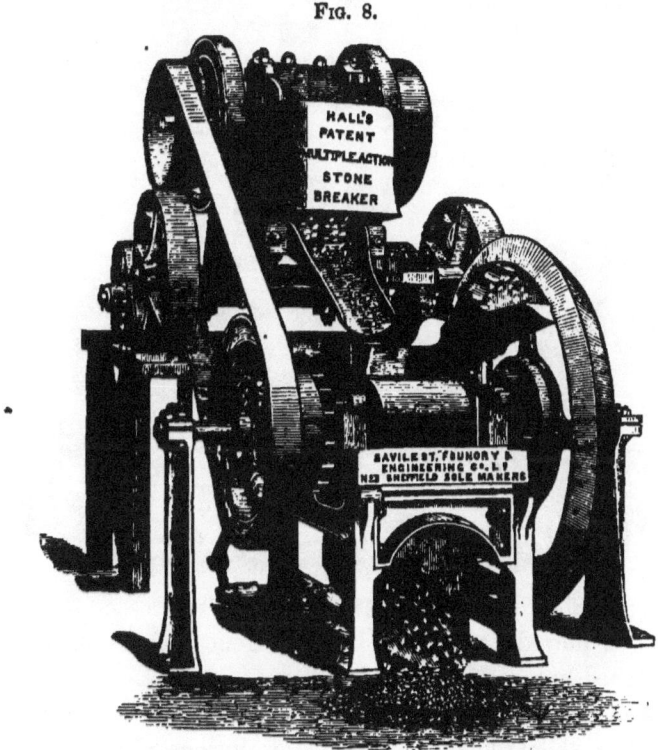

HALL'S COMBINED BREAKER AND CRUSHING ROLLERS.

relative tasks can be accurately apportioned. For some purposes, where for instance a sufficient quantity of fine material is desirable, this double crusher will be found both convenient and inexpensive.

In some operations it would be an advantage to have

K

machines of a portable character in combination with the necessary driving power. For this purpose the "self-moving engine and stone breaker" offers great facilities, and in harbour or other similar works it would be found useful.

Fig. 9 is an illustration of this combination, with boiler, engine, and stone breaker arranged in a compact form. It

FIG. 9.

PORTABLE ENGINE AND STONE BREAKER.

can be moved as desired by the agency of the specially designed link-motion reversing gear.

The normal speed at which these multiple-action stone breakers should be driven is 250 revolutions of the driving shaft per minute.

Fig. 10 is another improved Blake, called by its inventor and maker, Broadbent, "The patent drawback-motion stone breaker."

In the ordinary Blake's stone breaker the movable jaw when forced forward compresses a spiral steel spring fixed in indiarubber, and when not in accurate condition acts as a brake, involving thereby considerable loss of power. The

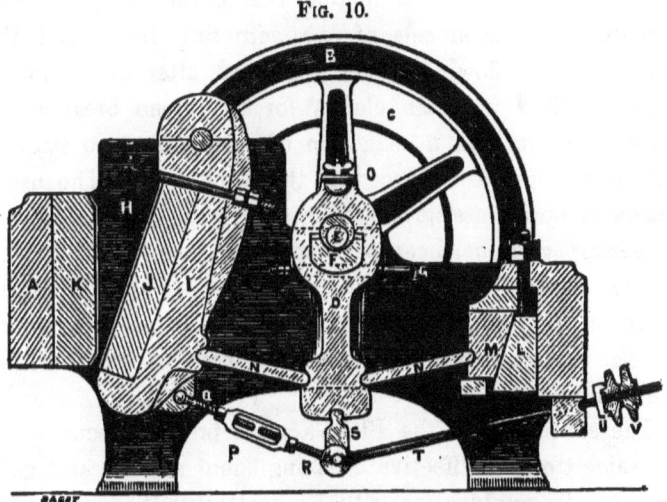

Fig. 10.

BROADBENT'S DRAWBACK-MOTION STONE BREAKER.

application of the drawback action is intended to remove this objection, and by its addition to the machine provide an accurate mechanical arrangement, whereby an uninterrupted and regular movement of the pendulous jaw is secured.

Fig. 11.

GIMSON'S DUPLEX STONE BREAKER.

Fig. 11 represents Gimson's double-acting or duplex stone breaker.

This machine derives its peculiar action from the rocking motion of the lever, actuated by the eccentric on the driving shaft. It is, in fact, a double Blake, the toggles being introduced on each side of the actuating lever, and the size of the outflowing stones regulated after the ordinary fashion. The patentee claims for his stone breaker an increase of work at a reduction of cost, while two sizes of aggregate may be produced at the same time. The usual drawback spring assists the withdrawal of the jaws after the squeezing action has been performed.

These types of the "Blake," as well as all the others based on its principle, are provided with toggles, by which the size and quality of the desired product from the machine are regulated.

The simplicity of the Blake's stone breaker, securing at the same time an effective crushing agent at comparatively low cost, in combination with its portability, has caused its almost general introduction for the preparation of aggregates for ordinary concrete purposes. In practice, however, some details of objection are frequently complained of, such as the high cost of renewal of jaws, and the difficulty of dealing with materials in a wet or slippery condition.

Fig. 12 shows the simple, single-action "Goodman's crusher," which differs from the "Blake" by being direct in action, without the intervention of the middle or motion-transferring jaw. The crushing movement is derived from the eccentric of the driving shaft impinging on the fixed jaw, as in the Blake. The adjustment for different sizes is effected through the agency of wedges regulated by attached screws. The speed at which this machine requires to be driven is much greater than that of the Blake, and objection is raised by some to the large amount of friction which this involves, but the writer has not in his experience found any insuperable difficulty from this cause. The simple character of the "Goodman's crusher," its comparatively low cost, and extreme portability, render it a most useful aggregate reducer.

Fig. 12.

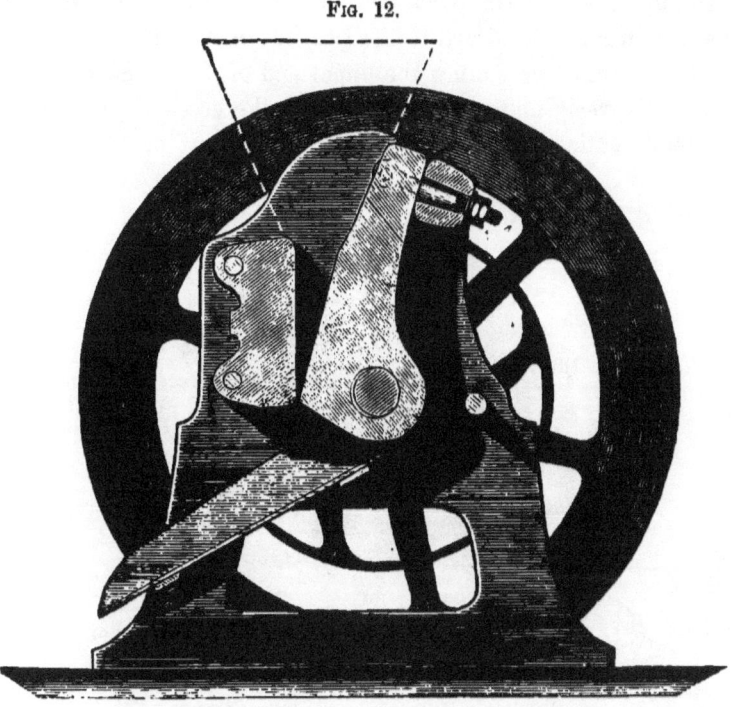

GOODMAN'S SINGLE-ACTION STONE CRUSHER.

Fig. 13 is an illustration of a double-acting "Goodman's crusher," into the scheme of which is introduced a toggle arrangement, differing, however, from the Blake, because it is not intended to regulate the size of the produce from the machine, but to facilitate the action of the movable or middle jaw; the desired size of reduced material being, as in No. 12 machine, controlled by the regulating wedges attached to the fixed jaws. The patent specification describes this somewhat novel machine as follows:—

"The extent of the motion of different parts of the block (movable jaw) is controlled by the fulcrum levers, each of which abuts on one of its sides in a cavity prepared for it in the upper end of the block, and on the other side each lever abuts in recesses formed in the two side-cheeks of the frame. The fulcrum levers allow the upper end of the block to rise and fall freely under the influence of the eccentric, but the lateral motion which they permit to the part of the block

Fig. 13.

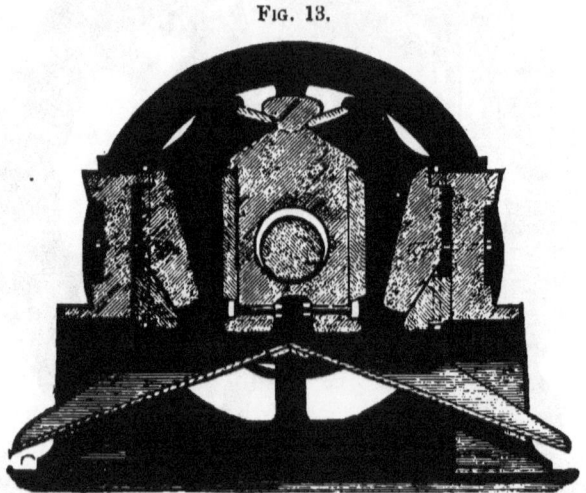

Goodman's Double-acting Stone Crusher.

which is between them is but very slight; but this lateral motion to and from the fixed crushing faces increases in passing downwards towards the bottom of the block, and it is greatest at the bottom, where a free escape of the material from the crushing cavity is required, and where the pieces of stone to be acted upon being of small size, a smaller crushing power will suffice. In place of the fulcrum levers, friction rollers or other equivalent means may be applied to guide the upper end of the block, but the arrangement shown is preferable."

In this crusher it is possible to reduce materials to a finer condition than by the ordinary Blake's stone breaker.

These references to machines of such well-known character are intended to illustrate the facility with which aggregates can be reduced to a moderate degree of fineness, but in the changing character of the concrete industry a reasonable desire is evinced for greater fineness of the materials for concrete purposes. As first agents in reducing the large pieces to a suitable state for their final reduction by a secondary or finishing machine, the variety of stone breakers is very suitable. When large-sized pieces were tolerated—and for large blocks they are not now unsuited—the ordinary "Blake" was competent to perform the whole of the work required; but modern experience on the concrete question generally indicates the necessity for finer aggregates than those hitherto regarded as most desirable. The frequent and, generally speaking, unsuccessful attempts to meet such requirements have at least resulted in a conviction that one machine on the Blake, or any modification of its principle, is incompetent to meet the desired want. Hence the modern necessity of a combination of machines, to which are duly allotted the several duties of the task requiring performance. Up to the present time the attainment of this desideratum through one machine has not been reached, but in the next, or third class of reducers, which we are about to describe, we shall find an approximate, if not a perfect, solution of the difficulty. For many years, we might almost say centuries, stamping machines have been used for the reduction of metalliferous ores in this and other countries. Their great original cost, ponderosity, and therefore unportable character, have limited their use to such purposes, and in those localities where the value of the products could alone warrant the continued waste of power and money.

136   A PRACTICAL TREATISE ON CONCRETE.

Third Division.

Impact Reducers.

The primary principle of this class of reducing machines is the impact force applied by a blow impinging on a strongly founded anvil.

The latest of these machines is "Patterson's Elephant

Fig. 14.

Patterson's Elephant Stamper.

stamper," represented by Fig. 14, which is at the present time creating some sensation in the mining districts of this country and abroad from the marvellous character of the invention and the beneficial results obtained. The drawing of the machine, as illustrated, shows a perspective view of this stamper, which differs widely in character from all other reducers of this class.

Its chief characteristic is the small amount of power required to drive it, and the entire absence of all waste of impact, any excess of which when inadvertently applied being conserved in the springs to which the hammer heads are attached. They are indeed the storehouses of the machine's energy, and neither more nor less of the impelling force is applied than is actually necessary to reduce the material under operation. The members of this novel machine are few and simple, avoiding thereby any waste of power from friction. The power is applied by a strap to the driving pulley attached to the shaft, on which are placed the two cranks, and to each of these is fixed a powerful semicircular spring. The springs are connected with two levers, made of the best hammered scrap iron, which hold the stamp heads, or hammers of the machine—the lever and stamp head together weighing about nine hundred pounds—balancing one another when in motion. The speed may be varied to any desired rate, and according to its velocity will be the value of the impacting blow of the hammers. This is a most desirable and valuable feature of the machine, for it permits of facile adjustment to meet the varying values of the mineral under treatment.

It will be seen on reference to the woodcut that the stampers perform their hammering in a receptacle called, according to the locality in which it is used, a "mortar" or "coffer." It is at this point where the output of finished produce takes place, and according to the size of the aggregates required the necessary perforated sides or screens must be regulated.

Owing to the peculiar action of the Elephant stamper, it is capable, without injury to any of its parts, of performing the reduction to a fine powder of the hardest granitic elvan or quartz with comparative ease and safety. The cost of performing this operation will vary according to the character of the material acted upon, but it is found from actual experience that at an expenditure of 25 lb. of coal for every ton of material operated on, it is able to reduce to the finest powder from twenty to twenty-four tons a day. There is practically no strain or jar on the machine, however powerful the impact, for the superfluous speed or blow is at once absorbed and stored up for future action by the powerful semicircular spring.

Although the machine and its coffer really constitute the machine proper, they are for all purposes of convenience or utility distinct and separate, so that while the movable and acting parts are equally uniform, any shape or form may be given to the coffer according to the purpose for which it is to be used. The materials to be reduced may either be treated dry or wet, a manifest advantage especially for concrete purposes, where it is so essential that all aggregates should be thoroughly washed and cleansed from the dust of fracture or other objectionable impurities with which the stone may be associated.

Although a massive machine, it is portable in character, and the facility with which its several parts may be taken down permits of its ready removal and re-erection at any desired point. This stamper appears to possess a great advantage over the pendulous jaw machine and to obviate the usual high cost for renewal of the wearing parts, for the changing of the stamp heads is easily effected at comparatively trifling cost.

For concrete, and indeed also for cement-making purposes, an arrangement for the regular withdrawal from the coffer of the finished product may be readily applied, so that no choking of the perforations in its sides can arise.

The following results have been recently obtained through the agency of the Elephant stamper in reducing copper ore and gold quartz in California, and now first published in this country.

No. 1.—30 tons of copper ore were reduced by the wet process to the fineness of a 150-mesh sieve in a day.

No. 2.—27 tons of gold quartz also with water reduced to a fineness of a 150-mesh sieve in a day.

No. 3.—15 tons of gold quartz by the dry process reduced to a fineness of a 1600-mesh sieve in a day.

No. 4.—10 tons of gold quartz by the dry process to a fineness of a 3600-mesh sieve in a day.

Dr. Oxland, of Plymouth, chemist, made during December of 1877 a series of approximate trials of the Elephant stamper with the view of testing its capacity for the reduction of phosphates, coprolites, and other minerals used in the manufacture of chemical manures. The machine was temporarily fixed in the works of one of the makers of the machine, Messrs. Willoughby Brothers, engineers, Plymouth.

The experiments were made with granite and limestone, 100 tons of each being operated upon with the following results:—

GRANITE.

No.	Tons.	
1	27·45	Fine enough without grinding.
2	16·40	Nearly so, would rapidly pass the millstones.
3	20·50	About the size of peas.
4	11·90	Size of beans and again returned to the stamper.
5	23·75	
	100·00	

LIMESTONE.

No.	Tons.	
1	37·90	Fine enough without grinding.
2	21·32	Nearly so, would rapidly pass the millstones.
3	24·90	About pea size.
4	15·88	Size of beans and again returned to the stamper.
	100·00	

The product returned to the stamps will contain a certain proportion of stuff as fine as No. 1.

Dr. Oxland remarks "that with stuff equal to granite in toughness, 40 per cent. will pass, requiring no grinding; and materials similar in character to Plymouth limestone, 47 per cent. will leave the stamp equal in quality to No. 1."

It was found in these experiments that Portuguese phosphate of lime realized a higher output from the stamper than the limestone.

From these experiments it is evident that much valuable assistance to the concrete-maker can be obtained through the agency of Patterson's Elephant stampers.

Figs. 15 and 16 represent another class of reducing machines, called "Sholl's pneumatic stamps," which are made either of a portable or fixed design. In the case of the former the stamper is reciprocated by a crank driven by a pulley from any available motive power. This crank carries with it a forked connecting rod, to the ends of which the pneumatic piston is connected by a steel pin, on which the whole of the motive power is centred. The stamper and pneumatic cylinder are in one piece, the cylinder itself being pierced with a double slot, through which the steel pin (working in the pneumatic piston) reciprocates. At the top and bottom of the "slot" are the air chambers, which remove by the compression and expansion of the air, the jar from the motive parts of the machine. In the latter case the stamper is driven by direct action from an inverted steam cylinder, the steam and pneumatic piston running together at the same speed. It is from this arrangement that the cushioning principle is obtained to counteract the violence of the descending blow of the hammers or stampers. The advanfage obtained in the Elephant stamper by the agency of powerful springs is in the case of Sholl's pneumatic stamps derived from the elasticity of compressed air.

As the piston descends in the air cylinder, it compresses the air beneath it, and thus provides an elastic agency competent to regulate the blow of the hammer to the required

FIG. 15.

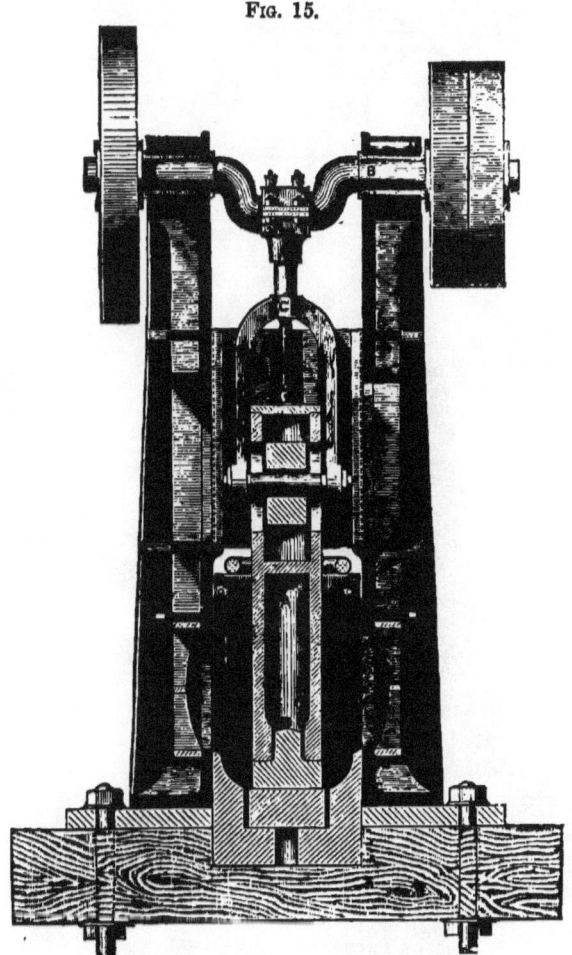

SHOLL'S PNEUMATIC STAMP.—Front Elevation.

stroke, and thus in effect converting it into a self-regulating action, capable of overcoming any danger which may arise from a negligent supply to the coffer.

142   A PRACTICAL TREATISE ON CONCRETE.

The hammer is guided through appropriate slides, and in connection with which is placed a metal box, from which

Fig. 16.

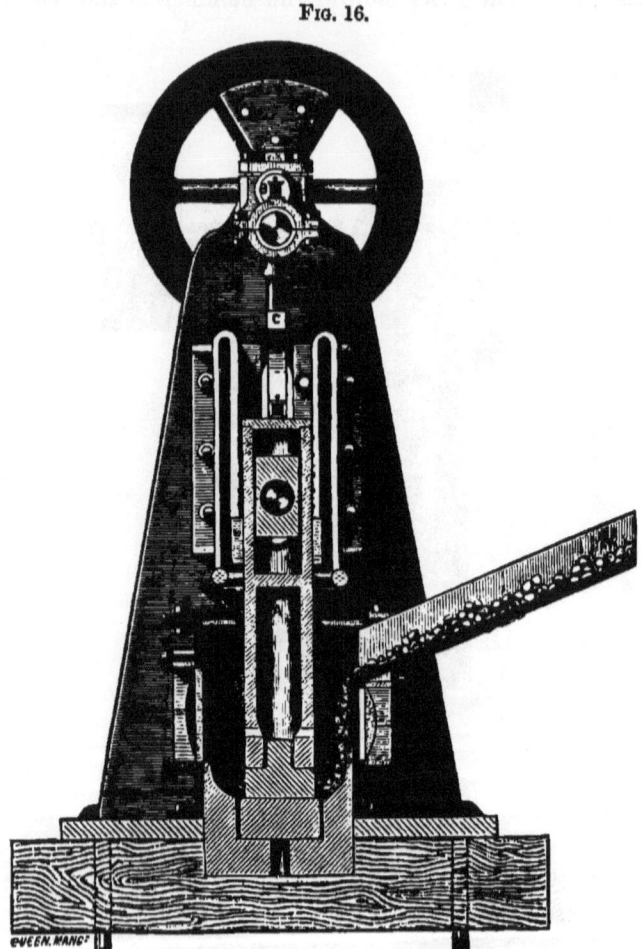

Sholl's Pneumatic Stamp.—Side View.

water is constantly supplied by an arrangement of pipes round the air cylinder, so as to cool it from the heating influence of the repeated compression of the air within.

The stroke can be regulated to any speed from 50 to 150 blows per minute, and it is claimed for this stamper that, owing to the peculiar character of its principle, much lighter stamps can be used, thereby making them particularly available for some descriptions of gold quartz.

The heat waste in compressing the air in the cushioning cylinder is equivalent to so much expenditure of fuel, but the object attained through the agency of this special pneumatic arrangement is a great improvement on the single steam cylinder. Before the introduction of this improvement the great difficulty encountered in the use of the steam hammer was the character of the dead blow, which rendered it necessary to introduce some elastic material between the base of the anvil and its foundation, so as to mitigate the violence of the shock. "Patterson's Elephant stamper" and "Sholl's pneumatic stamps" get over this difficulty, through the means which we have described and illustrated.

Fourth Division.

Abraders or Triturators, Millstones, and the ordinary Grinding Machines.

Motte's Patent Universal Crushing Machine.

This machine is of a somewhat novel construction, its action and influence being modelled on the principle of the pestle and mortar, in fact the "Mortarium" of the Romans. The members of the machine are few and simple in character, and the motion being rotary, secures an even wear of the working surfaces. A reference to Fig. 17 will show the mode by which the materials to be ground are treated. They are put into the machine at the opening where the indicating arrow denotes the points where they first become subjected to the grinding action of the revolving pestle, and are gradually reduced in their descending course,

Fig. 17.

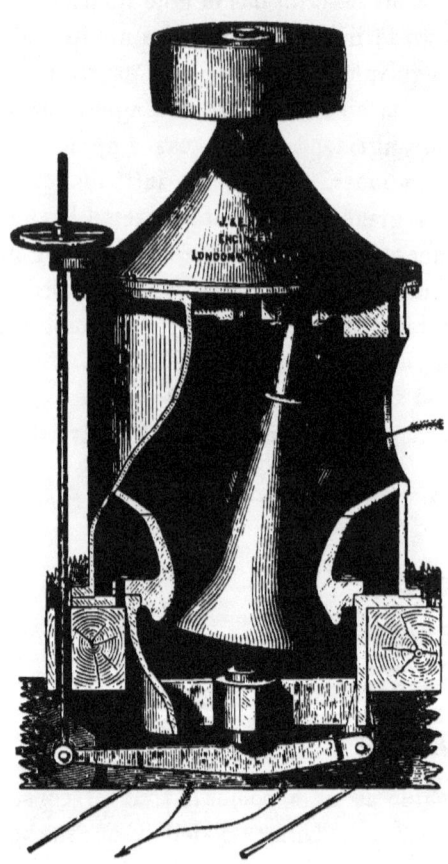

Motte's Patent Universal Crushing Machine.

where they issue in the direction of the outflowing arrows, as shown at the base of the machine. The lever to regulate the quality or fineness of output is on the principle usually applied to control the produce from horizontal grinding mills, and is guided by the vertical spindle working in a wheel screw at the top, as delineated in the illustration.

There can be no question of the simple character of this machine, and from the experience obtained during its use in Belgium it may be considered as a most useful one. At Liége, where it has been employed for some time in the reduction of limestone for road-making purposes, it breaks 300 tons into the desired size for "Macadam" in ten working hours. Another mill of the same (largest) size, working at Charleroi, broke or reduced 150 tons of hard Quanest porphyry for road metal in ten working hours.

Motte's machine has been introduced into this country by Messrs. Hall, Engineers, of Dartford, in Kent, and at their hands has received an almost complete overhaul, resulting in a machine much more efficient in character and better adapted, by its greater simplicity and strength of construction, to the hard and trying work of reducing obdurate minerals, and the equally hard and indestructible artificial slags and clinkers.

This crushing machine as now improved appears to possess the desired qualities which must commend it to concrete-makers, from the possibility of obtaining through its agency materials of all sizes. There is only one objection, which of course can be readily overcome if the produce from the machine is subsequently washed; we refer to the fractural dust which in concrete-making must, if the best results are desired, be eliminated, and the peculiar abrading action of the pestle is likely to produce a large amount of finely ground material. For other purposes, such as Portland cement-making, this dust difficulty would be rather advantageous than otherwise.

In some extensive and carefully conducted trials made

with this machine at Wetter, in Westphalia, the following results were obtained:—

Materials	Size of the reduced materials.	Production per ten working hours.
		Tons.
Calcined fireclay	3–4 mm. ($\frac{1}{8}$–$\frac{1}{6}$ in.)	15
Ganister stone	3–4 ,, ($\frac{1}{8}$–$\frac{1}{6}$,,)	15
Old firebricks	3–4 ,, ($\frac{1}{8}$–$\frac{1}{6}$,,)	30
Old crucibles	1 ,, ($\frac{1}{25}$,,)	21¼
Firebricks (Villeroy and Boch)	Fine	25
Blend	,,	9
,, in gravel	,,	27
Calcined clay	1–2 mm. ($\frac{1}{25}$–$\frac{1}{13}$ in.)	15
,, claystone	1–2 ,, ($\frac{1}{25}$–$\frac{1}{13}$,,)	10
Clay slate	3 ,, ($\frac{1}{8}$ in.)	12½
Hard cement clinkers	4–5 ,, ($\frac{1}{6}$–$\frac{1}{5}$,,)	30
,, ,,	Fine	6
Hard blue limestone	5–6 mm. ($\frac{1}{5}$–¼ in.)	15
Limestone	1 ,, ($\frac{1}{25}$,,)	9
Glass	1–2 ,, ($\frac{1}{25}$–$\frac{1}{13}$ in.)	12
White lead ore (hard conglomerate)	1 ,, ($\frac{1}{25}$ in.)	30
Blend with quartz	Fine	16¼
Pyrites	1 mm. ($\frac{1}{25}$ in.)	5
Quartz with blend	1–2 ,, ($\frac{1}{25}$–$\frac{1}{13}$ in.)	12
Galena in pieces (lead ore)	2 ,, ($\frac{1}{13}$ in.)	12½
,, in gravel	1 ,, ($\frac{1}{25}$,,)	13¼
Arsenic ore	Fine	3
Granite	1 mm. ($\frac{1}{25}$ in.)	6
Feldspar	Very fine	18
,,	A little coarser	30

The varied character of the materials operated upon indicate that both natural and artificial stones of any kind are easily reduced by the peculiarly effective action of this grinding principle, now for the first time introduced in a machine actuated by steam, and suitable for grinding purposes in general.

It is noted that the machine during these trials was controlled by inexperienced workmen, and that no special or skilled training had in any way beneficially contributed to the results obtained.

The dimensions of the reduced materials are in French measures, but the equivalents in English inches and their parts are worked out for the more ready comparison of inquirers in this country.

Disintegrators have been for many years more or less employed in the reduction of minerals for various purposes, and Carr's well-known machine of this class is employed in the reduction of hard materials required in the chemical and other industries. The general objection to high-speed disintegrators is their liability to accident and the large percentage of dust produced. This is unavoidable when the result of the percussive action of the rapidly revolving bars, and the continued and incessant attrition of the materials under operation against each other are considered.

FIG. 18.

HALL'S DISINTEGRATOR.

An improvement (as shown by Fig. 18) has been made in the mechanical arrangement of the percussive force by Mr. Hall, who has substituted for the bars of the original machine peculiarly formed beaters, which during their revolution incessantly attack the materials under operation, until reduced to the desired fineness. The beaters are fixed on the moving spindle of the machine in pairs, which rotate within a case of an irregular circular form. The parts of this machine are unusually strong, to resist the impacting force resulting from the heavy and repeated blows of the revolving beaters, as well as the violently agitated materials tossed about under their influence.

We have limited our observations to these two somewhat novel machines (more especially that of Motte's), but should state that there have been innumerable inventions of a cognate character designed for grinding purposes—conical and conoidal mills, triturators, and such like, some of which possess qualities of great use in dealing with the objects for which they were primarily designed; some of them being so adapted as to combine the use of the well-known French burrstone with such specially devised mechanical contrivances as were suitable for the object in view. Generally speaking, however, all these adaptations are not of such a character as to entirely dispense with the necessity of preliminary reduction of the materials before being fitted for the action of these abraders or triturators. Another disadvantage inherent to most of them, is the constant necessity of reparation, or dressing the grinding surfaces so as to maintain the beneficial action of the cutting surfaces of such machines.

The well-known horizontal millstones, as shown in Fig. 19, which represents one of the simplest portable machines of this class, is extensively used in the reduction of cement clinker and phosphates of lime, besides many other useful purposes. The framework provides for the driving shaft and pulleys, while the millstones are fixed on the top of the frame, the under or nether stone being fixed, and the runner or upper one being accurately balanced and swung on the top of the revolving vertical spindle.

When it is desirable to reduce materials to a very fine state there is perhaps no arrangement better adapted than this old principle of rotating surfaces of obdurate hardness acting against each other. In ordinary concrete operations, however, unless it is required to take advantage of existing machinery, cheaper and equally good auxiliaries may be found in one or other of those reducers we have described,

Fig. 19.

Horizontal Grinding Mill.

all of which are free from unavoidable high cost of dressing or sharpening the hard siliceous "burr."

It should not be forgotten that by all of those machines "dust" is produced, whether fine and almost impalpable powder from fracture, or the other equally objectionable quality resulting from the attrition of violently disturbed particles under the influence of reducing surfaces travelling at high speeds and acting against each other.

CHAPTER VI.

TREATMENT OF THE AGGREGATES.

HAVING selected the aggregate best suited for concrete purposes, either from the controlling influence of locality or otherwise, it is then necessary to treat it in the most approved way before mixing it with the cement. Generally speaking, few materials suitable for aggregates will be found in a naturally fit condition for concrete-making, and they therefore require a certain amount of preliminary treatment. Some river gravels and sea-shore sands are capable of direct mixture without running any risk of failure. In the majority of cases, however, it will be requisite not only to reduce the rock from which the aggregate is obtained by the proper mechanical appliances, but even after its reduction to be careful that all dust arising from fracture is effectually eliminated. The neglect of this apparently insignificant precaution has led in past days to much loss and dissatisfaction, but in carefully prepared and scientifically controlled concrete-making the presence of fine dust should no longer be tolerated. The reason for this precaution is, on careful examination, made obvious, and may thus be explained. Fine or impalpable dust produced from the mechanical and rapid fracture, especially of crystalline rocks, coats or covers the broken pieces with fine dust, thus interposing between their surfaces and the cement, preventing thereby the beneficial action of the latter. In some cases a stream of water is passed through the machine while the operation of crushing is being performed, but we do not regard this as sufficient to purge the aggregate, and the operation of cleansing should be subsequent to, and independent of, the reducing process. In granite especially this precaution is indispensable, as from its crystalline character the dust produced by fracture is more insidious, and indeed is in fact converted into fine clay on the application of moisture, and in that condition a dangerous ingredient in any concrete or mortar is compound.

While on this subject we may profitably consider the size to which the aggregates should be reduced. In Mr. Buckwell's original paving slabs this point was disregarded, and, in consequence, after a certain amount of wear the evils of too large pieces were apparent. So also in the early preparations of the Victoria stone paving slabs. Nature in this, as in many other matters, is a good guide, for if we examine rocks of all kinds it will be found that few except the comparatively imperfect pudding stones, millstone grits, and large-crystalled granites are coarse in grain or texture. Indeed, as we approach excellence in natural building or paving stones we find that the hardest and strongest are the finest grained. The object for which the concrete is to be used will influence the size of the aggregate, and pieces that are unfit for paving slabs or sewage or water pipes will do for large blocks or foundation work. It is a matter of indifference as to the size or shape in those rough operations of concrete blocks, such as are used in sea walls or similar structures, so long as sufficient small sand or gravel is used to fill up the interstitial spaces. Generally it may be said that aggregates from the size of a carraway seed to a hen's egg are suitable, provided the proper proportions are duly attended to. Sands and gravels, as commonly understood, are of a globular form, and not the best therefore for concrete purposes. Fractured aggregates produced by the proper machinery are invariably angular in character, and more likely therefore in the mixture to become compacted together, for the most careful treatment cannot possibly reduce the cavities formed by the rounded surfaces of natural sands or gravels. A brecciated concrete is in all cases more durable than one of a rounded or pudding-stone character. In every case it is preferable to use freshly broken aggregates, for after lying a short time they get dirty, and in some sorts chemical atmospheric action would produce deterioration in quality. Many concrete-makers, in ignorance of the advantages which a clean aggregate secures, and others from an utter disregard of its necessity, fail to produce the best attainable results from using unwashed materials. Those engineers in charge of the execution of important harbour, dock, and other works are alive to the imperative necessity of strict attention to this apparently insignificant condition. Sir John Coode, who has had perhaps the greatest experience in concrete works built in exposed situations, insists upon attention to the purity of the aggregate; and in the works of Douglas Harbour, in the Isle of Man, the rubble stone with which the large concrete blocks were built was carefully scrubbed with a brush and water before being used. It is all-important that the surface on which the cement is to operate should be clean and free from any extraneous matter that would interfere with its beneficial action.

Porosity or roughness of surface is an essential condition of the aggregate, and where it can be commanded without injury to the other necessary requirements all the better. For instance, a glassy material does not appear to offer a favourable surface on which the cement can operate, yet it is possible to unite two pieces of glass together with a good Portland cement. A glazed surface, however, produced by the chemical kiln action, is not always to be depended on, for such glazing is at best only a thin film, which may be eventually disturbed by the influence of the cement. In the various preparations of cement sold for mending glass, china, &c., we have a favourable example of the necessity of cleanliness of the surfaces to be cemented, for without this precaution an effective joint is impossible.

CHAPTER VII.

THE SILICATING PROCESS.

FOR many years various endeavours have been made to render stones impervious to water and proof against the insidious action of vitiated atmospheres. Silicates of soda and potash, singly and doubly in combination with chloride of calcium, have been applied in divers ways, but for so far without any very satisfactory results. These processes, however, indicated that stones, both natural and artificial, were sensibly improved in hardness by being impregnated with a prepared solution of silicate of soda. Indeed the double silicate of soda and potash applied in a highly concentrated form had been successfully used in preparing internal walls and ceilings for fresco paintings, and some eminent chemists have given the subject much attention.

The analysis of the compound used for this purpose was :—

Silicic acid	23·21
Soda	8·90
Potash	2·52
Water	65·37
	100·00

The specific gravity of the solution being unusually high.

This preparation was practically a coating of glass on a previously prepared mortar, having the smallest possible amount of lime in its constitution, and fit to receive the painter's colours without entailing ultimate damaging efflorescence. The mortar required careful preparation, and was composed of cleanly washed quartz sand, with the smallest amount of lime as a matrix.

Our business is, however, simply to examine the influence of silicates of soda or potash in improving the quality and character of concrete compounds generally. The silicates of soda and potash are more generally known as soluble glass or dissolved flints. As these chemical preparations now play an important part in the more advanced preparation of concretes, we shall shortly describe their nature and character.

Soluble glass is prepared in three different ways, being manufactured on a large scale, for in addition to its value as a concrete auxiliary, it is employed in glass-making and soap-making, indeed new uses are continually being found wherein its valuable properties are made available. One of the preparations is to calcine black flints to a white heat and then break them into moderately small pieces, which are placed in a wire cage and put into a jacketed iron digester. Between the outer jacket and the internal digester steam is introduced from a boiler under the desired pressure, and the temperature is thus raised to the required point, which ought to be much beyond that of boiling water obtained in the usual way. The water in the digester (the cover of which being first carefully screwed down) is then saturated with caustic soda, with which the silicic acid of the flint unites forming silicate of soda. Pure caustic soda should be used, for certain imperfectly made qualities contain sulphate, which in practice would render the silicate liable to degradation, owing to its liability to crystallize out on exposure to the moisture of the atmosphere. Mr. Ransome prepared the silicated alkalies used in the manufacture of his siliceous stone by dissolving broken flints in a solution of caustic soda at a temperature of 300°.

Increase of pressure raises the boiling point, and the celebrated Papin invented the digester (Fig. 20) known by his name for the purpose of obtaining results unattainable by boiling water at the ordinary temperature. This vessel is of great strength, covered by a lid fastened down by a powerful screw, to which is attached a safety valve to prevent accidents. The tension of steam thus controlled increases rapidly, and may finally acquire enormous power. Thus at 200° the pressure is that of sixteen atmospheres, or about 240 lb. to the square inch.

A more scientific process, however, is by heating the

Fig. 20.

Papin's Digester.

sand (silica) and alkali (either soda or potash) together in a reverberatory furnace, and as the glass forms it is raked out into water, when it is afterwards boiled in a suitable vessel. No sulphate if it existed in the alkali remains, as it would be dissipated in the form of sulphuric acid, through the flues of the furnace.

Gossage's process is performed by heating common salt to a high temperature, by which means it is volatilized, and in this state is put in contact with steam of a high temperature, when a compound decomposition takes place. The chemical composition of steam is oxygen and hydrogen, while that of salt is sodium and chlorine. The chlorine unites with the hydrogen and the oxygen with the sodium, forming by these combinations hydrochloric acid and oxide of sodium. If these two substances thus produced were allowed to cool together, the chemical process would be reversed, and each again would resume its normal condition. To obviate this and complete the process, the vapours are brought into contact in a strong chamber lined with fireclay, in which are placed masses of sand. The oxide of sodium unites with the sand and produces silicate of soda, which is thus almost completely removed from the action of the hydrochloric acid, and a larger yield of silicate obtained.

One of the formulas for making silicate of soda and potash is as follows:—

Fifteen parts of fine sand thoroughly incorporated with eight parts of carbonate of soda or ten parts of carbonate of potassa, and one of charcoal fused in a furnace, will produce a silicated alkali which is soluble in boiling water.

The late Dr. Graham made some very ingenious experiments to show that silicic acid or silica was soluble in water. These investigations led eventually to the discovery of the process of dialysis, which may very shortly be thus described.

The dialyzing apparatus consists of two guttapercha rings about a foot in diameter, which fit into each other and tighten (in tambourine fashion) a piece of vegetable parchment. It may be thus used to produce silicic hydrate.

In a quantity of water put 5 per cent. of silicate of soda, and in this solution as much hydrochloric acid as will make the liquid distinctly acid. Place this mixture in the dialyzing apparatus, when the chloride of sodium formed by the union of the chlorine of the hydrochloric acid with the sodium of the silicate of soda will pass through the vegetable membrane leaving hydrated silica behind in solution in the water with which the silicate was mixed. The solution should not be more than an inch deep in the vessel, which is floated on the surface of distilled water, contained in a larger vessel of suitable dimensions. The distilled water should be changed every day, until no precipitate can be obtained from it by nitrate of silver. When this point is reached all the chloride of sodium will have passed through the parchment into the larger vessel, leaving behind in the apparatus only silicic hydrate. On this liquid being allowed to stand for some time it will gelatinize, and afterwards contract, becoming very hard; so much so that eventually a piece when broken produces a fracture like that of flint.

The preparations of silicates of soda and potash, when in solution and concentrated, are thick and viscid, capable of uniting wood, paper, glass, &c., most effectually, and from being so used are sometimes called "mineral glue." In this condition, however, it is not suitable for applying to stone or concrete. In a dilute state, its application as a stone preservative has, we have already observed, been unsatisfactory, or at all events less successful than its advocates promised. In any case the carbonic acid of the atmosphere acts slowly upon the silicate and results in a protracted deposition of the silica in the pores of the stone, which consequently renders the application in this form practically useless. Besides, during the lengthened interval of the process, the silicates, from their being soluble, get dissolved out by rain and moisture. Compounds of silicate of soda and chloride of calcium, resulting in a silicate of calcium, have also been used, but the injurious natural action again interferes prejudicially before the desired precipitation can be secured. The colloidal or gluey character of the silicate prevents its penetrating to any depth in the stone, such tendency preventing the absorption of the chloride of calcium. There is greater penetrative capacity in the chloride of calcium, but when applied first it enters a depth to which the silicate cannot follow, precluding thereby the possibility of the required combination and joint precipitation.

The above remarks are intended to illustrate the question of silication, and to show the steps that have been made in reference to the preservation of stone. All, or nearly all, of these experiments were made on natural stones, and with the view of preserving them from injurious atmospheric action, or to check the tendency to decay incidental to their originally defective chemical constitution. A good illustration of this natural decay is seen in the Houses of Parliament. Some siliceous stones, after being quarried, possess the property of acquiring a surface hardness, due, in the opinion of some chemists, to the *sweating* of the inherent siliceous water, which produces a thin film of silica or of a nitrate. There may be some reason for arriving at this conclusion in the case of purely siliceous stones, but probably with as much accuracy as in the case of a supposed silication of lime from its combination with sand. Mr. Spiller, in a careful examination of Roman mortar from Burgh Castle, near Great Yarmouth, showed beyond question that the supposed property of silication from such a source was illusory. We have seen that in the artificial preparation of silicates heat of high temperature is an indispensable condition of the process of manufacture.

Concrete made with Portland cement as a matrix has been proved to become in the course of years as hard, if not harder, than the best natural stones or highly vitrified products from clay. Time, however, is a main factor in this realization, and it is with the view of anticipating the natural induration that the comparatively modern process of silication has been resorted to. In the case of paving and piping, as well as articles of ornate character for architectural purposes, it is inconvenient to wait on this natural process, and hence the desire to accelerate the hardening without prejudicing the eventual natural process of induration. The desired film or coating of the artificial silicate need not be of great thickness, for indeed it is found that under the best conditions its penetration seldom exceeds the sixteenth part of an inch—a mere protection while the concrete itself is reaching its final point of induration, and without any prejudicial influence on that desirable operation. By a process of this character, and with the beneficial results which experience proves it to possess, a new and valuable feature is introduced to the important industry of concrete-making in all its branches.

Although silicates are regarded with the greatest favour, and seeing also that the most durable rocks in nature are composed of various proportions of silica, yet there are other artificial preparations which have been found to possess considerable value as aids to the hardening of concrete and natural stones. Aluminate of potash is among the number, being obtained by fusing alumina and potash together. If used in a concentrated form, a certain quantity of the alumina will be precipitated, but when used in a dilute state no such difficulty occurs. If aluminate of potash of 1·12 specific gravity is mixed with a solution of silicate of potash of 1·2 specific gravity, the precipitation is somewhat protracted, and will not gelatinize until some hours have elapsed. If, however, solutions of greater density are used, an almost immediate jelly-like substance is formed, which on being dried slowly will, in the course of one or two years, be hard enough to scratch glass, and resembles in chemical composition felspar, being quite insoluble in ordinary mineral acids. If a dilute solution of this mixture be applied to the surface of stone or concrete it will fill up the interstices, and being in accurate combination, will jointly penetrate to the same depth. This artificial felspar, when thus impregnated, will act as a binding agent between the aggregates and thoroughly protect them from further wear or decay by the action of the atmosphere, however vitiated its character may be. In using this mixture of aluminate and silicate of potash, care must be taken that it is in a sufficiently dilute state, otherwise it will gelatinize and set before it can penetrate, then only forming a surface coating without the necessary bond or penetration, and can be easily rubbed off.

The subject of silication must in the future be one of very great importance, and we have thus alluded to it so that its details may be considered by those interested in the question of concrete generally. The increasing desire to use concretes in a variety of portable forms for house-building affords excellent opportunity of improving their value by the addition of these undoubtedly valuable chemical agents when required for early use.

It should be remembered, however, that the application of silicate to a concrete requires to be carefully performed, and although under the varying circumstances different rules have to be observed, it may be generally stated that until the setting of the mass is complete no beneficial result can be secured. Without porosity and capacity of absorption of the silicated liquid, no beneficial influence can be imparted. There are very few preparations of concrete, however, so dense as to be impervious to absorption, unless it be some of the best examples made from well-selected materials and put together by well-directed and violent impact. The best quality of concrete at present made, either by hand or machine, contains a large amount of moisture, which requires to be evaporated, and the space which it occupied filled as far as possible by the silicate. The penetrative power of the silicate under ordinary circumstances is, however, as we have shown, very limited, but it is enough to protect the

M

external surface of the concrete and enable it to be used with advantage before the eventual natural induration has been reached. A protective coating, when carefully applied, of so hard a material as silicate renders it practically air and water tight. In other than Portland cement concrete mixtures this would be attended with a certain amount of disadvantage, and especially in the case of lime preparations it would preclude the passage of the carbonic acid so necessary to ensure their ultimate induration. Portland cement requires no such adventitious aid, for it sets without the assistance of carbonic acid, and possesses the valuable property of cohesive capacity in a high degree. In these remarks concerning silicates it is not intended to insist that Portland cement concrete of the best kinds is eventually benefited by silicate applications, for the cement itself is really a double silicate of lime and alumina, and can reach by its own inherent value the highest attainable point of induration. It is only in such cases as where an expeditious use of the concrete is desirable, that advantage may be taken of this valuable auxiliary. Generally speaking, however, for pavements and pipes the process of silication is almost indispensable, as it would not pay to keep the slabs long enough to render them equal to the condition which the silicate quickly secures, and its cost is well repaid by the advantage thus realized.

In describing the manufacture of paving and pipes, further details of this process will be given in Chapter IX.

CHAPTER VIII.

ESTABLISHED PROCESSES OF CONCRETE MANUFACTURE.

THE most celebrated of the artificial stone compounds is that made under Coignet's system of "béton aggloméré," which, especially with French engineers and architects, has obtained a high reputation. Many important structures have been erected with its aid, and notably the blocks which were used in the construction of Port Said, at the Mediterranean entrance to the Suez Canal. The employment of this doubtlessly valuable composition requires to be conducted with great care and skill, for any careless departure from the prescribed formula for its preparation is attended with much risk and danger.

In England we have not had much experience of the constructive value of this concrete or béton, but its introduction was attempted some years ago, during the construction of the Thames embankment in front of St. Thomas's Hospital. An archway under the stairs leading down from Westminster Bridge was formed of the "béton aggloméré," and the whole of the preliminary mixture as well as the construction was examined by the author while the work was in progress. It was afterwards used in some sewer works on a small scale, on the south side of London, but no further progress except in this introductory manner was made, and for so far the works in England may be regarded as of a purely experimental character. A concrete of this kind, the manipulation of which requires to be performed by skilled workmen, possesses too many ingredients of risk in the English labour direction to warrant its general adoption.

The machinery used in preparing these early examples of "Coignet's béton aggloméré" was a specially made pug mill, in which the mixture of hydraulic lime, cement, and sand were put, after having been previously well mixed by hand in their dry state, the subsequent careful application of the least possible amount of water being made by a fine jet to the pug mill whilst in motion. This process of preparation was rather protracted, but it resulted in the production of a thoroughly homogeneous mass, somewhat incoherent in character, but which, on the completion of the impinging or ramming process, produced a compact mass of even texture and uniform density.

In preparing the matrices great care is necessary and considerable fineness is absolutely imperative, otherwise the desired results are unattainable. The lime (preferably hydraulic) is mixed with certain proportions of Portland or other good cements, and the incorporation of this *blend* with the sand requires to be carefully and accurately performed. Mechanical means are resorted to for this operation, and on its extent and character will pretty much depend the value of the resulting concrete compound. However desirable it may be to use the best qualities of Portland cement or lime, in some localities their great cost necessitates the employment of common limes, and in such cases the greatest care must be exercised owing to the want of cohesive value in matrices of that class. The mixing process under these circumstances must be done carefully and in as dry a condition as possible. The mechanical arrangement designed by M. Coignet for the perfect amalgamation of the materials is of an ingenious character, and as its employment in the preparation of some of our concrete

ESTABLISHED PROCESSES OF CONCRETE MANUFACTURE. 165

mixtures may be found desirable, we will shortly describe "The Malaxator."

This machine consists of twin screws, whose blades interlock each other, travelling in the same direction.

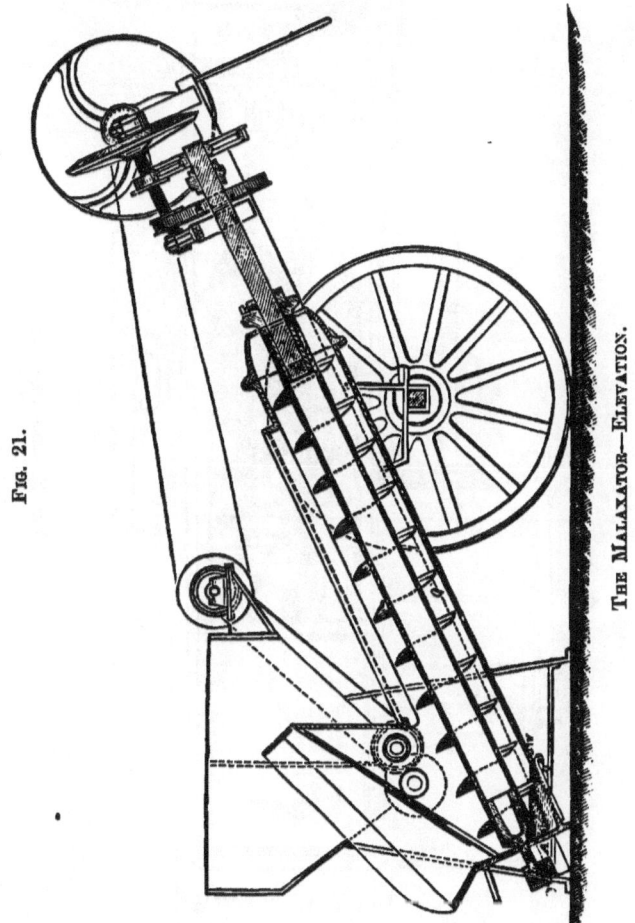

Fig. 21.

THE MALAXATOR—ELEVATION.

Figs. 21 and 22 represent elevation and plan (partly in section) of the "Malaxator."

166   A PRACTICAL TREATISE ON CONCRETE.

The machine is fixed on a frame at the upper end of which is placed the necessary gearing to actuate the shafts of the screws.

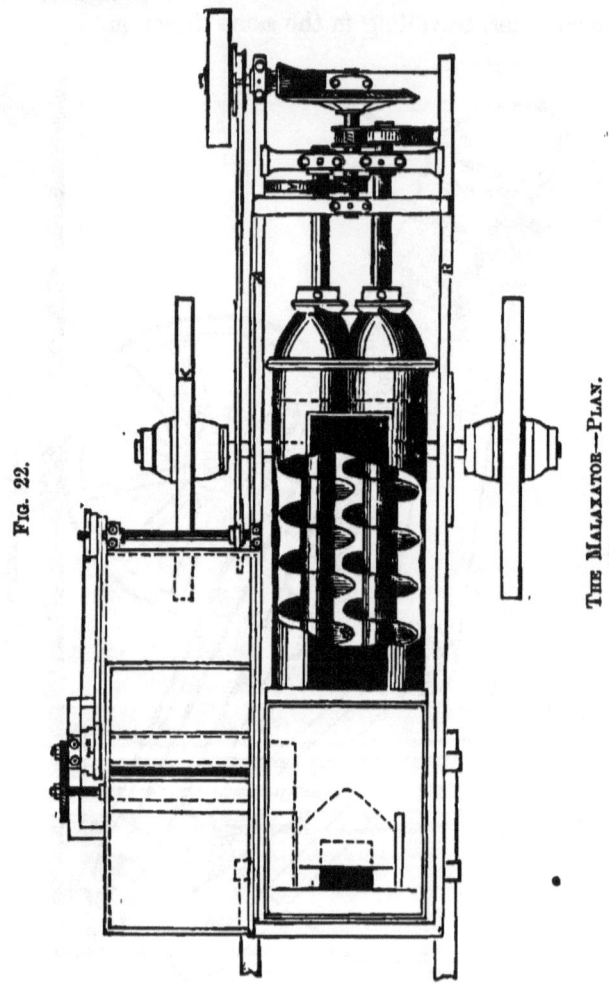

Fig. 22.

The Malaxator—Plan.

The wheels and axles on which the frame is placed permit of the ready transfer of the machine to the required working points, the desired angle, about 25°, being secured by this arrangement.

The hoppers at the side, in which are placed the matrices and sand, have regulating adjustments, whereby the exact proportions are secured; and at the bottom (at the point of delivery of the duly proportioned materials to the screws) the water is introduced in the required quantities. All these arrangements are automatic in character, and when the machine is started the proportions cannot be varied or any error arise in the operation, if the feeding of the hoppers is correctly performed.

The materials then pass up between the screw blades to the upper end of the machine, and at any desired point the stone paste (for it may now be so termed) is dropped into pails or barrows and wheeled or carried off to where it is to be used.

A careful examination of the woodcuts will show that there is not much complication in this machine, and it secures a thoroughly proportionate mixture of the materials. By this mechanical arrangement not only is the proportion properly adjusted, but the succeeding admixture is also most effectively secured. This latter operation may be varied by increasing the pressure at the point of delivery on the top of the machine, thereby retarding the speed of the output.

When Portland cements are used the side hoppers are generally dispensed with, and in such cases the materials, first mixed in a dry state, are at once put into the bottom of the mixing machine, or "Malaxator."

There is nothing very novel in this machine except it may be the idea of placing the screws at an angle. This to some extent will tend to make the mixture more dense in character, but we do not see enough advantage in this inconvenient position to recommend its adoption. For washing aggregates its elevation would in some measure facilitate the ejection of the rejected loam or dust.

This machine can only be used with advantage when sand or very fine gravel is employed, for aggregates of large size would interfere with its beneficial action, and in fact it would become clogged and useless, for the larger pieces would jam between the feathers of the revolving screws and stop or break the machine.

The "stone paste" thus prepared is now spread in thin layers in the required form and position, being carefully and evenly beaten by rammers formed of hard wood, weighing from 15 to 20 lb. The attached handles are from 3 to 4 ft. in length, and the operator has to be careful that he applies the impact truly and evenly.

It is obvious that the béton so treated for the construction of monolithic masses should be of such consistency as to be benefited by the impacting blow of the rammer. If too moist, the ramming would be useless, as no resistance would in that case be offered to the blow of the operator. The mass therefore should be slightly incoherent in character and somewhat viscous in texture.

Very considerable works have been successfully accomplished through the agency of the "béton aggloméré," far exceeding in originality and extent anything attempted in this country through the means of Portland cement concrete. We shall shortly refer to one of the well-known French structures which has proved beyond dispute the suitability of this béton for works of any extent on land.

In the Vanne Aqueduct, which supplies the city of Paris with water (traversing the whole length of the Forest of Fontainebleau), there are nearly three miles of arches and eleven miles of tunnels, nearly all of which have been constructed of "béton aggloméré." Several of the aqueduct arches are nearly 50 ft. in height, and many bridges crossing rivers, canals, and roads are of considerable span, some being as much as 125 ft. wide.

In the more important bridges the piers have been carried

up to springing height in béton aggloméré, and the arches built in hydraulic lime and Portland cement masonry.

These works were prosecuted without cessation during the winter of 1868–69 and the following summer, but did not suffer in the slightest from the extremes of temperature.

The spandrils of the arches were built up to the level of the crown, and upon this arcade was made the monolithic pipe (6½ feet in diameter), the whole being firmly and carefully compacted together in one solid jointless mass.

In the spring of 1869 water was let into a portion of this conduit pipe, and M. Belgrand, Inspector-General of bridges and highways, and director of drainage and sewers of the City of Paris, certified that "the impermeability appeared complete."

In the architectural direction, equally satisfactory results have been attained, as is shown by the erection of a church at Vesinet, near Paris.

This church is built in the Gothic style, and the steeple is 135 feet high, the founder certifying thus:—"During the two years consumed by M. Coignet in the building of this church, the béton aggloméré in all its stages was exposed to rain and frost, and that it has perfectly resisted all variations of temperature." The floor of this church is paved with the same material, being composed of various beautiful designs, having agreeable and diversified contrasts of colour. The crushing strength of "béton Coignet" used in building this church was only 2634 lb. per square inch, that strength being regarded as sufficiently strong for such a building.

We refer only to these well-known examples of advanced important engineering and architectural edifices, although there are others of a less imposing character, such as the sewers of Paris, many miles of which have been formed of the same material. There is, however, one other example we think worthy of reference, viz. the arched ceilings of the cellars under the Municipal Barracks of Notre Dame, Paris. The spans of the arches varied from 22 to 25 feet in width, the rise of the arch being one-tenth of its span, and the thickness at the crown 8·66 inches.

To prove the strength of these arches the following tests were made. The vault arch selected for the experiments was 17 feet 6 inches in span :—

1st Test. A pyramid of stone, weighing 36 tons (2000 lb. each), was placed on the centre of the vault.

2nd Test. · A mass of sand, 13 feet thick, was spread over the surface of the same vault.

3rd Test. Carts loaded with heavy materials were driven over it.

In no case was the slightest damaging result noted or observable. It is not stated what was the age of the béton thus proved, but we may presume that the interval between building and testing was not long under the circumstances.

The proportions used in the béton are variable, and subject to the character of the work to be performed, as well as the relative cost of the materials used.

The mixture used in the béton for the Paris sewers was as follows, by measure :—

> Sand, five parts;
> Hydraulic lime, one part;
> Paris cement (considered as good as Portland cement), one-fifth part.

The saving of cost in the case of these sewers was 20 per cent. less than the lowest estimated price of any other kind of suitable masonry.

For external work of good average quality, the following proportions may be used, by measure :—

	1.	2.	3.	4.
Portland cement	1 part	1 part	1 part	1 part
Common lime powder	$\tfrac{4}{10}$,,	$\tfrac{1}{4}$,,	$\tfrac{3}{4}$,,	$\tfrac{1}{4}$,,
Coarse and fine sand	6 ,,	$6\tfrac{1}{4}$,,	7 ,,	$7\tfrac{1}{4}$,,

And for coarse work, such as foundations, where a fine or smooth surface is not required, the following proportions (by measure) may be used with advantage:—

	1.	2.	3.	4.
Portland cement	1 part	1 part	1 part	1 part
Common lime	$\frac{4}{12}$ „	$\frac{1}{3}$ „	$\frac{2}{3}$ „	$\frac{6}{14}$ „
Gravel and pebbles	12 „	13 „	13 „	14 „
Coarse and fine sand	6 „	6¼ „	7 „	7½ „

Many large city houses have been built in Paris of béton aggloméré, one of which, six stories high, having a Mansard roof, has the exterior walls of the following thicknesses:—

Cellar, 19·7 inches; first story, 15·7 inches; second story, 13·8 inches; third story, 12·8 inches; fourth story, 11·8 inches; fifth story, 10·8 inches; sixth story, 9·8 inches.

The cellars under houses of the above character are usually bisected in their length or street front by a mid wall, from which spring flat arches, the usual proportions being a rise of one-tenth of the span, the crown being 5½ inches thick, and at the springing 9 inches.

It must be evident that results of so satisfactory a character as those we have described could only have been realised by the greatest care in the manipulation of the béton. Thorough mixture of the matrices and aggregates, the smallest amount of water, and the perfect compacting of the whole together by the blows of the rammer; the force applied in the last and final operation not being violent, and the thickness of the layers limited to the exact depth which could be beneficially acted upon. The strict adherence to these conditions would produce a dense, and therefore non-porous mass, free from contained air or water spaces.

The successful operations carried out through this material in France led to its introduction in the United States of America, where for some years the system has been carried out with much and deserved success. In that country there is at the present time at least one disadvantage in using the béton, owing to the necessity of importing foreign Portland cement, as the manufacture of this invaluable concrete matrix has not yet made much advance in that country. The materials for its manufacture must exist in great abundance, but cost of fuel and labour probably has, in some districts at least, acted as a deterrent in attempting the production of Portland cement on any extensive scale.

General Gillmore, however, takes a hopeful view of the subject, and estimates the cost of manufacture of Portland cement in the United States according to the following statement—the cement works being located at the point of supply of the limestone:—

"Estimated cost of making Portland cement in the United States by the dry process, using hard limestone and a Hoffman kiln."

"Annual capacity of works, 30,000 barrels, or 6000 tons." All costs in dollars.

Interest on $40,000 capital	2,800
9000 tons raw limestone, delivered at 1·50	13,500
2000 tons clay, at 1·40	3,920
1100 tons peat and dust coal, at 4·30	4,730
Salary of superintendent	2,000
Salary of foreman	1,200
Thirty-eight labourers, at 45 per month average	20,520
Two burners, at 75 per month	1,800
Contingencies, &c., &c.	9,534
Total cost	60,004
Equal to per ton Or 2*l*. English.	10·000

Even that cost would be from 12 to 14 per cent. less than the wholesale market price of the famous American Rosendale cement, the cost of package in both cases being excluded in the calculation, that item being in the case of imported European Portland cement about 10s. per ton;

the relative weights of Portland and Rosendale cements being, the former 410 lb. per barrel, and the latter 310 lb.

With the advantage shown in the calculation made by General Gillmore (a high American authority on the question of limes and cements), there must be, sooner or later, large manufactories of Portland cement established in the United States. The introduction of the béton aggloméré and other similar systems of concrete manufacture will accelerate the introduction of home-made artificial cements.

The make of natural cements of high quality already exists to a very considerable extent in various districts of the States.

Since these remarks were written the author has been supplied with a prospectus of the "National Portland Cement Company, of Kingston, New York."

The manufactory of this company is situated on the banks of the Hudson river, at Kingston, Ulster County, State of New York.

The materials used are fuller's-earth, kaolin, and lime, which are treated by the semi-wet process of manufacture.

The National Portland Cement Company, in their prospectus, state that:—

"The cement made by them is from an artificial mixture of different qualities of clays with lime, in a regular and certain manner, in the proportions of the different elements required to give the highest grade of cement, as found by careful testing; and the manufacture is thus maintained daily without variation, by mechanical devices especially constructed to accomplish a certain, regular, and uniform product."

In calling attention to their produce, the Company also say that:—

"The foreign makers depend for the quality of their manufacture upon a mixed species of alluvial clay found in certain deposits in Europe, and as every layer of this clay is of a different chemical constitution, so the cement made from it varies in quality with the different layers mixed. Cement of the grade manufactured by this Company can only be furnished by the foreign makers in small lots at special prices."

The foregoing claims to such excellence are warranted from the results of tests at the Philadelphia Exhibition, 1876, as shown by the following table:—

CRUSHING AND TENSILE STRENGTH OF THE HYDRAULIC AND OTHER CEMENTS AT THE PHILADELPHIA EXHIBITION. TESTED IN THE MANNER EXPLAINED IN THE FOLLOWING NOTES BY GENERAL GILLMORE.

NAME OF EXHIBITOR AND PLACE OF MANUFACTURE.	Crushing Strength.		Tensile Strength.	
	Average strength per square inch.	No. of trials.	Average strength per square inch.	No. of trials.
PORTLAND CEMENT.	lbs.			
National Portland Cement Co., Kingston, N. Y.	1482	20	213	5
Toepffer, Grawitz & Co., Stettin, Germany	1439	12	216	3
Hollick & Co., London, England	1330	10	216	3
Wouldham Cement Co., London, England	1140	12	199	3
Saylor's Portland Cement, by Coplay Cement Co., Coplay, Pa.	1078	8	184	2
Wampum Cement and Lime Co., Newcastle, Pa.	968	12	163	3
Pavin de Lafarge, Teil, Canton of Viviers, France	931	12	158	3
A. H. Lavers, London, England	926	6	192	2
Francis & Co., London, England	907	14	163	3
Wm. McKay, Ottawa, Canada	882	10	141	3
Borst & Roggenkamp, Delfzyl, Netherlands	826	12	132	3
Longuety & Co., Boulogne-sur-Mer, France	764	12	108	3
Riga Cement Co., by C. X. Schmidt, Riga, Russia	693	5	134	2
Scaman Cement Co., Lomma, near Malmo, Sweden	606	14	112	3
Bruno Hofmurk, Port-Kund, Esthland, Russia	580	6	154	2

Even making full allowance for the unavoidably interested commendation of home-made wares, the Transatlantic cement-makers thoroughly understand the position of their English competitors, and make no secret of their intention to enter on a keen contest for ultimate superiority. The prospectus from which we have taken the above extracts also says that the National Portland Cement Company "are manufacturing Portland cement of the very best quality, for engineers' use, by a process which enables them to turn out every barrel of a uniform quality, and of standard test grade, which it is impossible to obtain in the best imported article, an advantage which will give the American article eventually the preference over the best of the foreign brands."

General Gillmore, one of the judges at the Philadelphia Exhibition, also certified to the accuracy of the above tests by giving an ex-official certificate of the first four in the list.

The method practised in making the above tests is thus described:—

" All the cements exhibited at Philadelphia were carefully tested before awards were recommended by mixing them dry in each case with an equal measure of clean sand, tempering the mixture with water to the consistency of stiff mason's mortar, and then moulding it into briquettes of suitable form for obtaining the tensile strength on a sectional area $1\frac{1}{2}$ inch square, equal to $2\frac{1}{4}$ square inches; the briquettes were kept in the air one day to set, then immersed in water six days, and tested when seven days old. After thus obtaining the tensile strength in each case, the ends of the specimens were ground down to $1\frac{1}{2}$-inch cubes, which were used the same day for obtaining the compressive strength by crushing. The results averaged from a number of trials with each sample of cement, and divided by $2\frac{1}{4}$ in order to get the strength per square inch, are recorded in the following table. It may be stated in further explanation, that although the table shows beyond question, under the conditions named, the strength of the several specimens exhibited, it may not correctly indicate the relative merits of the customary productions of the several manufactories represented. Some of them may have used especial care in preparing the article exhibited, while others may have sent average samples from stock on hand."

There can be no doubt that the Americans are alive to the very great importance of securing a good native Portland cement. With Portland cement works in the United States, Buenos Ayres, Rio de Janeiro, Madras, &c., our English makers will be wise if they look to the possible contingency of some of their best foreign markets being shut against them.

The American makers are not very wide of the mark when they say that chalk and fine clay cements are unreliable, owing to the variable quality of the latter ingredient. The difficulty of obtaining the finest and best-suited river muds or clay is becoming every day greater, and there must be eventually some change in the English mode of manufacture to meet this difficulty. Should coal gas be superseded by the electric light, the London cement-makers will lose another advantage in the increased cost of fuel now got on such favourable terms from the gas works.

Béton aggloméré, if accurately proportioned, and when all interstitial spaces are filled up, is more impervious to water than the best qualities of sandstone, some granites, limestone, and marbles. The proper proportions to ensure such a quality are one of matrix to from two and a half to three of suitable clean sand.

In a chemical examination made by Dr. Isidor Walz, of New York, it was found in two small specimens weighing about $2\frac{1}{2}$ grammes, the specific gravity of which was $2 \cdot 305$, that after fifteen minutes' immersion in water, and a subsequent exposure of four days in air saturated with moisture, that no increase of weight occurred in one specimen, and the other only absorbed $\frac{16}{100}$ of 1 per cent. of moisture.

To attain such excellence in this manufacture the exact degree of the water of mixture must be secured, for an excess would prevent the required profitable result from the ramming or impacting operation. If the mass is too moist the blow would only shake the mixture, and if too dry it would break up and cling to the rammer like so much clay.

Another machine used for mixing the materials in the manufacture of "béton aggloméré" is represented by Fig. 23, and is called the "Greyveldinger mortar mill." This machine is equal, if not superior, as a mixer to the best of the pug mills now generally used for that purpose. Within a cylindrical trough is fixed an Archimedean screw, which is rotated by a pinion and bevel wheel connected with driving pulleys (fast and loose) fixed on the framework attached to the wheels and axle of what may be called the carriage. The materials are roughly mixed in a dry state, and placed in the hopper at the lower end of the machine (the angle best suited in practice being the same as that adopted in the case of the "Malaxator," or 25 degrees), and the mixture is passed up by the revolving action of the screw to the point of outlet. The time occupied in this operation is about twenty seconds. It is dropped into buckets fixed on a revolving table, so that no interruption in the progress of the machine arises, as there is always a bucket beneath to receive the outflowing mass. A mortar mill of this construction is competent to turn out, when used horizontally, from 30 to 40 cubic yards of mixed material in the course of a working day of ten hours with a half-horse-power engine, the labour and superintendence absorbed in tending it being eight ordinary labourers, foreman, and engineer.

Particulars of the compressive values of the béton aggloméré will be found at page 189, and at pages 188, 190, and 191 are recorded the several values of some of those concrete and other compounds which we will now

178   A PRACTICAL TREATISE ON CONCRETE.

proceed to consider. We should state, however, that the references now made to those foreign-made concretes are

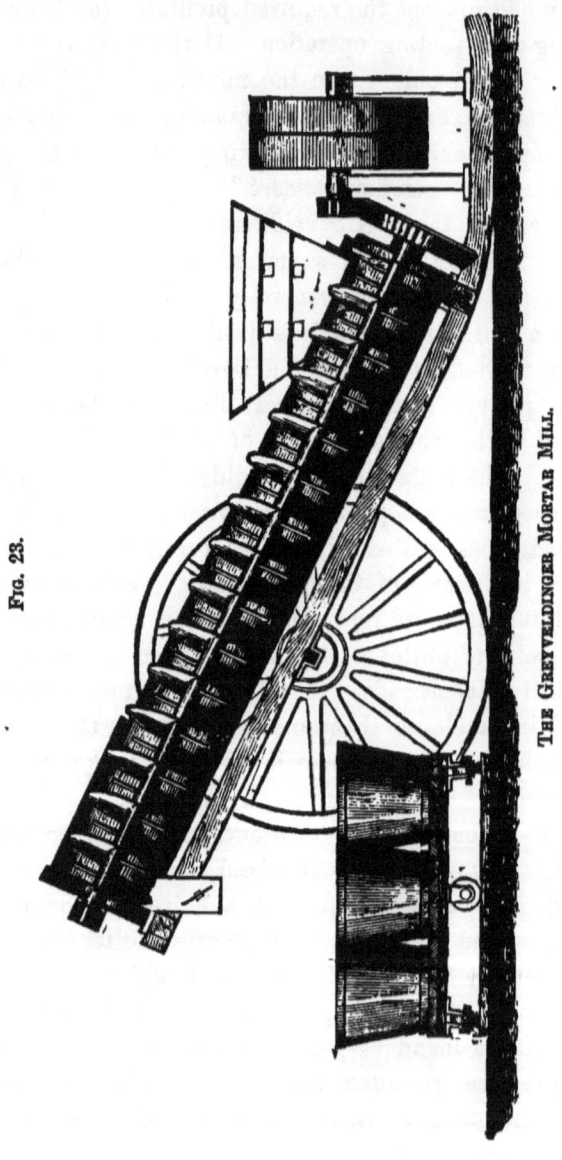

Fig. 23.

THE GREYVELDINGER MORTAR MILL.

not intended to preclude the necessity of discussing the various processes in use in this country, but merely to show the progress of the concrete art elsewhere.

Another artificial concrete, much used in the United States of America, is called the "Frear Artificial Stone."

The novelty of this invention consists in the application of gum shellac for the purpose of hardening any mortar mixture with which it is incorporated.

The hydraulic cement is first mixed in a dry state with the sand in the proportions of one of the former to two and a half of the latter, such an admixture securing as nearly as practicable a resulting stone free from porosity. This mixture is then moistened with a solution of gum shellac of the following quality:—One pound of the gum is dissolved in from 2 to 4 ounces of concentrated alkaline solution. The gum thus treated is mixed with as much water as will secure about an ounce of it to each cube foot of stone. The mortar or stone paste is then thoroughly mixed and placed in strong moulds of the required form, in which it is subjected to a pressure of from 15 to 25 tons, according to the size and character of the block. There are special machines designed by the inventor for this purpose. Almost immediately after the pressure has been applied the stone blocks are removed, and when exposed to the air for two or three days and dried, are afterwards damped when the hardening process continues. After three or four weeks ordinary blocks, such as sills, steps, &c., may be used.

Blocks of the Frear stone used for constructive purposes in Chicago and other western districts have not resisted with advantage the damaging influence of such climates. The failures have not been of any great extent, and were due, it is believed, to the imperfect character of the hydraulic cement employed. There is no Portland cement required for this manufacture.

The use of gum shellac is, to the mind of the writer, a doubtful advantage, and if mixed with the best quality of Portland cement would, in his opinion, be attended with a deterioration of any concrete mixture in which it should be so applied.

There are two other systems of artificial stone-making in use in America, promoted by the "American Block Building Company."

One made under Foster's process is simply a mixture of slaked lime with wet siliceous sand. The smallest quantity of water is used, and the theory is that each particle of sand shall be covered with a thin coating of lime paste, preventing by this supposed accurate arrangement the presence of any void spaces in the mixture. The blocks are subjected to considerable pressure.

In the process thus described (of a very simple character) the inventor claims that by his invention silicate of lime is formed by the close junction of the matrix and sand. We have elsewhere shown that this is illusory, for no such advantage can, in our opinion, be secured under the circumstances.

Another process, regarded as an improvement on the above, is that devised by Van Derburgh, who uses ground quicklime instead of slaked lime, taking advantage of the heat evolved from the lime in that state when placed in contact with moist sand. Steam is likewise used to hasten the slaking of the lime.

Professor Horford, in reporting on this improvement, says:—

"Van Derburgh by his process, as at present carried out in practical working, intimately mixes *finely ground unslaked lime* with moist sand, in a close chamber, kept in constant agitation. The affinity of the unslaked lime for water causes the lime dust to adhere wherever it touches the

ESTABLISHED PROCESSES OF CONCRETE MANUFACTURE. 181

surface of the moist sand. Slaking instantly commences and is aided by the introduction of steam into this confined space. Under these circumstances the heat evolved in slaking the lime, as well as the heat due to the steam admitted to the interior of the continuously stirred and kneaded mixture, is brought to bear on the silica, at the surface of the sand grains, in contact with the moist hydrate of lime. After continuing in this condition for a suitable time, it is subjected to great pressure, imparted by successive percussions in metallic moulds. The pressure results in a block, the surface of which rapidly becomes hard, and the hardness extends from the surface toward the heart of the mass."

Blocks made under the Van Derburgh system were used in the construction of the Howard University and Hospital buildings at Washington, district of Columbia, in 1868–69, and during the progress of building, a portion of the walls fell. This accident led to an examination of the materials by experts appointed for that purpose.

The blocks were 10 inches long by 5 inches wide and 4 inches deep, having a hollow space in the centre 6 inches long by 1 inch in width. The solid area therefore under compression was 44 square inches.

A block from twelve to thirteen months old was crushed under a pressure of 443 lb. per square inch.
A block six months old, at 295 lb. per square inch.
,, eleven months ,, 384 ,, ,, ,,
,, ten years old ,, 1455 ,, ,, ,,

It was found that the blocks improved in strength by age, and at the rate of about 1000 lb. per month, which divided by 44 would give an increasing value per square inch in that time of 22·7 lb.

These experiments were made on metal and plaster beds, the latter of which gave much better results than the former. The average crushing weight was found to be much in excess of that required. The weakest block was 173 lb. per square inch and the strongest (10 years old) 1455 lb., a somewhat unsatisfactory comparison, seeing that blocks of such an age could not be readily forthcoming. It, however, proved that a continuous and prolonged induration was possible through the agency of the process in question.

The above system of block-making has much resemblance to the Ranger concrete process, adopted in England more than forty years ago, and some important buildings were constructed with these blocks, amongst others the College of Surgeons, in Lincoln's Inn Fields, and a school at Lee, near Blackheath.

Mr. Ranger's patent was for making artificial stone blocks, in moulds of any form and size, of gravel or other aggregate with lime as a matrix. The lime was used in the shell state, as received from the kiln, mixed with the aggregate, the water being heated in making the concrete.

Mr. Ranger obtained his first patent in 1832 for "a cement or composition," which he denominated "Ranger's artificial stone."

This artificial stone was prepared by mixing siliceous materials (in combination with brick or stone) into blocks, and using as a matrix Dorking or other limes in their pure or caustic state, mixing the mass with hot or boiling water. Occasionally a portion of sulphate of iron was put in the water "as well also as gaseous or other matters when thought desirable."

The proportions ordinarily used were 30 lb. of aggregate with 3 lb. of powdered lime and boiling water. It was necessary to be careful that only so much of the concrete mass should be mixed at a time as would fill the mould used, its initial set being exceedingly rapid. The use of rammers was resorted to, for increasing the density of the artificial stone and to assist in filling up the void spaces.

In 1834 another patent was obtained by Mr. Ranger to use the aggregate hot, thus securing a more perfect combination of the mass and increased capacity of crystallization.

The use of Ranger's artificial stone was, as before mentioned, applied successfully in buildings on land, but when subjected to the action of running or tidal waters it succumbed to their dissolving action and became disaggregated.

We have here introduced these remarks on Ranger's artificial stone to show how much similarity exists between the process used in its manufacture and that adopted by Van Derburgh, in both cases heat being used, one from water and the other from its vapour.

Another artificial stone-making process in much favour in the United States of America was discovered by M. Sorel, a French chemist, and the manufacture is carried on extensively by "The Union Stone Company." This process of M. Sorel's consists essentially in the adoption of a valuable cement, obtained from a careful and judicious treatment of magnesian or dolomitic limestone.

The production of hydraulic lime from such rock has been proved in this country, and the late Professor Calvert established the practicability of making a first-class hydraulic cement from a magnesian limestone found in the Island of Anglesey, North Wales. The magnesian or dolomitic limestones are used extensively in the United States of America in the manufacture of building blocks under what is called the "Sorel process."

The production of a reliable hydraulic lime or cement from the dolomitic limestones is attended with much difficulty unless the accurate degree of heat is applied in calcination. If the temperature is so high as to cause the expulsion of the carbonic acid from the limestone the magnesia will be overburnt and the hydraulicity of the resulting lime destroyed.

Carbonic acid can be expelled from the carbonate of magnesia at a temperature of from 550° to 700° Fahr. (or a cherry-red heat), whereas its expulsion from the carbonate of lime requires from 1500° to 1700° Fahr. It is evident therefore that when the object is to convert the carbonate of magnesia into oxide of magnesium, the carbonate of lime in the stone cannot be much altered in character.

Any magnesian limestone having an analysis of 60 per cent. or more of carbonate of magnesia may be converted into a good hydraulic cement if not overburnt.

M. Sorel discovered that oxy-chloride of magnesium was an hydraulic cement of much value, attaining great hardness. It is formed by adding chloride of magnesium to oxide of magnesium obtained by calcining the carbonate of magnesia.

Deposits of this mineral exist in several localities in the United States of America and elsewhere, having as much as 95 per cent. of carbonate of magnesia in their analyses. In this condition, when submitted to a temperature of about 650° Fahr., the resulting product will be quite white, which after being finely ground is called by the "Union Stone Company" "Union cement." The purity of colour of the cement depends very much on the character of the carbonate of magnesia, for if the analysis gives any appreciable quantity of oxide of iron, the cement has a reddish tinge, which does not, however, injuriously affect its cementitious value.

The burnt magnesite (oxide of magnesium) is used for various concrete purposes, and according to the purpose for which it is required is the quantity of powder regulated to mix with the aggregates.

For building blocks, sills, lintels, and such like forms, from 6 to 10 per cent. of the oxide of magnesium (finely ground) is mixed with properly prepared aggregates, such as marble powder, siliceous sand, or other suitable materials, while for ordinary work, such as walls, 5 per cent. is found sufficient.

The above proportioned ingredients are first mixed in a dry state by hand or machine, and then moistened with chloride of magnesium, for which is sometimes substituted bittern water—the refuse of seaside salt-works. When thus moistened, the mortar or concrete is passed through a mixing or pug mill and submitted to a process of trituration. When ejected from the mill the mass should resemble the condition of clay when used by the brickmakers in moulding bricks.

The stone paste is then rammed or tamped into the necessary moulds, and the block may soon thereafter be moved. The blocks thus made speedily harden and can be used in three or four days' time.

Besides building blocks the Union Stone Company make a variety of emery wheels and hones, which have acquired a high reputation for their excellence, and it is also used in the manufacture of billiard balls.

The durability of this stone has been submitted to severe tests, in Boston the blocks having withstood the degrading influences of several winters.

Dr. C. T. Jackson, State Assayer of Massachusets, reports on this stone as follows:—

"I find that the frost test (saturated solution of sulphate of soda) has not the power of disintegrating it in the least. The trial was made by daily immersion of the stone in the sulphate of soda solution for a week, and allowing the solution to penetrate the stone as much as possible and then to crystallize. From this test it is evident that your stone will withstand the action of frost more perfectly than any sandstone or ordinary building stone now in use. I see no reason why it will not stand as well as granite."

A resistance during one winter to the freezing and thawing of a climate like Boston may be taken as evidence of the Union stone being able to resist the ordinary destructive action of frost, while such magnesian concretes are supposed to withstand the action of sea water better than lime compounds.

The following extract from a Report of the Committee of the Middlesex (United States) Agricultural Society is interesting:—

"Union Stone Company's articles, building stones and bricks; soapstone sinks and tubs without joints; tiles, curbstones, posts, emery wheels, grindstones, mantle-pieces, &c. &c. This stonework is a new enterprise, and in the opinion of the Committee promises to become one of great general utility. The Committee learn that this Company has several modes of manufacture, and have worked out and recorded more than two thousand formulæ, using in all cases the same cementing agent, *magnesia*, for which they have several letters patent. The large stone upon which their goods were exhibited, and upon which the spectators stood while examining them, was made upon the spot, of *magnesian cement*, prepared at the manufactory and brought to the ground, where it was mixed and moistened with the earth and gravel found there! The whole mixture was then pounded down, and in the course of a week was transformed into stone as hard as granite!! The manufactured article partakes of the nature and colour of the mineral substances used; that is, broken soapstone is used to make soapstone, marble to make marble. The amount of cement (magnesia) used is so small a proportion of the whole mass as to make no perceptible change in colour or quality.

"The Committee were greatly interested in the articles made by the Company—such as beautiful soapstone stoves, white stoves, hones, medallions with a surface as smooth as polished ivory, emery wheels, and numerous other articles of value. It does not seem improbable to the Committee that this device may be carried so far as to furnish coverings for buildings, tiles, walls, outside chimneys, and even underpinning, where stones cannot be had short of heavy cost of transportation."

The cost of the magnesite is rather high, and in America is dearer than English Portland cement, which costs about one dollar per 100 lb. weight. The advantages of this doubtlessly excellent process must be contingent on the cost of the carbonate of magnesia in the required pure state. In England we are not favoured with any known deposits of this mineral, and to import it from Greece, Canada, or other distant sources would add so much to the cost as to prevent its use for other than the higher priced articles.

For making window caps, sills, steps, &c., the following proportions are used, and the cost given as under for the preparation of one cubic foot of artificial stone :—

		Cents.
100 lb. beach sand cost		·05
10 „ comminuted marble		·02½
10 „ Union cement (oxide of magnesium)		·50
10 „ chloride of magnesium in solution (20° Baumé)		·02
130 lb.	Total cost	·59½

To the above cost would have to be added the labour of forming the blocks, which runs from 20 to 25 cents per cubic foot, and in the case of rough concrete where large stones are used the cost of labour would range from 10 to 15 cents per cubic foot.

These four different artificial stone preparations are thus lengthily referred to for the purpose of showing what is being done in the United States of America in the direction of concrete-making, and it must be evident that this subject receives much more serious attention in that country than in England.

We shall now proceed to examine the various recorded

compressive values of these several compounds, beginning with the Sorel stone by reason of its being the best and strongest. The question of expense, except in so far as already referred to, we shall not further examine, as the cost of all these preparations must be influenced by the local values of the cement, labour, and aggregates employed.

In some trials made at the Boston Navy Yard with 2-inch cubes, the following results were obtained (four samples were tested):—

		lb.
No. 1 crushing strength per square inch	.. .. ..	7,187½
„ 2 „ „ „ „	.. .. ..	11,562½
„ 3 „ „ „ „	.. .. ..	21,562½
„ 4 „ „ „ „	.. .. ..	7,343½

In none of these samples was a larger proportion of the oxide of magnesium used than 15 per cent. by weight. The ages of the cubes are not given, but from the great difference of recorded values it is to be presumed some cause was known for so wide a divergence in the results. The lowest, however, exhibits great strength, and the highest exceed considerably anything obtained through the ordinary agency of a Portland cement concrete.

In another series of experiments made with different aggregates, and the blocks varying in age, the results obtained were as exhibited in the following table:—

TABLE OF TRIALS OF SOREL STONE.

Nos.	Character of the inert material.	Proportions by weight of oxide of magnesium.	Age of blocks.	Size of blocks.	Total crushing strength	Crushing strength per square inch.
		Per cent.		inches.	lb.	lb.
1	Coral sand	12	1 year	2 × 2¼ × 1⅞	26,500	6,235
2	Pulverized quartz ..	12 to 15	1 „	1⅜ × 2 × 1⅜	20,000	7,272
3	Washed flour of emery (piece of hone)	not known	2 years	1⅜ × 2 × 1⅜	54,000	19,636
4	Fine marble	15	3 „	1½ × 1½ × 1½	26,000	11,555
5	Mill sweepings	12 to 13	9 months	1⅞ × 2 × 1⅞	23,000	6,133
6	Marble and sand ..	12	2 years	1⅝ × 2 × 1¼	16,000	4,923
7	Marble with coloured veneer	not known	not known	1¼ × 1¼ × 1	12,000	7,680

ESTABLISHED PROCESSES OF CONCRETE MANUFACTURE. 189

It is not stated in what manner the aggregates were prepared before being mixed with the cementing agent. The mixture of emery indicates a high value, as might have been expected from its favourable character as an aggregate.

The following table shows a series of tests made by General Gillmore (during December, 1870, and January, 1871), of Coignet béton blocks.

The dimensions of the blocks were $3\frac{1}{2}$ in. × $5\frac{1}{2}$ in. × 3 in., the area under compression being $19\frac{1}{4}$ square inches, and two blocks of each compound were tested, with the results as under:—

No.	Proportions of dry ingredients, by volume, loosely measured.	Crushing strength in lb.	
		Total.	Per sq. in.
1	Boulogne Portland cement, 1; common lime powder, 0·4; sand, 5·6	18,000 16,000	935 831
2	Boulogne Portland cement, 1; common lime powder, 0·8; sand, 5·6	15,500 19,000	805 987
3	Boulogne Portland cement, 1; common lime powder, 0·4; sand, 7·5	8,000 10,000	415¼ 519
4	Boulogne Portland cement, 1; common lime powder, 0·8; sand, 7·5	10,000 11,000	551 571
5	Boulogne Portland cement, 1; common lime powder, 0·4; sand, 5·6; gravel and pebbles, 5	12,500 13,100	649 681
6	Boulogne Portland cement, 1; common lime powder, 0·4; sand, 5·6; gravel and pebbles, 13..	13,000 16,000	675 831
7	Boulogne Portland cement, 1; common lime powder, 0·8; sand, 5·6; gravel and pebbles, 5	12,500 12,000	649 623
8	Boulogne Portland cement, 1; common lime powder, 0·8; sand, 5·6; gravel and pebbles, 13..	12,500 14,500	649 753

The whole of the above blocks were only two months old.

General Gillmore states that "Bétons and mortars, in which the matrix is Portland cement *alone*, acquire, during the first two years, fully nine-tenths of the strength and hardness which they ultimately attain in process of time."

During the early period of the introduction of the "Coignet béton" in France, various tests were made by competent authorities, and in July, 1864, M. P. Michelot, Chief Engineer of the "Ponts et Chaussées," obtained a crushing strength of 7495 lb. to the square inch, and in fourteen tests made by the same authority he obtained an average strength of 4670 lb. per square inch.

Frear Stone.

Next in importance to the béton aggloméré. General Gillmore, in a test of several 2-inch cubes, obtained the following results:—

				lb.	Per sq. in.
No. 1. Four weeks old, crushed at	..	..	18,000	=	4,500
„ 2. „ „ „	..	..	18,500	=	4,626
„ 3. Three „ „	..	..	9,000	=	2,250
„ 4. Six months „ „	..	..	8,000	=	2,000

The composition of the above blocks, as reported by Mr. Frear, was one measure of hydraulic cement, two and a half measures of sand, moistened with an alkaline solution of gum shellac of sufficient strength to furnish 1 ounce of the shellac to 1 cubic foot of the finished stone. Portland cement was used in Nos. 1, 2, and 3, and Louisville cement in No. 4. A 4-inch cube of the composition of Nos. 1, 2, and 3, and four weeks old, sustained a pressure of 57,000 lb., equal to 3562½ lb. to the square inch, and was not crushed.

Foster and Van Derburgh System.

The previous remarks on the compressive strength of some of these blocks at page 181 may be supplemented with advantage by the table on page 191, furnished by General Howard.

All the blocks were crushed on metal beds except No. 3, which was tested on a plaster bed.

The introduction of a ten-year old block into these experiments is not quite reasonable, for it lifts up the averages to a misleading height.

These tests show, however, that there is considerable progressive value in this material.

ESTABLISHED PROCESSES OF CONCRETE MANUFACTURE. 191

Size of blocks, 10" long by 5" wide by 4" deep. Hollow space in centre, 6" × 1". Area under compression, 44 square inches.

No.	Remarks.	Age of blocks.	Crushing strength of blocks.	Strength per square inch.
			lb.	lb.
1	From Howard University 1st story	12 to 13 months	19,500	443
2	Ditto 2nd „	not known	20,000	456
3	Ditto 3rd „ A	6 months	13,000	295
4	.. ..	11 „	16,900	384
5	.. ..	10 „	24,800	564
6	.. ..	9 „	15,500	352
7	Red block	8 „	15,900	361
8	.. ..	7 „	12,000	273
9	.. ..	6 „	9,000	205
10	.. ..	5 „	8,700	198
11	Damp	4 „	7,600	173
12	.. ..	3 „	7,900	180
13	Angle off block	15 to 16 months	18,760	426
14	.. ..	about 12 months	15,600	355
15	.. ..	ten years	64,000	1,455
16	Made of yellow and loamy sand, refuse block	not known	8,200	186
	Average strength of blocks 1, 2, and 3	.. ..	17,500	398
	Ditto of the whole 17 blocks ..	.. , ..	17,353	394

In point of cost, these four examples of American manufacture of artificial stone stand in the following order of cheapness:—

1st. Foster and Van Derburgh stone.
2nd. Béton aggloméré.
3rd. Frear stone.
4th. The Sorel stone.

If we except the Sorel stone, it will be found that the others are in a great measure indebted to the admixture of Portland or other good hydraulic cement for their excellence, and when it is omitted in any of these preparations a marked lessening of indurative value arises.

The blending of common or any other limes with Portland cement is not, from our experience, of much advantage in saving of cost, and we feel satisfied that the danger of inaccuracy in the mixture (which must be made before the blend can be added to the aggregate) is more than counterbalanced by any fancied advantage in cheapness. The lime must be either finely ground or slaked, and in the latter condition it is difficult to mix. In all the tests made by us in this direction we never have obtained good strong briquettes, and frequently their hydraulicity was greatly impaired by the addition of the lime. In "béton aggloméré" the process of compacting is so good and accurate, that possibly the disadvantages of the combination are not apparent when so compounded.

In the experiments made by Pasley during 1836 on Ranger's artificial stone in comparison with Yorkshire natural stone, the following results were arrived at:—

The blocks under examination were 3 feet long by 18 inches wide and 15 inches deep, the age of Ranger's stone being two years.

The Yorkshire stone bore a weight of 13,512 lb., applied in the centre of the supported beam.

Ranger's artificial stone was tested in a similar manner with three beams: 1st, breaking with 6285 lb.; 2nd, 5141 lb.; 3rd, 2930 lb.

Pasley took the average of 1 and 2, rejecting 3, owing to his belief in its imperfect manufacture.

It must be remembered that these artificial blocks were made with unslaked lime, and consequently liable to imperfect combination with the gravel with which it was incorporated.

The same experimenter found that the artificial stone so made was, when compared with Yorkshire stone, as thirteen to one, in small prisms of 2 inches wide by 2 inches deep supported on points 3 inches asunder.

With such treatment of the materials employed by Ranger, it is indeed surprising that even such favourable results as those recorded were obtained.

At the time of these experiments an accurate or technical knowledge of concrete preparations did not exist in this country, and the most advanced engineering intelligence of that day could not even reach the excellence attained during the Roman concrete constructive period.

We have thus lengthily referred to the American progress in block or artificial stone-making for the purpose of comparing it with English practice in the same direction. With the exception of the "Coignet béton" system, all these processes have in view the moulding of suitable forms for constructive purposes, and they ignore, or, at all events disregard, the assistance of movable frames. The béton, as generally used, is of course rammed or impacted (by gentle and frequently recurring blows), and when so treated the result obtained is an almost compact mass, from which the comparatively temporary guiding boards may speedily be removed without endangering the stability of the structure. This result is primarily obtained by the accurate character of the blended compound and the absence of any excess of moisture. In fact, under this intelligent process the character and condition of the paste or béton is regarded as of the first importance, and the frame or guiding arrangement as quite subsidiary in the operation of building. Compare this accurate application of sound technical knowledge with the concrete building operations in this country. Since the introduction of the frame building system much attention has been directed to the suitability of concrete for building purposes generally. From 1836 (when Ranger first introduced his lime concrete blocks) until now the subject has received much consideration, but from that inventor's ambition to develop his system, and also probably from a disregard of the chemistry of the question, he executed works for which his artificial stone was unsuited. In the wharf wall at Woolwich Arsenal the lime-dissolving action of the water

o

soon proved that it would be dangerous to use a concrete of that character for such a purpose.

The failure of this work checked the use of concrete, and it was not again in vogue until ten years later, when improvements in cement encouraged further examination of the question, and a quasi-scientific commission was appointed to obtain engineering evidence as to its suitability for the construction of the harbour works at Dover. Since then, notwithstanding the opposition of some engineers, concrete is now almost universally adopted in hydraulic engineering, and with marvellous success. Some of the most important harbour works at home and abroad have been built of this valuable constructive material, and it may be safely asserted that many of these indispensable refuges for storm-tossed vessels would have been impossible without its aid.

In the description and particulars we have given of the more prominent foreign concrete or artificial stone preparations, we have limited our attention to what may be strictly characterized as the simple compounds obtained from the associated ingredients readily commanded in almost every locality. From the ingenuity displayed in nearly every one of these preparations, it must be evident that the subject of block-making especially commands much more attention in America than with ourselves. While houses and such moderate-sized structures as come within the range of assistance from frames are being erected in various parts of this country on the monolithic principle, much progress is also being made in the manufacture of paving slabs, sewer pipes, and other articles. The importance of these industries, however, will warrant our noticing them in a more lengthened and particular manner, and for that purpose we shall devote Chapter IX. to their special consideration.

Ransome's Siliceous Stone.

A work on concrete would, we think, be incomplete if it did not refer to this well-known material.

The processes by which the Ransome stone is fabricated are essentially chemical in character, and require the exercise of previously trained or skilled labour. The materials employed are of the simplest kind and obtainable in abundance in almost every locality.

Flints (converted into "soluble glass" or "liquor of flints" by the agency of caustic soda and heat) are mixed with sands naturally in a clean state, or rendered suitable by washing; the mixture afterwards in a concrete form being placed in solution of chloride of calcium. In fact the cementing agent used in this preparation is silicate of soda, instead of cement or lime, and the character of the resulting stone depends on the quality and amount of silicate used.

The silicate may either be made in the manner described in Chapter VII., or purchased direct from the chemical manufacturers. Both plans have been adopted, but it is perhaps under certain circumstances better to manufacture your own.

The silicate of soda is used about the consistency of treacle or syrup, and in that condition has a specific gravity of 1·7.

The proportions used are about one gallon of the silicate to a bushel of sand or other suitable aggregate. This compound is carefully mixed in a pug mill or similar machine, and becomes by such treatment a viscous, putty-like paste. The process of mixing must be done quickly, being usually performed in about five minutes. The paste is then moulded into the desired forms and may be readily shaped, for in that state the material is capable of facile conversion into any form. The manufacturing process is then completed by drenching the blocks or slabs with a solution of chloride of calcium (cold). They are then, and as soon as possible, placed in suitable cisterns, containing a solution of chloride of calcium of 1·4 specific gravity heated to boiling point. In this bath the chemical reaction is perfected, resulting in obtaining an almost indestructible silicate of lime throughout the concrete mass. The chloride of sodium (common salt) developed in the process by the combination of the liberated sodium and chlorine is carefully washed out; the artificial stone is then complete, and when accurately prepared is exceedingly hard.

The Ransome stone process is no doubt highly ingenious, being based on correct chemical principles, the disregard of, or inattention to which must lead to disappointment and loss. It is this necessity of careful manipulation that has prevented the general adoption of this undoubtedly valuable material in the production of blocks for decorative purposes, for which it is admirably suited. The manufacture of this stone cannot be entrusted to such a class of labour as would be competent to make Coignet's béton and the other concrete preparations having for their basis of strength good hydraulic limes or Portland cement.

Dr. Frankland in December, 1861, made elaborate experiments with the object of ascertaining the relative value of Ransome's artificial and other well-known natural building stones. We give the full report as beyond the question of Ransome stone; it illustrates the chemical properties of building materials in most general use in this country.

Dr. Frankland says:—

"I have submitted to experimental investigation samples of stone forwarded to this laboratory (Royal Institution, London), and have now to report as follows:—

"The experiments were made in the following manner; the samples were cut as nearly as possible of the same size and shape, and were well brushed with a hard brush. Each sample was then thoroughly dried at 212°, weighed, partially inserted in water until saturated, and again weighed; the porosity or absorptive property of the stone was thus determined.

"The sample was then boiled with water until all acid was removed, and again weighed. Finally it was dried at 212°, brushed with a hard brush, and the total degradation or loss since the first brushing was ascertained. The following numbers were obtained."

The specimens of Ransome stone experimented on were fourteen days old.

The object of these experiments was to determine the porosity of those stones examined, and also to value the amount of degradation to which they are liable under varied circumstances.

DR. FRANKLAND'S EXPERIMENTS.

Name of Stone.	Porosity. Percentage of water absorbed by dry stone.	Percentage alteration in weight by immersion in dilute acid.						Total percentage, loss by action of acid, and subsequent boiling in water.	Further loss by brushing.	Total degradation from all causes.
		Of 1 per cent. acid.		Of 2 per cent. acid.		Of 4 per cent. acid.				
		Loss.	Gain.	Loss.	Gain.	Loss.	Gain.			
Bath	11·57	1·28	..	2·82	..	2·05	..	5·91	·26	6·17
Caen	9·86	2·13	..	4·80	..	·67	..	11·73	1 60	13·33
Aubigny	4·15	1·18	..	4·00	..	..	1·04	3·56	·29	3·85
Portland	8·86	1·60	..	1·10	..	1·35	..	3·94	·24	4·18
Anston	6·09	3·52	..	3·39	..	3·11	..	11·11	·27	11·38
Whitby	8·41	1·07	..	..	0·53	none	none	1·25	·18	1·48
Hare Hill	4·31	·75	..	..	·60	none	none	·98	·15	1·13
Park Spring	4·15	·71	..	..	·10	·15	..	·81	none	·81
Ransome's	6·53	..	·95	none	none	none	none	·63	·31	·94

Another series of experiments by Professor Ansted on the transverse and tensile strength were as under.

Transverse Strength.—A bar of Ransome stone, 4 inches square, supported on an iron frame (on which it was bedded an inch on each end), having 16 inches clear, broke by a weight of 2122 lb. suspended from its centre. A bar of Portland stone, of the same dimensions and similarly treated, broke with a weight of 759½ lb.

Tensile Strength.—These tests were made on pieces of stone having a sectional area of 5½ square inches at their weakest points, and notched so as to receive the clamps by which they were held during the operation of testing.

	Lb.	Per square inch.
Ransome stone broke at	1,980 =	360
Portland „ „	1,104 =	201
Bath „ „	796 =	145
Caen „ „	768 =	140

A 4-inch cube of Ransome's stone bore a crushing weight of 30 tons or 4200 lb. to the square inch.

These may be regarded as the highest values obtained in the examination of Ransome stone. In other experiments it was found that considerable differences in quality resulted, the variability ranging in the tensile strength from 97 lb. to 533 lb. per square inch.

In this process the utmost strength was obtainable at the final stage of the manufacture, and unlike the Coignet béton and other purely cementitious compounds, the strength was stationary in the faultless blocks, and probably would deteriorate by age. The best specimens of Coignet béton at two years old gave a crushing strength of 7500 lb. to the square inch.

Mr. Ransome took out another patent for an improvement in the manufacture of artificial stone, dispensing with the use of chloride of calcium, avoiding thereby any difficulty which might have arisen when the common salt had not been effectively eliminated from the stone. The new process is much more simple and less expensive.

The Ransome stone process may be regarded as a purely chemical one, and in the operation of the manufacture success was only possible when the rules were accurately attended to. A process so hedged in with technical details, unusually precise in character, necessitated the employment of intelligent labour, which coupled with a dependence on outside help for the proper quality of some of the chemical agents, led to many mishaps in the prosecution of this interesting industry; so much so indeed, as leading ultimately to a partial discontinuance of the business. The many existing examples of architectural decoration in Ransome stone indicate that the material is a fairly reliable one, capable of being used for ornamental details, and it is to be regretted that its prosecution on an extensive scale in London is for the present in abeyance, for it is well suited for internal use.

We will now proceed to consider a valuable process of artificial concrete, called by its inventor (Mr. Buckwell) granitic breccia stone. Unlike Ransome's siliceous stone, this preparation is produced from the simplest ingredients, and its value really depends on the peculiar manner of their treatment. Portland cement of the best quality as a matrix, and oolitic or magnesian limestone as an aggregate, are brought together in the closest contact in iron moulds, by the impact blows of a rammer, directed by manual labour. This industry was introduced at a time when the manufacture of Portland cement was carelessly performed, and some of the results obtained were unsatisfactory, owing to the doubtful quality of the matrix used.

The operation of compacting the mass by impacting blows enabled the process to be accomplished with the smallest quantity of moisture; indeed, the amount used was almost imperceptible, giving the mass an appearance of incoherency which a few days' setting converted into a hard, granitolike stone. The quantity of water now generally employed in concrete work would prevent any beneficial action from the impact blow, as the percussion, instead of driving the materials into close contact, would merely keep them in a state of unprofitable disseverance. The amount of mixture put into the mould at one time was small, this process resembling the treatment adopted in 'making "Coignet béton," except that in the one case blocks are only attempted, and in the other, the monolithic operation is carried out.

There are, however, many examples remaining to prove that the process of impact was a sound one and capable of advantageous employment. In 1845 a considerable surface of pavement, in slabs 6 feet long, 3 feet wide, and 3 inches thick, was laid at Lewisham, in Kent, and they continue until now in good condition, apparently equal to another thirty years' wear. The aggregates used in the "granitic breccia" were too large, and in some cases a mixture of rough gravel was employed, which in those slabs most worn develops an uncomfortable and unsightly roughness on the surface. Buckwell's "granitic breccia" was used for many purposes besides that of paving, testimonials of a satisfactory character having been given as to its suitability for a variety of engineering operations.

The experiments made to test the value of "granitic breccia" stone in comparison with natural stones of established reputation, resulted in the following figures:—

	Granitic breccia stone landing.			Calverley wood stone.	Ottley Bramley Fall.	Trickett's Bramley Fall.	Yorkshire stone.	Cheesewring granite.
	Twelve months old.	Six months old.						
	6" × 6" 3¼" thick.	6" × 6" 3¼" thick.	6" × 6" 3¼" thick.	6-inch cubes.	6-inch cubes.	6-inch cubes.	6-inch cubes.	6-inch cubes.
Pressure in tons to crack..	73	73	56	73	62	..	..	..
Pressure in tons withstood before breaking	104	88	81	81	73	52	66	45

These experiments were made by Mr. Andrews at the London Docks in March, 1858, to test the value of this artificial stone for constructive purposes in connection with these works under his control.

From the experience obtained in the preparation of at least two of these artificial stones we have referred to, other important industries have sprung, the Sorel or magnesite process, and the Ransome stone. From these inventions great advantages have resulted in the manufacture of siliceous and emery wheels now used most extensively for dressing and polishing metals of every kind in this country and America.

For the purpose of testing the relative values of the Newcastle (natural) grinding stone and the Ransome artificial stone (sand hardened by the chemical treatment), Messrs. Bryan, Donkin, and Co., in the year 1867, made the following experiments:—

By an ingenious method of fixing a bar of steel $\frac{3}{4}$ of an inch in thickness, to which a spring fixed in a tube was attached, and which pressed against the stone under examination, the result obtained was thus:—

A quarter of an inch of the steel was ground by Ransome's stone in sixteen minutes, the same length of steel taking eleven hours to grind by the Newcastle stone. The speed of the latter stone was at least 20 per cent. greater than the former, which is still further in favour of the artificial grinder.

These wheels are manufactured in England by Messrs. Bateman and Company, of East Greenwich, near London, who have kindly furnished materials for the tests and experiments recorded in Chapter XVII.

The magnesite, or rather oxide of magnesium, wheels made under the Sorel process have a high reputation in America, where they are well known as the "Union Wheels."

They are exclusively manufactured in England by Messrs. Hodges and Butler, of East Greenwich, near London.

The proportions used are from 10 to 15 per cent. of oxide of magnesium with emery. These wheels are of high tensile strength, as is shown by the high velocity with which they can be driven. They are quite safe at a speed of from two to three miles per minute of circumferential velocity, and do not usually break until the speed is increased to from four to five miles per minute.

The materials for the tests and experiments in Chapter XVII. were furnished by Messrs. Hodges and Butler.

Messrs. Williams and Son, of Hales Cliff (near Liverpool), carry on the manufacture of Ransome's stone on a considerable scale, and some extensive works, notably that of the St. George's pier head approaches to the landing stage, Liverpool, have been made by them. Although greatly improved of late years, this stone does not, in all cases, resist with advantage the action of severe frost. In our Chapter of experiments at page 363 will be found tests from materials furnished by Messrs. Williams and Son, which indicate a high tensile capacity.

CHAPTER IX.

ENGLISH CONCRETE INDUSTRIES.

IN the last chapter we described the various manufactures of concrete and artificial stone making carried on in France and the United States of America, and we now proceed to notice similar processes in this country.

The beginning of scientific concrete-making may be said to have originated with Messrs. Ranger and Buckwell, who had, during their early efforts at least, the difficulty of inferior matrices to contend with. The difficulties encountered by the former gentleman we have elsewhere referred to, and in their description (quoted from the writings of Pasley) ample proof is shown that none of the details of the Ranger process can be advantageously adopted by modern constructors. Although in a modified form, some parts of the specifications of Ranger's patents are adopted in the Van Derburg process in America; we do not gather, however, after a careful examination of the practical results obtained, that any very satisfactory progress has been made by such adaptations.

The principle of impact, however, first applied by Mr. Buckwell, in addition to a careful selection of the materials, reaches a point of excellence unattainable by any of the other manufacturers of artificial stone, either in this country or abroad. The primary object in the application of the impact process is to drive into the closest contact the matrices and aggregates. This can only be done with advantage in the absence of a superfluity of moisture, and indeed the small amount of water used in the operation is calculated to create a doubt in the minds of the uninitiated as to the possibility of obtaining a durable concrete at all by such means. But the advantage of the impact treatment reduces the porosity to a minimum, and almost succeeds in getting rid of the more or less objectionable interstitial vacuities. A concrete of such a character cannot, according to Mr. Buckwell's theory, be benefited by the application of any silicate, for to use these chemical aids with advantage, a certain amount of absorptive porosity is necessary in the stone to which they are applied.

Mr. Buckwell made blocks of 90 tons weight and upwards, more than thirty years ago, and at that time they were undoubtedly the largest concrete monoliths attempted by any engineer. A block of this size used for mooring purposes was placed in the River Thames at Pickle Herring Wharf, and during the first tide after its deposition, a vessel struck her keel against it and remained in that position until the next tide; no damage was done to the artificial mooring block. The materials used in making this block were Portland stone chippings and Portland cement, the aggregate being damped before mixture with dilute silicate of soda.

Many blocks of a similar kind were used by the River Thames Conservancy at different places, having been found less expensive and more reliable than those blocks made from granite or Portland stone. In fact, the cost and difficulty of obtaining such large natural stone blocks of a sound and reliable character was very great, and the artificial blocks made under Mr. Buckwell's process were regarded with much favour by Mr. Leach, Engineer to the Board of Conservators of the River Thames.

Another difference in the practice of "granitic breccia" construction was the use of the heaviest slow-setting Portland cement obtainable, the difficulty of contending with its tardiness of initial set being overcome by the use of dilute silicate of soda instead of water. This application of the silicate increased the setting energy of the cement, and obviated any objection to a protracted induration of the resulting concrete.

Pipes of all sizes, ranging from 6 inches to 12 feet in diameter, in 4 and 5 feet lengths, were also made by the impact process we have described.

Although Mr. Buckwell does not approve of the silicate bath (since the time of his operations prosecuted with much success under the Victoria stone patent), he resorted to a modified form of silicate treatment in his mixtures before applying the impact force of combination, and, from the examples of his "granitic breccia" stone we have examined, succeeded in producing monoliths of surpassing excellence.

The force applied in the "granitic breccia" process differs essentially from what is generally understood as pressure produced either by hydraulic agency or screws. These forces should be regarded as simple squeezers or compressors, for the thickness of material capable of successful compression by them is but limited. In many of the machines now recommended, and in some systems of blockmaking employed, the difficulty of profitably compacting concrete masses is imperfectly understood, otherwise a continuance of the use of such means would soon cease. The difficulty surrounding these pressing machines is the impossibility of getting rid of the contained air, which, from its elasticity, resists the pressure applied, acting simply and wastefully as a cushion, which the utmost efforts fail to expel. This defect is common to some brick presses, and in their case is well illustrated by the distortion of the plastic brick on the withdrawal of the applied force of compression on leaving the mould.

In the production of "béton aggloméré" the impact force is applied in a limited, and therefore effective, manner, so that the contained air in the paste as received from the "mixer" is eliminated by repeated and gentle impact blows of a moderately weighted rammer.

Neither "granitic breccia" stone nor "béton aggloméré" have any appreciable quantity of voids or unoccupied spaces when carefully made, and both the inventors (Messrs. Buckwell and Coignet) therefore consider their artificial stones so perfect as to dispense with any chemical aid from soluble silicates, as from their point of view there is no space which they can profitably occupy.

A reference to the tables in Chapter XVII. will show the relative values of the various artificial stones we have referred to in this work.

There is a little bit of interesting history in connection with the early efforts of Mr. Buckwell to introduce his system of stone-making into London.

The late Dean Buckland was particularly impressed with the value of an artificial stone of the character of the "granitic breccia," and we believe that name was originally given to it by the eminent geologist. A short time before his death he purposed giving a lecture on the "granitic breccia" and "Ransome's siliceous stone" at the Royal Institution of Great Britain; his last illness, however, prevented his doing so, and Faraday, the great philosopher, lectured on behalf of his friend the Dean.

The lecture was delivered on the 26th of May, 1848, and in a condensed report of it (furnished by Professor Faraday) in the 'Athenæum' of the 17th June, 1848, the following information was given:—

"Owing to the absence of Dean Buckland, who had arranged to give a lecture on Ransome's stone and Buckwell's concrete.

"As the artificial stone of Mr. Ransome is chiefly applicable for ornamental purposes, so Mr. Buckwell's invention, termed by him artificial granite, appears exclusively designed to supply the place of blocks brought from the quarry for large works, whether walls of houses, or of aqueducts, sewers, &c. Mr. Buckwell uses the following simple process. Fragments of suitable stones (Portland stone for example) are gauged and sorted into sizes. These are cleaned and carefully mixed on a board with cement, in the proportions of five parts of large fragments, two of smaller ones, one of cement, and a portion of water, but the water is in no *greater quantity* than will bring it to the dampness of sawdust. This being done, the materials are put into a strong mould to the depth of about 1½ inch at a time; they are then *driven together by percussion;* more materials are now put in; these in turn hammered together, till the water has escaped by holes pierced for that purpose in the moulds, and this process is continued until the block or pipe has attained the required magnitude. It is then taken out of the mould, and now found to be so hard as to ring when struck, and in ten days is fit for service. It is affirmed to harden under the influence of moisture, to bear, when moulded in the form of girders, a greater transverse pressure than any rock except slate, and to be nearly one-sixth of the cost of brickwork. It will be noticed that the process is characterized by the use of fragments, by the quantity of cement employed (not one-fourth of the proportion used in common *grouting*), and by water instead of fire being made the means of bringing the fragments into close union. Mr. Faraday then noticed two scientific principles, on the success of which Mr. Buckwell's process depends.

"1st. *The use of water in effecting the approximation of the particles, and the exclusion of air.*—It has been ascertained by Dr. Wollaston (Bakerian Lectures, 1828) that, in order to bring the particles of platina into close contact, it is best to bring them together in water. When a freshly made road is watered to make the materials bind together, the same principle assists in the result. Having filled a measure glass with sand, Mr. Faraday showed that when the glass was first filled with water, and then the sand added with agitation, it occupied less space than when dry.

"2nd. *The effect of percussion in bringing particles together.* —Mr. Faraday noticed that simple pressure will not displace interstitial air or water, but that a blow will. Water contained in a small cylinder of wire gauze was shown remaining in the open network when subject to the pressure of a column of the same fluid, though it freely ran through the meshes when the cylinder was gently struck; on the same principle moistened sand on the sea-shore gives way and leaves a footmark under the limb that strikes it. In conclusion, Mr. Faraday noticed the remarkable fact that the sedimentary matter in sewers, &c., does not accumulate on Mr. Buckwell's granite as it does on glazed pipes."

These particulars and advantages claimed by Mr. Buckwell as peculiar and essentially salient features in the "granitic breccia" stone have, unfortunately for its reputation and that of its inventor, failed to gain what can be regarded as a permanent position in the constructive sense. A similar fate has all but overtaken Ransome's siliceous stone, as an external aid in constructive operations.

Both these ingenious processes designed and, in their early stages at least, prosecuted under the personal attention of their respective inventors, were surrounded by circumstances of more or less favourable encouragement. Indeed it is seldom that we find any novel and untried invention receiving so much considerate attention as that given to the "granitic breccia" and "Ransome's siliceous stone" by two such illustrious scientific men as Professor Faraday and Dean Buckland. Yet these two undoubtedly valuable applications of artificial concrete-making have, even under such favourable auspices as those which we have recorded, failed in acquiring a permanent position of acknowledged usefulness.

Mr. Buckwell's retirement from the prosecution of his "granitic breccia" process, in this country at least, may in a great measure account for want of progress in the use of his material in England, and probably the limited employment of "Ransome's siliceous stone" is due to the difficulty of securing the required technical or skilled labour so necessary for its successful manufacture.

On the lines, however, of these two original discoveries, and partaking more or less of their distinctive peculiarities, new concrete industries have been established which command public confidence, due to an appreciation of their excellent and now well-tried advantages. We will now proceed to describe these manufactures, beginning with the Victoria patent stone process.

Victoria Patent Stone Process.

This industry, limited, for the present at least, to England and Scotland, has obtained a well-earned reputation for the excellence and durability of the paving slabs and other articles of constructive utility produced by the manufacture having the above title.

For many years, and more especially since a reliable chemical appreciation of the origin and causes of the premature decay of natural rocks and the numerous products derived from the various geological formations, a great desire has existed to avert the dangers incidental to their incipient and inevitable disaggregation. At first various chemical solutions were applied in a number of forms, but in the end, and after the most persistent efforts of well-intentioned and equally well-informed experimenters, these had to be abandoned owing to the want of the desired success.

P

From want of success in the direction of the surface solution and the indurating bath, chemical ingenuity was finally directed to the possibility of meeting nature half-way by an interference with her apparently most beautiful and indestructible productions. The granites, porphyries, sandstones, and marbles, in their variety of colour and texture, were found by chemical investigation to have in their constitution inherent blemishes, which on the exposure consequent on their utilization for constructive purposes became developed into hitherto unseen and unavoidable (in their natural condition) dangers.

The use of concrete in all ages of recorded history indicated that stones of all sizes and qualities had been used in conjunction with cementing agents of varied quality with considerable success. These ancient remains, still open to our examination, show that they were generally concreted in irregular monolithic masses by the agency of lime and various natural and artificial compounds, such as puzzolana, lava, and burnt clay. In almost all cases it is found that according to the care bestowed in the reduction of the lime to a fine condition, so was the success realized in a strong and firmly bound concreted mass; the cement apparently indestructible while the aggregates, according to their mineralogical constitution, exhibited, where externally unprotected by the binding agent, a varying amount of decay.

Metallic oxides and soluble alkalies in all their variety of forms and developments, with which the aggregates in various degrees were associated, eventually became the inherent causes of their disaggregation when they were exposed to the action of air or water.

The experience gained from an intelligent examination of the best known building and paving stones, developed the fact that granites (more especially those with large crystals of quartz or felspar and an excess of alkalies) became speedily degraded by the dissolving out in moist situations of the cementing paste by which they were apparently indissolubly bound together. Soda and potash, with various proportions of other less soluble chemical compounds, loosened their hitherto beneficial grip when subjected to continuous or repeated states of moisture.

Sedimentary rocks in their various fine laminæ of original mechanical deposition, in like manner became degraded, not from any inherent chemical cause, but simply owing to the wearing out, where exposed, of the thin leaves of (in some cases almost invisible) siliceous and carbonaceous partings to which in some of the best known flag or paving varieties the coherence of the mass is due.

From a careful examination of these causes, and assisted in his inquiries by the experience obtained from such concrete pioneers as Messrs. Ranger and Buckwell, and doubtless also in a great measure guided by his own chemical knowledge, the late Mr. Highton hit upon the now well-known process of "Victoria stone" making. Conscious of the weak points inseparable from the granites in their original massing, he reduced them to small pieces, added the best binding agent (Portland cement) at his command, and finally completed the concrete thus produced by impregnating it with, under the most favourable conditions, a specially prepared solution of flints or silica (silicate of soda).

Mr. Highton established his industry in London, and found within a comparatively easy range of his manufactory the necessary materials for his purpose; granite from Leicestershire, Portland cement from the Thames or Medway, and the natural silica from Farnham in Surrey. The only ingredient for which he had to go a distance (Lancashire), was the caustic soda with which he treated the silica to convert it into a silicate. This was, however, a small matter, as the caustic soda required in the process was very limited in quantity, and did not by its extra cost of carriage appreciably increase the cost of the manufactured concrete.

Like all other novel industries the "Victoria stone" has had to contend with the prejudice and ignorance of those whose appreciation of its merits could alone ensure its early profitable reception. And in addition to these usual concomitants of stubborn opposition, much disfavour in the recognition of the value of the artificial stone arose from errors of judgment on the part of the inventor himself, by placing a too confident reliance on the virtues of his silicating process. He unwisely, and before becoming familiar with the best properties of Portland cement, believed that lias lime and other less valuable cements would suffice to produce a good concrete, and that whatever deficiencies arose from their being used would be more than counterbalanced by (to his mind) the all-powerful panacea of silication. To these avoidable causes a degree of evil repute surrounded the use of the "Victoria stone" in its infancy, but this perhaps not unreasonable prejudice and the causes of its origin gradually disappeared when new lines were adopted in the details of the process of manufacture. The use of all inferior or doubtful cements was abandoned, the broken granite was washed and rendered scrupulously clean, the best available Portland cement insisted upon, and the concrete itself ultimately saturated with the silicate in its most efficient and profitable condition. All these stages of the process were regulated and controlled by the best technical rules, nothing being left to the accidents and dangers of hap-hazard or rule-of-thumb treatment.

With these preliminary observations we will now proceed to describe the particulars of this daily increasing and important industry.

The granite is at present obtained from Leicestershire, and after being reduced to the desired size it is carefully washed, either at the works or the quarries, and sometimes a small proportion of clean gritty Thames sand is added, according to the quality and character of the work for which it is destined.

ANALYSIS OF LEICESTER GRANITE FROM THE GROBY QUARRIES.

Silica (soluble)	0·55
Ditto (insoluble)	65·26
Alumina	13·06
Lime	4·55
Magnesia	1·01
Oxide of iron	9·81
Carbonic acid	0·03
Soda	2·34
Potash	2·85
Water, &c.	0·54
	100·00

An examination of this preliminary treatment of the aggregate wherever it may be performed, whether washed at the source of supply or in the "Victoria Works," will show the most heedless concrete-maker that even the best aggregates should be purged of their impurities under all and every circumstance. The Victoria Stone Company cannot afford to use unclean materials, and, very wisely in the writer's opinion, leave all impurities behind at the quarries, as well as, by the present arrangements, the responsibility of crushing and cleansing to the quarry owner. By so doing they reduce their labour and eliminate any cause of anxiety consequent on the obtainment of not the least important ingredient in their manufacture.

The Leicestershire granite quarries are on the Midland Railway and thus command favourable facilities of transit. This granite, tested by Mr. Kirkaldy in 1863, was found to bear a crushing or compressive strain equal to 20,742 lb. per square inch, and in the same series of experiments it was proved that a cube of Guernsey granite, under like conditions of testing, stood a pressure of only

15,062 lb. per square inch. In reference to this somewhat unexpected result, Mr. Etheridge, F.G.S., of the Museum of Practical Geology, London, thus observed, "And it is not a little singular that the specimen from Guernsey, which possessed the highest specific gravity, should have stood the least pressure."

The crystals of the Leicestershire granite are regular in character, being of moderate size, and well cemented by a paste which the analysis shows is unusually free from the destructive alkaline ingredients.

The small size of the pieces of granite used in this manufacture renders the presence of the alkalies practically innoxious, because the artificial cement used surrounds these pieces with a protective coating, and prevents the possibility of any dissolving action by the air or moisture.

A reference to Plate 2 will illustrate the size of the broken granite (being exactly the natural size) now used in making the Victoria stone. Fig. 1 on Frontispiece shows a section of Victoria stone, nine years old (also the exact natural size), when much larger pieces were used. Since that time the advantage of small-sized aggregates has been fully recognized, and is now adopted in all cases, whatever the character of the manufacture, and whether it is required for simple paving slabs, or the most ornate mouldings as aids to architectural embellishment.

Fig. 2 on Frontispiece shows a piece of Victoria stone, six years old, which exhibits smaller aggregates than those of Fig. 1.

Pavements have been laid with slabs made according to both systems, one of which, of the quality of Fig. 1, may be seen at Piccadilly, near Devonshire House, and the other, of the mixture of Plate 2, in Holborn, between the Fire Brigade Station and Chancery Lane on the same side of the street.

Plate 2.

BRIQUETTE SECTION OF
PATENT VICTORIA STONE
NATURAL SIZE AND COLOUR.

AS AT PRESENT MADE WITH FINE GRANITE
THREE MONTHS OLD.
TENSILE VALUE 740 LBS PER SQUARE INCH.

The above references to works of easy inspection are given for the purpose of affording those desirous of information on the subject an opportunity of personal examination.

A considerable mileage of paving has been laid in London as well as in other parts of the country.

Having secured the preparation of the aggregate, the next ingredient required is all-important, and no precaution is neglected to secure its being of undoubted excellence.

The Portland cement is at present obtained from the Medway or Thames, and when possible weighs upwards of 112 lb. per imperial bushel. Under existing circumstances the fineness of the cement is not equal to the requirements of the Victoria Stone Works, and to bring it up to the wished-for condition it is put through a sieve so as to eliminate the coarsest particles of unground clinker.

ANALYSES OF PORTLAND CEMENTS USED BY THE VICTORIA PATENT STONE COMPANY.

FROM THE RIVER MEDWAY.

Potash and soda	1·90
Lime	61·20
Magnesia	1·30
Peroxide of iron	2·80
Alumina	9·35
Silica	20·80
Sulphuric acid	1·85
Phosphoric acid	0·10
Carbonic acid	0·40
Water	0·30
	100·00

FROM THE RIVER THAMES.

Silica	22·05
Alumina and oxide of iron	12·15
Lime	58·80
Magnesia	0·91
Combined water, alkalies, &c.	6·09
	100·00

The cement is taken out of the sacks in which it is received from the manufacturers, carefully stored away in lots of about 60 tons in the cement shed, well protected from the weather. From the heaps so stored sample lots are taken by the superintendent of the works and made into briquettes having a breaking surface of 2·25 square inches. These briquettes are exposed to the air for twenty-four hours and then immersed in water, where they remain for six days, when they are tested by a single-lever testing machine, and if the average results of tensile value realize 350 lb. per square inch, the cement is regarded as fit for the further operations at the works. This is what may be regarded as the works test, performed by the superintendent and under his most vigilant control, but further precautions are observed by having other samples of the cement sent to the offices in London, a different treatment being there pursued.

Instead of the ordinary and, to the writer's mind, somewhat wasteful and not always reliable test by the single-lever testing machine, the double-lever machine, described and illustrated at page 59, Fig. 6, with briquettes of one inch square section, is used.

This is the final or crucial test, and if it is sufficiently fine and passes this last ordeal the cement is considered safe and fit to use. Until, however, this double, and by no means unreasonable, testing is observed and proved satisfactory, not one ounce of the cement is allowed to be used in the manufacture of "Victoria stone."

It may appear that the above-described care in testing the cement is too complicated in character, but it is the practical result of long experience, and doubtless arose from damage and loss sustained in past times from a too confiding reliance on the cement-maker and his failings.

For upwards of six years some of the original cement

briquettes have been preserved, and they on examination indicate a high indurative value. Since they were broken (when taken out of the water after six days' immersion) they have been kept in the testing room of the works, and no indication of surface or internal disturbance is apparent in any of them.

The aggregate and cement having become in a manner guaranteed by the treatment and precautions we have described, are mixed together in the desired proportions. Paving slabs being the largest item of Victoria stone industry, we shall select that department of the manufacture for our present description of the process.

Three parts of aggregate (including, if found necessary, a limited proportion of the cleanest and purest Thames sand) are thoroughly mixed in a dry state by hand, and the water then added in a careful manner, so as to avoid the danger of washing out any of the finer and more soluble portions of the cement. The mixture is again carefully turned over by hand, and before any initial set of the crude concrete mixture can arise it is put into the moulds, in which it is carefully worked with the trowel so as to fill up the angles and sides, thus ensuring accurate arrises all round. To prevent adhesion to the sides and bottom of the moulds, they are rubbed with an oleaginous preparation, which effectually secures this desirable object. The moulds are made of wood, being lined internally with metal, not only to secure accuracy of form, but also to render them durable and proof against the liability to distortion incidental to the varying character of the work they have to perform. Fig. 24 shows a part of the room with the men at work, moulding the slabs.

Slabs are made of various sizes, but it is found after many years' experience that the most useful sizes for London paving work are 2 feet 6 inches in length by 2 feet wide, and 2 feet square by 2 inches thick. Paving slabs of these sizes weigh

26 lb. to the foot super, and are convenient to handle, and can be made at least cost, for when the surface dimensions are increased a proportionate addition to the thickness is necessary.

The amount of water used in the mixture of aggregate and cement is fluctuating and depends very much on the state of

FIG. 24.

·FILLING THE MOULDS·

the former, for if freshly washed a considerable surface moisture is unavoidable.

In making cisterns, sills, sinks, or other more ornate forms, a much less quantity of water is used than that required for the slabs, but in such cases the moulds are filled more carefully and rammed or tamped by a specially prepared hammer-headed tool.

Fig. 25.

Fig. 25 shows some of the many ornamental and useful forms made under the Victoria stone process.

One of the most elaborate architectural productions turned

out by the Victoria Stone Company is the clock tower of Messrs. Peek, Frean, and Company's biscuit factory, at Bermondsey, a panel of which is shown in Fig. 25.

Before the adoption of the small aggregate it was customary to make the surface and bottom of the paving slabs of finer stuff, introducing in the middle coarse and irregularly sized stones. That practice no longer prevails (unless in exceptional cases), for it is found that the more homogeneous and regular combination of small aggregates secures the best qualities of density and good wearing texture.

The moulds filled in the manner thus described are allowed to remain on the benches of the moulding sheds until the concrete has sufficiently set, and so much of the water of plasticity evaporated as will permit the slabs to receive the beneficial influence of the silicating operation. This indurating process is one of absorption, and the best practice is that which provides a reasonably porous mass to which may be introduced an accurately prepared liquid silicate of the desired specific gravity. Unless due regard is had therefore to these two essential conditions—the necessary porosity of the slab on the one hand and a suitable penetrative silicating agent on the other—the desired accuracy of result is impossible.

The slabs when sufficiently dry are relieved from the surroundings of the moulds, which being made in pieces can be readily detached by unscrewing the fastenings. The slabs are then taken to the tanks in the silicating yard (unprotected from the weather) and placed one upon the other or side by side, covered by the silicate solution, where they remain until the proper beneficial influence has been duly imparted. The period of time required to complete the silicating process is not of a fixed or arbitrary character and depends on the condition of the slab and its capacity of absorption. About fourteen days under ordinary circumstances is regarded as sufficient to secure the desired advantage of the process. Fig. 26 shows a view of the silicating tanks.

The slabs after being taken from the tanks are stacked in the store yard, where they remain to season and are taken away in the order of their age. The Victoria Stone Com-

Fig. 26.

pany find it most advantageous to have in hand a stock of paving slabs sufficient to cover eight or nine acres.

The mode of preparing the silicate is simple, and consists of treating the natural silica of Farnham with caustic soda under the influence of heat, thus producing a silicate of soda.

222   A PRACTICAL TREATISE ON CONCRETE.

The silicate so prepared is conveyed by wooden shoots or conduits to the various tanks, which are maintained at the normal value of specific gravity.

Fig. 27.

The use of natural silica from Farnham secures considerable advantages in this industry owing to a large proportion of it being soluble, effecting thereby a saving in cost of

FIG. 28.

the silicating ingredient, and to some extent immunity against the fluctuating value of the chemical equivalent obtained from the silicate manufactu

Fig. 29.

The machinery required for the conversion of the crude silica into silicate is of a very simple character, consisting of a pair of iron edge runners to reduce the stone, and a series of jacketed boilers to which is supplied the steam of the required temperature. The caustic soda is obtained from the best sources and of the purest quality, because the presence of sulphur, which sometimes exists in carelessly manufactured soda, has a most prejudicial influence on the silicate.

It will be observed that all the operations of the Victoria stone process are the result of manual labour, and with the exception of the steam power for grinding the silica stone, no mechanical assistance is called in to aid the manufacture. In fact the Victoria paving slabs are hand made, and the continuance of this practice is almost unavoidable in order to overcome the usual prejudice attaching to any departure from an established industry. The Victoria Stone Company are conscious of the extra cost of their existing process of manufacture, but hesitate to adopt machinery until all prejudice is completely removed. Their stedfast adherence to the original practice has enabled them to overcome the prejudice of some of the authorities controlling the paving of our cities and towns. They do not despair, however, of being able at the fitting time to introduce such machinery as will, while maintaining all the advantages of hand labour, secure a considerable reduction in the cost.

The analysis of a piece of Victoria stone paving slab is,—

Silica	50·35
Alumina	11·87
Oxide of iron	7·33
Lime	18·33
Magnesia	2·03
Potash	1·78
Soda	3·81
Carbonic acid	1·80
Water, organic matter, &c.	2·70
	100·00

Fig. 27 is a medallion made for the Post Office authorities.

The Victoria Stone Works are favourably situated with regard to railway and water carriage, thus commanding the most economical means of receiving the raw materials, and for the purpose of delivery of the manufactured article.

In reference to the continued practice of making the Victoria paving and other slabs by hand, it may be presumed that the advantages secured by such means are appreciated in the same way that hand-made paper is still preferred to that made by machine. Fig. 28 shows a base and pedestal; Fig. 29, another design for a similar purpose.

Rock-Concrete Pipe Manufacture.

The Victoria stone industry may be regarded as of English origin, having been originated and prosecuted to its present condition without any extraneous or foreign assistance. That of concrete pipe making is not, however, similarly circumstanced, for that industry has in its present shape been imported into this country from the United States of America. The cost of stone-ware pipes (formerly imported from England) used in the drainage of American towns led to inquiries as to a less costly and equally efficient substitute. The native Rosendale (natural) cements in conjunction with English and French Portland cements were used as matrices in the preparation of concrete pipes for sewerage purposes in many populous American towns. In the cities of Boston and Brooklyn extensive drainage works had been executed with concrete pipes before and during the year 1873, when Mr. J. W. Butler visited the States, and appreciating the advantages of pipes so made, entered into negotiations with the patentees of the machinery for making them, and ultimately purchased the English patent. On his return to this country he first experimented with the machinery at the pottery works of Messrs. Henry Sharp, Jones, and Co., at Bourne Valley (near Poole), in Dorsetshire, and in consideration of their having made these initiatory experiments granted them the exclusive right to use the patent machine in the counties of Dorset, Hants, Somerset, and Wilts. This was in the year 1875, and since that time Messrs. Henry Sharp, Jones, and Co. have continued this concrete pipe industry in addition to their extensive works for the manufacture of stoneware and terra-cotta.

It soon became apparent to these gentlemen that for the purpose of providing pipes of large size (those exceeding 15 or 18 inches in diameter) the concrete was admirably adapted, and they accordingly introduced the "rock concrete" pipes of large sizes for general sewerage purposes. In so doing they had due regard to the utilization of the unavoidable waste in their more important and established industry of glazed stoneware pipe making. In all the best and more reliable experiments made for the purpose of testing the relative values of various concrete compounds, it had been found that broken pottery realized high and satisfactory compressive results, and was therefore considered the best suited as an aggregate. Under ordinary circumstances, the quantity of broken pottery available for concrete purposes must necessarily be limited in quantity, and therefore unattainable where concrete works of any considerable extent were contemplated. In extensive pottery works, such as those existing at Bourne Valley, a considerable amount of broken stoneware waste is unavoidably produced, and it was for the purpose of absorbing or utilizing this that the rock concrete pipe making was introduced in Dorsetshire.

Broken pottery when of a semi-vitreous or lava-like character offers perhaps the most favourable surfaces whereon a good binding agent can advantageously adhere. The pieces of the damaged or rejected stoneware pipes are crushed (from pieces of the size of peas to that of fine sand) with a Blake's stone-breaker and heavy rolls. This produce, after being carefully washed so as to eliminate all fractural dust, is mixed with a small proportion of pure siliceous white sand found on the ground near the works. A small portion of fine clean yellow pit sand is sometimes added to the broken pottery.

The districts of Poole and Bournemouth are celebrated for the quality of the clays obtained from the deposits in the tertiary geological formation found abundantly within a wide circle of which these localities may be regarded as the industrial centre.

An analysis of the broken pottery-ware used in making the concrete pipes gives the following results:—

Insoluble siliceous matter	97·60
Oxide of iron and alumina	1·97
Magnesia and loss	0·43
	100·00

The Portland cement used in the rock-concrete pipe manufacture is obtained from the River Thames, near London, and is especially selected, being at least 112 lb. per imperial bushel in weight. Until recently the cement was tested at the manufactory where it was made, the results of breakings being forwarded regularly to Bourne Valley. The increasing business, however, has led to a system of testing at the works by the double or compound lever testing machine described at page 59. In addition to this careful testing of the cement before it is mixed with the concrete, briquettes are made daily from the mixtures of the concrete in process of conversion. By this means not only is the proper quality of cement secured, but a constantly recorded tensile history of the actual strength and quality of the concrete itself.

In fact these apparently trivial details are scrupulously attended to, and no pipe is made without a reliable and accurate test of its value, so that at any future time, on referring to the date and number on the pipe, its tensile value is readily ascertained. The proportions now used are three of well washed and properly balanced aggregate with one of the best Portland cement. The use of machinery for mixing is resorted to, thus ensuring accuracy of amalgamation, including the necessary moisture, before entering the moulding machinery. The mixer used for this purpose is as shown by Fig. 30. The water is imparted to the hand-proportioned

FIG. 30.

MIXING MACHINE.

mixture of cement and aggregates by a carefully arranged jet under the immediate control of the workman in charge of this machine. It is most important that the proper and exact quantity of water should be carefully imparted to the mass, otherwise the desired quality of density in the concrete is unattainable.

Experience in the use of the American moulding machine

and its various mechanical appliances indicated that the best quality of work was not reached by its aid, and accordingly from time to time improvements have been effected, culminating in a new machine patented by Messrs. Henry Sharp, Jones, and Co. Two of these machines have now been at work for some time, and the improved quality of work which they turn out clearly indicates a considerable advantage over the original American one. The moulding machine is shown by Fig. 31, and consists of a double hollow cylinder, into which the concrete mixture is fed with constant regularity from the hopper above. As the speed of the machine working the tamp or beater can be increased or diminished at will, only just enough of the concrete mixture is introduced as can be well and effectually rammed by each single blow at a time. The mould is placed on a revolving table, which rotates by properly regulated mechanical appliances, so that the speed is balanced in accordance with the quantity of mixture supplied and the velocity of the beaters or tamper. Unless this combination of supply of the material and its conversion by the machinery is regulated with almost automatic accuracy, it is obvious that irregularity in the tamping or overfeeding of the revolving mould would result in an imperfectly impacted mass, and any pipes so fabricated would be faulty in character and deficient in strength. Besides this disadvantage, another equally dangerous one in its results is an overdose of moisture. An excess of water would render the beneficial action of the impinging tamper inoperative, as the blow therefrom could not consolidate an overwetted mass, owing to its spongy character. For two reasons, therefore, it is imperative that great care should be bestowed on this part of the process, for too much water would result in a porous pipe and the imperfect tamping weaken its tensile and compressive value.

The moulds are allowed to stand for a certain time (from

Fig. 31.

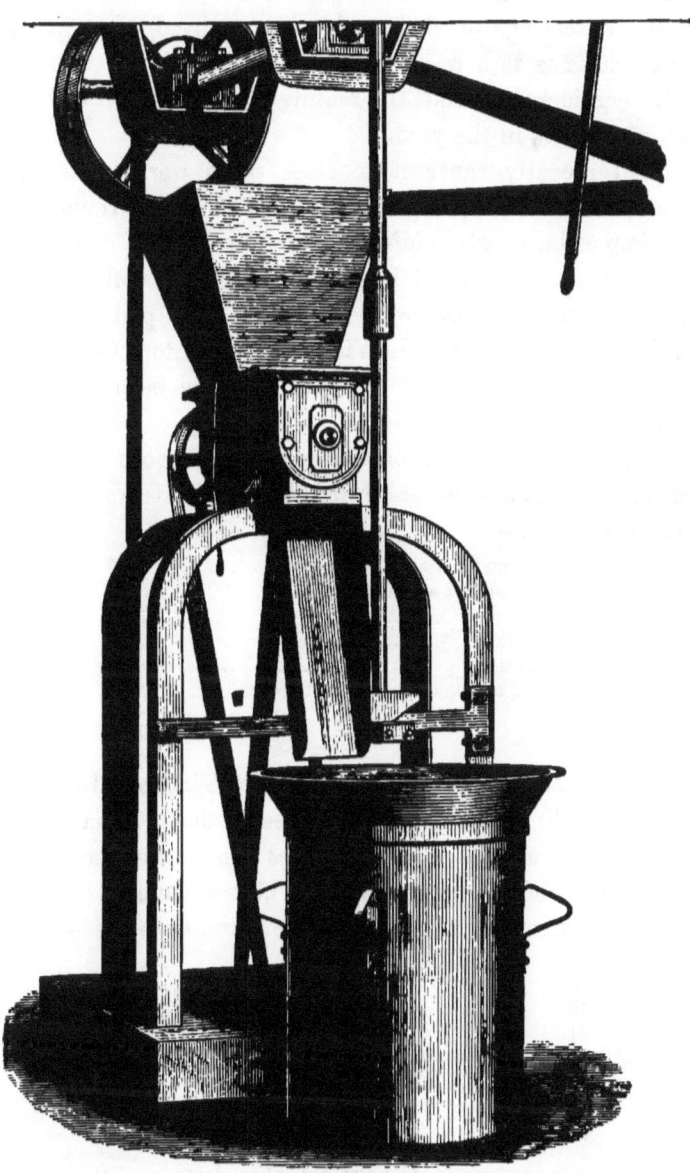

Pipe-moulding Machine.

one to three days, according to the state of the weather), and when the casings of the moulds are removed, the pipe is taken at once to a reservoir of water, in which it remains until considered sufficiently matured to be stored without risk of fracture in the yard.

Until recently the rock-concrete pipes were treated as thus described, but owing to the pressure of the trade, and the impossibility of holding them in sufficient quantities until the proper natural induration had been realized, the process of silication is now resorted to. The plan adopted for this purpose is that of the Victoria Patent Stone Company, from whom a license to use their process had been obtained by Messrs. Henry Sharp, Jones, and Co.

Figs. 3 and 4 on Frontispiece show sections of rock-concrete material, the following being an analysis of its chemical value:—

Water of combination	6·80
Oxide of iron and alumina	5·15
Lime	17·02
Soluble silica	12·95
Insoluble silica	56·20
Magnesia, carbonic acid, &c., not determined	1·88
	100·00

The sizes of pipes at present made are 18 inches, 21 inches, 24 inches, 30 inches, and 36 inches in diameter, as it is found by the makers that these sizes can be made cheaper than corresponding diameters of the best stoneware.

The weights of pipes of each of the above sizes are:—

Diameter. in.	Thickness. in.	Weight. lb.	
18	1½	175	—2-ft. lengths.
21	1⅝	215	,,
24	1¾	250	,,
,,	2	320	,,
30	2¼	406	,,
36	2½	500	,,

The pipes are made without rolled sockets, and therefore,

unlike the stoneware pipes, having only a rebated joint O G in form, permitting by such arrangement an even surface inside and outside of the pipe when laid and jointed.

In a series of experiments made by the author to determine the tensile and compressive value of the concrete used in making these pipes, the following results were realized.

Table No. 1.

On pieces of pipes 2 feet in diameter, two years old, having been laid in a sewer trench for nearly a year, and fractured in consequence of improper laying.

A. Experiments.

Tensile Strength.

No. 1	..	..	..	..	..	..	$339\frac{1}{8}$ lb. per sq. in.
No. 2	..	..	..	..	..	..	$430\frac{1}{2}$,, ,,
No. 3	..	..	..	..	..	..	$431\frac{1}{3}$,, ,,
No. 4	..	..	..	..	..	..	355 ,, ,,
No. 5	..	..	..	..	..	..	529 ,, ,,
No. 6	..	..	..	..	..	..	385 ,, ,,
No. 7	..	..	..	..	..	..	368 ,, ,,
No. 8	..	..	..	..	..	..	480 ,, ,,
No. 9	..	..	..	..	..	..	610 ,, ,,
No. 10	..	..	..	..	..	..	340 ,, ,,
							4268

An average equal to 426·8 lb.

B. Experiments.

Tensile Strength.

No. 1	..	..	..	..	..	..	700 lb. per sq. in.
No. 2	..	..	..	..	..	..	560 ,, ,,
No. 3	..	..	..	..	..	..	560 ,, ,,
No. 4	..	..	..	..	..	..	505 ,, ,,
No. 5	..	..	..	..	..	..	475 ,, ,,
No. 6	..	..	..	..	..	..	562 ,, ,,
No. 7	..	..	..	..	..	..	473 ,, ,,
No. 8	..	..	..	..	..	..	475 ,, ,,
No. 9	..	..	..	..	..	..	554 ,, ,,
No. 10	..	..	..	..	..	..	393 ,, ,,
No. 11	..	..	..	..	..	..	491 ,, ,,
							5748

An average equal to $522\frac{1}{2}$ lb. per sq. in.

C. Experiments.
Tensile Strength.

No. 1	540 lb. per sq. in.
,, 2	440 ,, ,,
,, 3	405 ,, ,,
,, 4	570 ,, ,,
	1955

An average equal to 489¼ lb. per sq. in.

D. Experiments.
Tensile Strength.

No. 1	213 lb. per sq. in.
,, 2	239 ,, ,,
,, 3	283 ,, ,,
,, 4	456 ,, ,,
,, 5	290 ,, ,,
,, 6	344 ,, ,,
,, 7	420 ,, ,,
,, 8	450 ,, ,,
,, 9	273 ,, ,,
,, 10	392 ,, ,,
	3360

An average equal to 336 lb. per sq. in.

A general average of these experiments gives the following, viz.:—

A.—426·8 lb. per sq. in.	
B.—522·5 ,, ,,	
C.—489·2 ,, ,,	
D.—336·0 ,, ,,	
1774·5	

Being on the whole of the tests, 443·6 lb. per sq. in.

Table No. 2.

A. Experiments.
Compressive Strength.

No. 1	2684 lb. per sq. in.
,, 2	3672 ,, ,,
,, 3	3955 ,, ,,
,, 4	2825 ,, ,,
,, 5	4340 ,, ,,
,, 6	3250 ,, ,,
	20726

Or an average of 3454·3 lb. per sq. in.

B. Experiments.

No. 1	5650 lb. per sq. in.
,, 2	5650 ,, ,,
,, 3	8390 ,, ,,
,, 4	3107 ,, ,,
,, 5	5390 ,, ,,
,, 6	5390 ,, ,,
	28577

Or an average of 4762·9

8217·2

The general average being 4108·6 lb. per sq. in.

The above experiments were undertaken by the author for the purpose of ascertaining the value of the concrete of which the rock-concrete tubes used in the drainage of Bournemouth were made. The pipes were condemned by certain engineers and the surveyor to the Board of Commissioners of that town, in consequence of which Messrs. Henry Sharp, Jones, and Company published a "Vindication," and with other engineers the author was engaged in the preparation of reports in furtherance of the arguments advanced by the manufacturers in their justification.

The experiments in Table No. 1 were with briquettes cut out of pieces of the rejected concrete pipes. Those of Table No. 2 were obtained from the same sources.

A. tensile strength experiments were with briquettes obtained by the author from broken pieces of pipes taken out of a trench in which they had lain for a period of ten months.

B. experiments from a similar source, but taken at an interval of a month.

C. experiments from pieces of pipes which had been submitted to hydraulic test (an internal water thrust) by the Bournemouth Commissioners' advisers.

D. experiments from pieces of pipes which had been broken under external hydraulic pressure by the Bournemouth Commissioners' surveyor.

Table 2 compressive tests were made on 2-inch cubes cut from pieces of the broken pipes, and comprise various sorts which had been previously examined for tensile strength.

These explanations are considered necessary, for they show under what disadvantages the tests were made, being from broken pipes out of the pieces of which briquettes and cubes were cut by hand.

These examinations, with the varied tests and experiments undertaken, go far to show that the blame of failure in that part of the Bournemouth drainage scheme in question was due rather to oversight on the part of the controlling authorities than to defects in the concrete tubes or pipes.

The Bournemouth Commissioners, alarmed at the position which their idle dissensions and fears had created, called in the aid of two engineers to assist their own adviser in this difficulty, and the outcome of such a combination is certainly amusing and worthy of being recorded.

This scientific triad agreed most unanimously in the condemnation of the concrete pipes, but on totally different grounds: Mr. A, because it was his opinion they could not readily be tested before use; Mr. B, because the water of Bournemouth was capable of dissolving the concrete; Mr. C, from the pipes being too thin "for all the exigencies of practice."

Yet the *experienced* surveyor of the Board (Mr. C), who in his published report says of himself, "touching sewers and the like sanitary questions, I am not a novice," found by actual experiment that the "thin pipes" were 13 per cent. stronger than the best glazed stoneware pipes of a well-known maker, and notwithstanding their proved superiority by his own tests, unhesitatingly condemned them. As these reports and the examinations which their preparation involved will no doubt be long remembered in connection with concrete progress and sanitary drainage, we will give just one more series of tests, being those voluntarily undertaken by the Commissioners' surveyor to test the relative compressive values of concrete tubes and those made of salt-glazed stoneware of the best quality made in the locality.

1st. Experiment with five concrete pipes taken from the sewer trench in which some of their fellows had been broken:—

No. 1 .. .. 2520 lb. the 2-ft. pipe.
 „ 2 3360   „   „
 „ 3 3192   „   „
 „ 4 3024   „   „
 „ 5 3360   „   „
         Or an average of 3091 lb.

2nd. Experiment with five concrete pipes which had not been used, and therefore not so old as the above:—

No. 1 .. .. 1848 lb. per pipe.
 „ 2 2772   „   „
 „ 3 2520   „   „
 „ 4 2856   „   „
 „ 5 2688   „   „
         Or an average of 2536 lb. per pipe.

Being on the two experiments a general average of 2813·5 lb. per pipe.

3rd. Experiment on five best glazed stoneware pipes, selected for the purpose of examination by the experimenter himself:—

No. 1 .. .. 2352 lb. per pipe.
 „ 2 .. .. 2553 , „   „
 „ 3 2301   „   „
 „ 4 2940   „   „
 „ 5 2258   „   „
         Or an average of 2480·8 lb. per pipe.

These three last experiments were made by the pipe-testing machine, Fig. 35, page 240.

Thus even by an engineer who cannot be convinced from his own tests of the superiority of concrete over stoneware pipes, a marked advantage is established of the former over the latter. And if the present value is higher we may confidently assume from all past experience that the future will still further add to the concrete pipe's superiority.

From these experiments made on defective concrete pipes, or at least presumably weak, we will shortly refer to experiments made on sound pipes by Mr. Ellice-Clark, a gentleman who was consulted on the Bournemouth disagreement. In his report he says, addressing the manufacturers of the rock-concrete pipes:—"Ten of the 24-inch tubes, 1⅜ inch in thickness, made by you last March, were placed in a trench 14 feet in depth to the crown of the sewer, the shoring timber removed, the tubes covered, and the trench filled in with clay in the ordinary manner. This gave a load of 6000 lb. to each tube, assuming that the whole of the superincumbent weight was borne by the tubes. This was successfully resisted. The trench was then saturated with water, the weight on each tube being increased to 8000 lb.: five of the tubes cracked, took their bearings, and sustained the weight; the other five remained perfect. I also had a concrete tube 2 inches in thickness tested, and found the cracking weight to be 7245 lb."

Mr. Ellice-Clark, in his position as surveyor to the Hove Commissioners, had used considerable quantities of the concrete tubes, and from his experience in their practical use may be regarded as a reliable authority on the subject.

In addition to these tests we have referred to, Mr. Ellice-Clark made other experiments, which, from the careful manner in which the fractures in the pipes were observed, are worthy of being recorded. In describing the manner of experimenting, the experimenter says:—"To determine what pressure the three tubes would bear under the same conditions as exist in practice, I had an excavation made the exact contour of the seating of the tubes; the sides were packed in clay, the compressive force was distributed uniformly over the tubes, as shown in the accompanying drawing, the mean of the cracking weights was 4393 lb.; the variation, as you will see from the accompanying statement, trifling."

Figs. 32, 33, and 34, and the figures of observations appended, are as follows:—

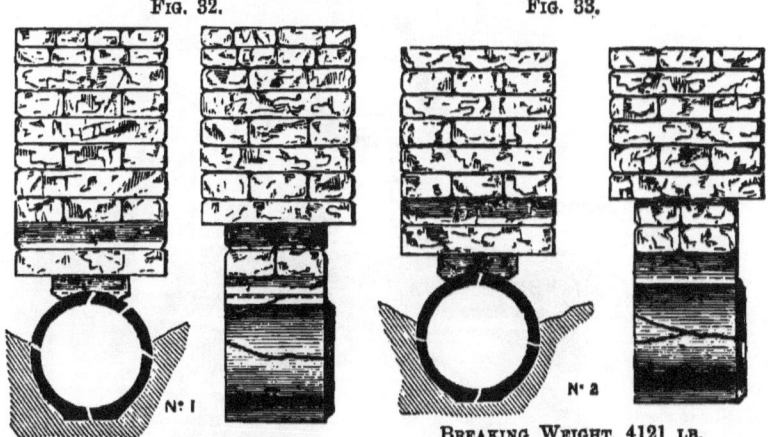

Fig. 32.
Breaking Weight, 4938 lb.

Fig. 33.
Breaking Weight, 4121 lb.

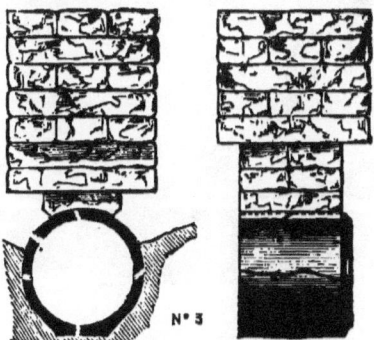

Fig. 34.
Breaking Weight, 4121 lb.

In addition to the carefully arranged system adopted at Bourne Valley to ensure the best quality of concrete and its advantageous conversion into rock-concrete pipes, the manufacturers have also introduced a hydraulic apparatus for

240 A PRACTICAL TREATISE ON CONCRETE.

testing the strength of the pipes, as shown by Fig. 35. Looking to the ultimate purpose for which these pipes are destined, it is desirable to know their capacity to resist, when

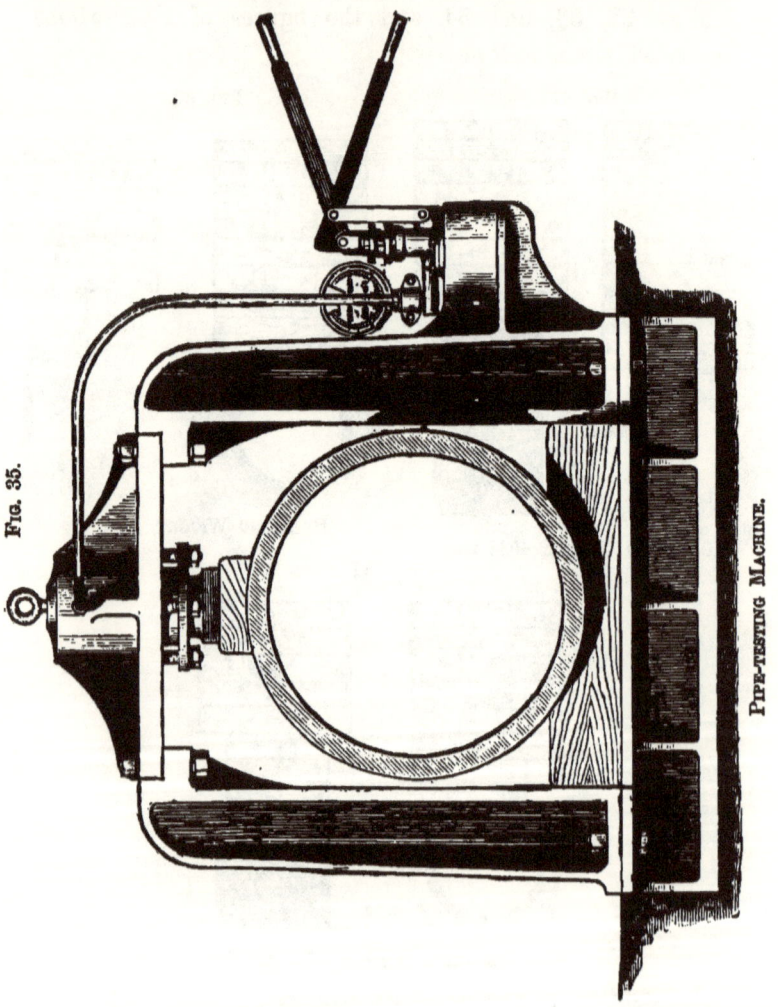

Fig. 35.

Pipe-testing Machine.

placed in position, the compressive action of the superincumbent mass of the filled-up trench in which they may be laid. The fluctuating character of this duty required of the pipe, owing to the careless and ignorant mode of treatment by some engineers, requires that it should at least be capable of sustaining a reasonable strain of this kind according to its size. The primary object of a sewer or conduit for the conveyance of the sewage of towns is to secure the perfect and uninterrupted passage of the objectionable waste of its inhabitants, and the first aim of the engineer entrusted with this responsible duty should be to secure the constant maintenance of the conditions which will best secure that object. His duty is one of responsibility, and he ought to consider the nature and extent of the strain to which any one pipe will be subjected to when placed as a factor in the scheme of drainage. Regardless of its constituent ingredients, whether of stoneware or concrete (although these questions will in the future obtain a more careful consideration), the pipe must be liable to external and dangerous influences, from which it should be carefully protected. The resistance to chemical action is a matter very properly left to the pipe manufacturers, but the mechanical dangers of the pipes' surroundings when placed, should be provided against by the engineer. An engineer who, after failure of his work, turns round and blames the maker of the pipe, realizes but a poor estimate of his honourable position, for to him have been entrusted duties of an almost sacred character. Nothing that an engineer uses should be beyond his capacity of testing and proving before it is accepted, and if his ignorance or supineness interferes with the accurate observance of that duty, he is incompetent and unworthy of being entrusted with the execution of the most trivial works.

Under any circumstances, either with the silicating process or without it, pipes exceeding 15 inches in diameter should be at least four months old before being used. In moving the pipes after moulding, great care is taken that they will not be subject to any jar or shock likely to disturb the setting process. Mechanical means are adopted to secure this important object, and the utmost attention is bestowed in this direction until the pipes are stacked away in the storing ground, where they remain increasing in strength and value until sold.

Mr. Baldwin Latham, the well-known sanitary engineer, recommends that the 24-inch concrete pipes should be made at least 2 inches thick, and, in deference to the opinion of so good an authority, the makers are now producing them of that thickness. For those purposes, however, where the tubes are merely used as internal conduits in concrete masses, pipes of less thickness are still manufactured.

Silicated Stone Manufacture.

Mr. Butler, from the experience gained by a personal examination of the extensive drainage works in the United States of America, more especially of the city of Brooklyn (acknowledged to be the best sewered city in America), and satisfied with the pioneer productions at Bourne Valley Pottery, determined to establish extensive concrete works in or near London. Mr. Hodges in conjunction with Mr. Butler established at East Greenwich, in the premises formerly occupied by "Ransome's Siliceous Stone Company," a manufactory for making pipes of all sizes and descriptions for sewer and cognate purposes. In addition to operations in connection with the pipe industry, they also manufacture paving slabs and other suitable concrete articles for a variety of uses.

Messrs. Hodges and Butler do not limit their manufacture to the use of one aggregate only, but, according to the objects for which their concrete products are required, select those most suitable for their purpose.

The works at East Greenwich, being situated on the bank of the river Thames, command favourable facilities for unloading the materials from river or seaborne sources, Thames ballast, Kentish ragstone, flints, granites, and slags from the iron-making districts, as well as many other natural stones, such as the oolites from Portland and elsewhere. The cheapness of the river ballast, and its suitable character for ordinary paving-slab making when judiciously prepared, give it a preference over other aggregates. This material is put through a Blake's stone breaker, and reduced to the required size. The broken aggregate from the Blake machine is raised to the mixing-room by the dust eliminating apparatus, and is passed through a series of screw elevators. On reaching the destined point it is completely purified and free from fine dust and other objectionable ingredients. The ragstone, slag, and other aggregates are treated in the same manner, regardless of their natural condition, for it is found that all stones and minerals of whatever kind are unfit for rational concrete purposes unless first subjected to this treatment of purification.

Iron slag is used for both pipe and slab making, from which good cement is obtained; the analysis of the slag in its fabrication being as under:—

LINCOLNSHIRE SLAG.

Silica	38·25	Protoxide of iron ..	1·09
Alumina	22·19	Manganese	trace
Lime	31·56	Calcic sulphide.. ..	2·95
Magnesia	4·14		

Superior slabs are made from Kentish ragstone, which gives the following analysis:—

KENTISH RAGSTONE.

Carbonate of lime	92·6
Earthy matter	6·5
Oxide of iron	0·5
Carbonaceous matter	0·4
	100·0

In the experiments made by Pasley, ragstone realized some of the best compressive test results, and although not obtained in large blocks, its durability for building has been well established, more especially in ecclesiastical architecture, for which purposes it is well adapted and very extensively used.

The special advantages obtained from using slag aggregate are derived from the fact of its approaching in value some of the qualities possessed by heavy, slow-setting Portland cements; the only really objectionable ingredient in the analysis being the sulphur compound, the damaging influence of which is, however, nullified in a great measure by its thorough and isolated dissemination in the concrete mass. The natural disintegration of waste heaps of slag is due primarily to the destructive character of the sulphur, its presence anticipating by a considerable period of time the ordinary disaggregation incidental to air exposure.

The slag used at East Greenwich is in a condition of fine gravel or sand, and from which in that state all, or nearly all, of the sulphur has been purged by atmospheric exposure. In experiments made by Professor Roscoe at Manchester some years ago, it was found that slag direct from the furnace, when ground very fine, possessed many of the properties of Portland cement, and was tried as an internal plaster in several cases, but the protracted character of the setting qualities prevented its being used with profitable advantage. Its use as an aggregate, and in the manner pursued by Messrs. Hodges and Butler, indicates an approach to the utilization of waste heaps in many districts of this and other countries to an almost unlimited extent. Messrs. Hodges and Butler have been using a Belgian granite gravel, which produces concrete of great density. The following is an analysis of a piece of this concrete after having been submitted to the silicating process for fourteen days: —

ANALYSIS OF SLAG CONCRETE.

Silica (insoluble)	31·95
Ditto (soluble)	8·02
Alumina	9·97
Oxide of iron	5·98
Lime	32·12
Magnesia	1·45
Potash	1·52
Soda	2·59
Carbonic acid	1·71
Water, organic matter, &c...	4·69
	100·00

The machinery (adapted and patented by the manufacturers) is of two kinds. First, the original scheme of impact as practised in America for making pipes and other cognate forms. Second, the slab paving-making machines, invented and patented by Messrs. Hodges and Butler, and, for the present at least, solely used by them at their manufactory.

We will first proceed to describe the pipe-making operation, and the machinery by which it is carried on.

The materials (aggregate and matrix) are first carefully proportioned and mixed dry in the "mixing machine," of which a sketch is shown by Fig. 30, page 229. This mixer (first introduced into England by Mr. Butler) consists of a circular iron trough, having part of the bottom cut away, which is replaced by an iron plate put underneath, and arranged that it can be moved at will so as to open or shut the aperture through which the mixed stone paste falls out. A vertical spindle passes through the centre of this trough, on the top of which are fixed the usual bevel wheel and pinion, keyed to the driving shaft, on the pulley of which the strap is applied. Upon the vertical spindle, and near the bottom of the trough, are fixed two paddles or arms, which in their revolutions thoroughly mix the materials.

The "stone paste" thus prepared is now taken to the moulding machines, of which there are two kinds, one being

246   A PRACTICAL TREATISE ON CONCRETE.

employed in making pipes of 12 inches diameter and under, and the other for pipes exceeding that size and up to

Fig. 36

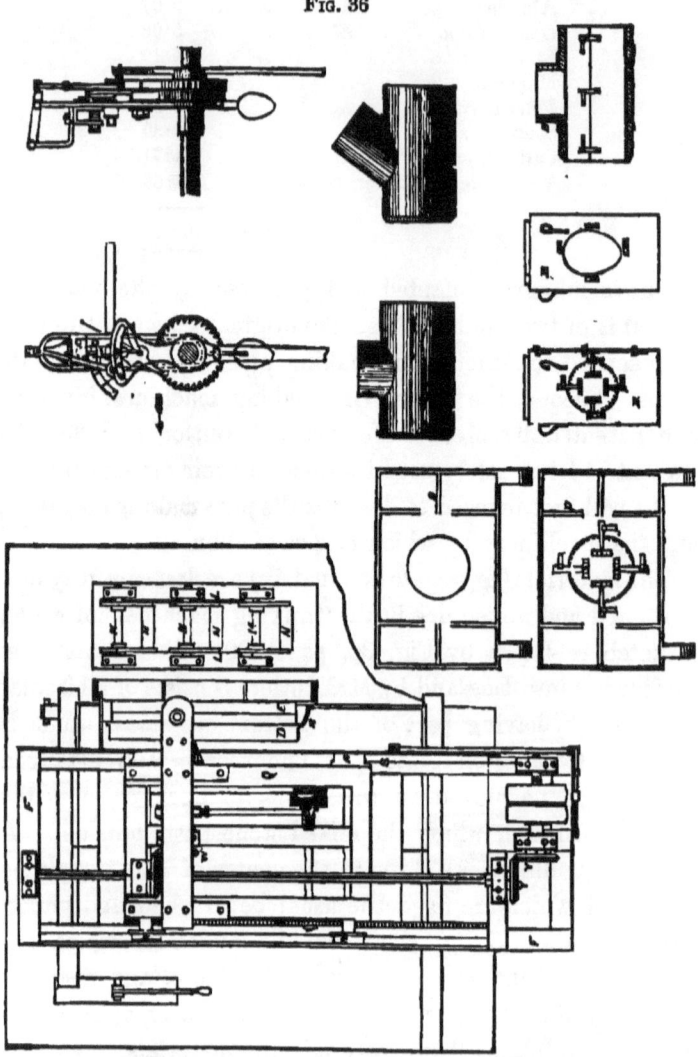

36 inches in diameter. The former machine, as shown in Fig. 36, consists of a wooden frame, carrying at a convenient

height the moulding table. In the middle of this table a circular opening is made, in which is placed a conical hopper cut to the required diameter of the pipe to be moulded; the moulds being placed in pairs on a board, which runs along a series of rollers, constituting a kind of tramway arrangement, for facilitating the placing of the moulds under the hoppers from which the supply of paste is delivered.

The tamping or impacting operation may be thus described.

A cylinder, of the diameter that will occupy the space between the external and internal sides of the mould (being the measure of the thickness of the pipe desired), to which is fixed a vertical spindle having the usual bevel gear driving arrangement. In this spindle an accurate key-way is cut from top to bottom, in which a key fastened to the top of the cylinder runs, securing a rotary motion from the spindle and the facility of rising along the feathered "key-way" as the process of filling the mould proceeds.

At the bottom of the cylinders are four projections (of the thickness of the pipe to be moulded), extending a considerable distance, but not touching the cylinder, so that between each a space is left sufficient to allow the ready fall of the concrete into the mould, to facilitate which the projections are bevelled off at one end on the under side. The top of the cylinder is formed with a projecting rim, over which is placed a loose bar, through which the vertical shaft or spindle passes; a lug at each end of the bar being placed on its under side so as to catch the projection around the top of the cylinder, securing by this arrangement its facile raising or lowering at pleasure.

The process of moulding is begun by feeding some of the "stone paste" into the upper part of the hopper, the machine being then set in motion. The spaces between the projections around the bottom of the cylinder allow the stone paste to fall, and the rapid motion of the cylinder, combined with its weight, assists in the compression of the mixed paste in the mould. Gradually, as the material is fed into the mould, it becomes compressed, and the cylinder rises up the vertical spindle until the mould is filled and the pipe all but completed. At starting, it is necessary to raise the cylinder by means of the cross bar about a foot, so as to permit it to fall upon the paste and thus compress it. When completed, the pipe and its casing are removed by the roller or tramway arrangement, another mould taking its place.

The second moulding machine for the larger pipes is differently arranged. The framework is similar in character to the other, but the moulding table has a hole in the centre the size of the pipe required, while over the inner part of the mould a cover is placed to prevent the concrete from falling through. The tamper, or rammer, consists of a vertical spindle, having an enlarged head corresponding with the thickness of the pipe to be moulded. The tamper is attached to the end of a horizontal bar above the level of the table, being supported in a socket at the extremity of a lower vertical shaft. The bar is held by a set-screw in the socket, and can be shifted to and fro to suit the several diameters of the pipes to be formed. Reciprocating and revolving motions are imparted to the vertical shaft, and through it to the tamper, by very simple and ingenious means. A horizontal timber in the frame above the table supports the bearings in which the pulley shaft runs, the vertical spindle also passing through it, and thereby securing a guide. At the end of the pulley shaft is placed an open cam with an internal pathway. A bar, the end of which is formed with an eye, is passed over the vertical spindle, the end entering the cam, which drives it for a certain part of each revolution and then allows it to drop. Within the eye at one end of the bar are two blocks, the inner faces of which are curved to the diameter of the vertical spindle which passes between them. These blocks are free to turn, being mounted on pins passing through them as well as through the sides of the loops in the bar. The cam, in driving the latter, lifts it, and causes the blocks to raise the vertical bar, which is carried partly round, and raised at the same time, but when the cam ceases to drive, the eye-bar drops, the grip on the tamper bar is released, and the latter falls, to be again raised and partially rotated by the next revolution of the cam. The quality and extent of the impact blow can be reduced or increased at pleasure by using different sized cams, and blows ranging from 50 lb. to 500 lb. in value can be struck by the means described.

A description of the moulds will assist in rendering the above description more intelligible.

The moulds are simply sheet iron bent to the required circular shape, secured by three latch fastenings on the outside skin, and by two in the inner one. The base of the moulds is formed of cast-iron rings, shaped so as to form an ogee socket at the bottom end of the pipes.

Junctions and T pipes, or elbows, are formed by cutting holes on the outside of the moulds, which are covered with a plate, also secured by latch fastenings.

The finishing of the upper end of the pipe is thus performed :—A small quantity of concrete is placed on the top of the pipe as received from the machine, on which is put a cast-iron ring of the required form, and slightly hammered, after which the finished socket is neatly trimmed by hand if necessary.

To secure success in the operation of moulding these pipes, it is necessary to be careful about the quantity of water used, otherwise the value of the impacting operation will be much interfered with. The pipes, after remaining in the moulds a sufficient time, are taken out and placed in a bath of silicate of soda, where they remain until saturated (at least as far as possible), when they are removed and stacked in the outside yards of the works.

The author was recently employed, at the request of Mr. Baldwin Latham (who is always on the outlook for any improvements which are likely to lead to beneficial results in connection with sanitary matters), to test in various ways the quality of the pipes made by Messrs. Hodges and Butler. In an analysis made of a piece of pipe in which Thames gravel was used as an aggregate, the following result was arrived at, viz. :—

ANALYSIS OF THAMES BALLAST CONCRETE.

Silica in the form of pebbles and quartz	62·20
Soluble silica	10·80
Magnesia, alkalies, carbonic acid	3·15
Oxide of iron and alumina	3·47
Lime	15·23
Water and loss	5·15
	100·00

Comparative tests to ascertain the capacity of the pipes to resist an impact blow, resulted in proving that while it took 161 lb. to fracture a silicated stone pipe, a force of 75 lb. broke any ordinary clay or stoneware pipe.

Two pipes were cemented together, and after some time a weight of a ton and a quarter was depended from a bar passed through two holes in the lower one without breaking the joint.

Twelve feet of pipes were cemented or jointed together as they would be in the ordinary practice of laying, and supported at either end on a 6-inch bearing, leaving 11 feet between the supports. At the middle of the length a load of nearly a ton was placed, and one of the pipes was broken near the third joint. Their capacity to resist a pressure of 30 feet head of water was proved by filling that height of pipes without any appearance of fracture, or indeed the least signs of water on the sides.

ENGLISH CONCRETE INDUSTRIES. 251

The paving manufacture is carried on by the same means as that described in the pipe-making, and the machinery is of two kinds; one being on the principle of the impact, and the other a new mechanical arrangement, called by the manufacturers and patentees "the Trembler."

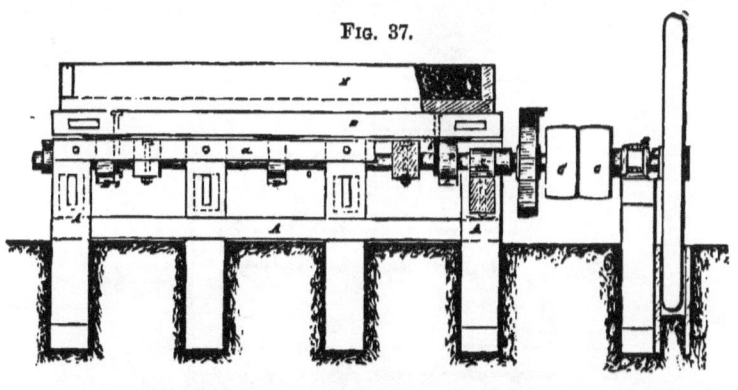

FIG. 37.

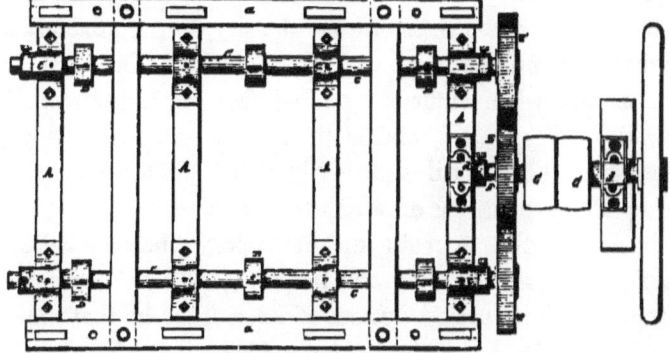

FIG. 38.

Figs. 37, 38, and 39 represent the drawings of this machine. Fig. 37 is the elevation of the table arrangement, showing partly in section the paving slab when filled in the mould. Two horizontal shafts passing under the table have a series of cams or eccentrics, which, in their rotation, alternately raise and lower the mould, imparting thereby a trembling motion, by which the concrete paste in the frame is gradually and effectually consolidated. Fig. 38 shows the plan of the shaft arrangement, the cams, and driving gear.

Fig. 39 represents an end view of the framework and

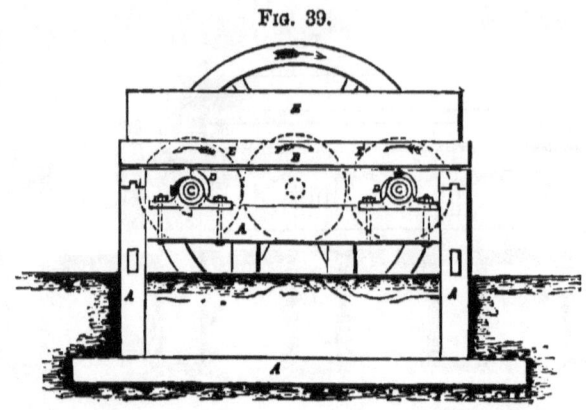

Fig. 39.

driving machinery. This machine is very simple in character, and produces slabs of great density, in the production of which a minimum amount of water is used. The movable slab mould, when sufficiently filled, has its upper surface trowelled off by hand, and is then taken away to a stack, where it is piled until sufficiently dried for the silicating tank. The foundations on which the machinery rests require to be substantial to resist the damaging vibratory action, which is necessarily rapid and frequent so as to secure the best results from such a system of *impact*, for by that name may its force be designated. It is evident that this principle is capable of application for the production of slabs of any size or thickness.

Fig. 40 represents an impact or tamping machine, by which Messrs. Hodges and Butler also produce paving slabs of any size. The end view of the machine shows F the framework and A that part of it by which the vertical tamper is guided as it traverses the length of the mould in which the concrete paste is filled. As the paste is added and

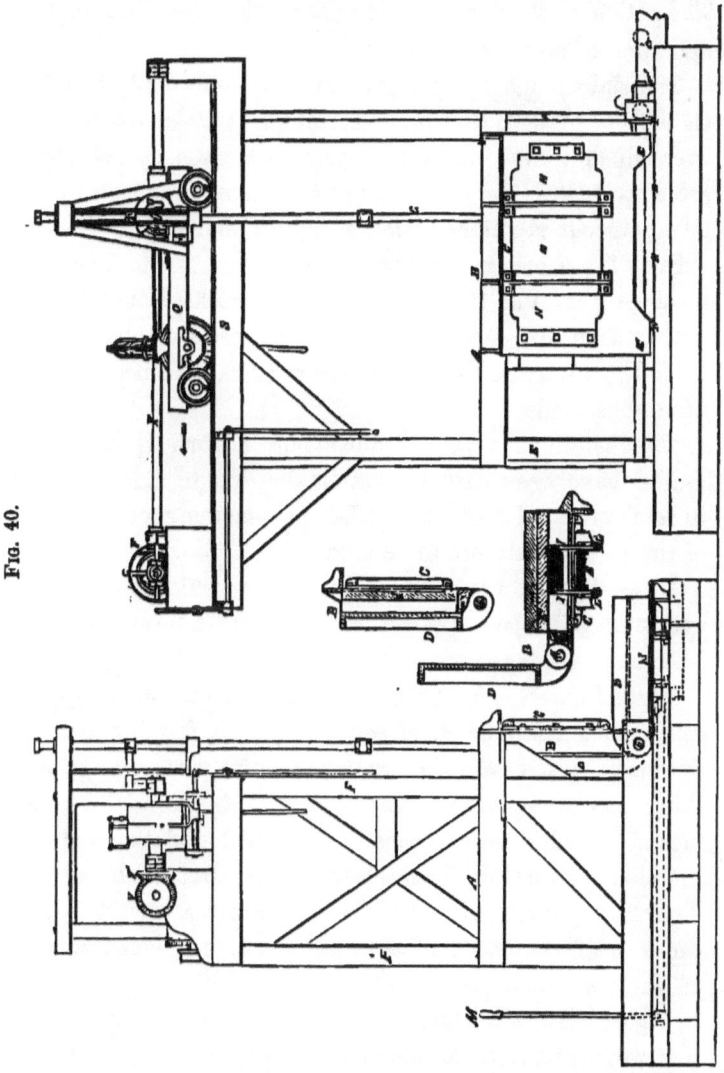

Fig. 40.

the tamping action performed, the tamper is raised gradually by mechanical means, and when sufficiently impacted the hinged side N is turned down and the slab removed, as shown by the enlarged drawings of the movable parts B and C. The action of the tamper is similar to that described in the operation of pipe-making.

The slabs made by this machine are more easily handled than those produced by the "Trembler." Its action is, however, much slower, and the paving slabs made by its means are consequently fewer in number than those produced by the advantageous vibratory action of the "Trembler."

Such is a somewhat imperfect description of the ingenious machinery used at the East Greenwich works, from which it will be seen that mechanical aid of a successful character is brought to bear on the production of silicated concrete forms of various kinds.

The sewerage pipes produced by Messrs. Hodges and Butler have been used in various districts in England, and in some of the London and other pavements good examples of the paving slabs are to be seen.

In addition to the ordinary pipes and paving slabs, other and more advanced applications of concrete forms for large sewers are produced.

Fig. 41 shows a front view of an egg-shaped sewer of large dimensions, constructed of separate pieces, forming with its apron, sides, and crown, a structure infinitely superior to anything reachable through the agency of brickwork or other means. Such a sewer, placed on a good foundation and surrounded by ordinary concrete as a protection against damaging thrusts of a horizontal character, would indeed prove itself competent to withstand every injurious influence of whatever kind.

Fig. 42 shows a perspective side-view of such a sewer as that we have described and shown in Fig. 41.

Fig. 41.

FIG. 42.

(257)

CHAPTER X.

CONSTRUCTIVE CONCRETE APPLICATIONS.

IN the preceding chapters we have discussed the various systems of artificial stone and concrete making in this and other countries. We will now proceed to describe and illustrate other manufactures which aim at an advantageous combination of concrete with various other constructive materials. In the more substantial and perhaps less accurately resulting products of blocks and other forms, such as those produced in "béton aggloméré," "Sorel stone," and "Victoria," and other silicated stone, we have shown the advantages derivable from their employment in such works as those for which they are suitable. The uses to which these several processes are applicable may be regarded as limited to the strictly substantial operations of the engineer and architect, and to which it is not desirable or necessary to introduce ornamentation, except in a comparatively limited degree. Time also, under such treatment of concrete, is not regarded as of primary importance, and indeed the best and most valuable results are obtained when it is possible, or convenient, to afford a large interval of time for advantageous induration. Paving slabs, sewer pipes, and other cognate articles, have to undergo, in their destined sites, wear and tear of an exceptionally trying character, with which they must be competent to contend, otherwise the advantages sought by their use would be unattainable. Hence resort to silicating and other careful preliminary treatment for the purpose of anticipating as much as possible the required and naturally resulting quality of hardness. Considerable

s

strength or resistive capacity to bear the strains of superincumbent weights or percussive and frictional wear is indispensable in those forms used for drainage and paving purposes.

In the industries, however, which we are about to describe, the quality of ponderosity, to resist abnormal pressure, although not altogether disregarded, is in some degree subordinated to that of ornamental display.

For many years various adaptations of concrete in conjunction with clay or terra-cotta products have been in vogue, but to a somewhat limited extent, owing to the difficulty of realizing any profitable result either in a pecuniary or decorative manner.

The most novel and undoubtedly important outgrowth of such lengthened experimental efforts in the purely "dwelling" constructive direction is the system so successfully prosecuted by Mr. Lascelles, and which we shall here designate the

" Lascelles System of Construction."

The principle on which this system of building is based, consists in the combination of timber and concrete. In fact interweaving, as it were, the best and most ingenious efforts of the carpenter and joiner with the no less valuable handiwork of the concrete-maker. And this combination of in a great measure controllable mechanical appliances produces in a rapid manner constructions of divers kinds and character, securing the erection of houses with great celerity. To show that we are not without some warrant in attributing to the Lascelles system an unusual capacity of speed, we will preface our more technical observations by mentioning a fact in connection with a building required under peculiar and pressing circumstances.

In one of the Government departments, owing to the

pressure of requirements of an unusual and exceptional character, a substantial building, 100 feet long and 30 feet wide, had to be built within a period of three weeks, one wall being formed by an adjoining building. Mr. Lascelles undertook this duty, and from the resources of his own ingenuity and the machinery of his works completed the task in such a manner as to satisfy the controlling authorities in the matter. The building so speedily built was composed solely of timber and cement, or concrete slabs put together with the necessary screws, the whole structure being externally clothed with slabs on all sides and across the roof, the floor being also formed of slabs. The manner of its execution will be more particularly described further on in these pages.

Mr. Lascelles' house or building may be regarded as partaking of a dual or twofold character, each separate operation being the work of different and distinct departments of labour. The building up of the skeleton or framework of the intended structure is accomplished by the aid of the most improved machinery, converting the well-seasoned timber into the required number of parts, which are readily put together from the accurately true method pursued in their preparation. Anyone at all acquainted with the character of the modern "labour saving" machines in use for the conversion of timber, can understand how easily and speedily such a framework can be produced. The proportions of the several members of the skeleton frame can be regulated to a nicety, according to the extent and character of the desired building and the work which it may be called upon to sustain. Unlike the protractedly formed skeleton of animal growth, this mechanically constructed creation has to be clothed with a covering of equally speedy production, which, when completed, renders the whole structure at once competent to afford shelter and comfort to those for whom it was fashioned.

The slabs used in covering the wooden framework are made from Portland cement and breeze or small coke, being faced with neat cement of any desired colour. There are several varieties of slabs, of divers sizes and design, according to the work required. The sizes and character of the slabs now generally used are as follows:—

1st. Fish-scale tile slabs, 3 feet long, 2 feet high, and 1½ inch thick.

2nd. Plain slabs, faced on one side, 3 feet long, 2 feet high, and 1½ inch thick.

3rd. Lining slabs, faced on one side, 3 feet long, 2 feet high, and 1 inch thick.

In addition to these slabs there are "studs," faced on three sides, 8 feet long and 4 inches square.

The first, or fish-scale slabs, are of the form generally used in a similar manner to the ordinary weather-tiling, as commonly practised in the erection of the common type of timber building, clothed with clay tiles. They are rebated on the lower edge, so as to render the joints waterproof.

The second, or plain slabs, having parallel sides, are made in a variety of ways, ornamental and otherwise, being used for party walls, chimney flues, ground floors, ceilings, and other numerous applications of an internal and external ornamental or useful character.

The third, or lining slabs, are only one inch in thickness, and capable of being applied to any of the purposes above mentioned, or used for facing the timbers of the higher-class structures, where timber boarding would be considered unsightly or otherwise objectionable.

The "studs" are simply blocks of cement concrete having an iron rod through their centre, three of their four faces being smoothed or polished. They are applied to many uses, as sills, joists, and such like purposes. Their general application is not recommended by Mr. Lascelles (as they are more expensive than wooden studs), except in damp situations, where timber would suffer, and therefore be less durable.

Tiles made in this way, by the agency of Portland cement, are not only capable of being moulded of large size, but possess the valuable property of maintaining their normal form in use under all circumstances. The ordinary kiln tiles produced from clay are necessarily limited in size and distorted in form, so as to render it impossible to make good and true work by such means. Cement concrete slabs, of 3 feet by 2 feet, require only four fastenings to the framework, whereas to cover the same surface with the ordinary tiles a great number of wooden pegs would be required; in short, a large number of joints of clay tiles against four only in using concrete slabs or tiles.

These few particulars will give some idea of the simple character of the materials which, when brought into the desired forms or combinations, produce what may be described as the carcass proper of the structure. In addition to these materials, there are manufactured under the same management "Patent concrete bricks," of all colours and every variety of form. They are moulded into quoins, pilasters, cornices, and other like designs of any desired pattern. Mr. Lascelles claims for these bricks, that they maintain their normal form and colour unchanged under the most trying circumstances, and, from the facility with which they can be produced, are competent to take the place of, or altogether supersede, the ordinary moulded or rubbed bricks. Their hard and durable character admit of their being sent any distance when properly packed without suffering damage.

In addition to the peculiarities we have thus shortly stated, the inventor places considerable value on their securing the following advantages:—

"Patent concrete slab cottages," he says, "require no brickwork, wood floors, excavations, window or door frames, tiling, lathing, or plastering.

"They are cheaper than a brick house, and one-fourth the weight.

"Can be built in any season, occupied as soon as built, erected very quickly, removed and re-erected, and can be sent by rail or road any distance, or shipped to any part of the world.

"If they are not affixed to the soil, they are tenants' fixtures; that is, they can be moved by the tenants and do not become the property of the landlord."

We can readily understand with what facility structures produced on the lines described can be erected at distant points and in the absence of skilled labour. But if we had any doubts as to the capacity of Mr. Lascelles or his manufactories to produce in the prompt manner and with the advantages claimed, such scepticism would be readily dispelled when we examine what has already been accomplished through the agency of the "Lascelles' system of construction." A group of ten workmen's cottages have been in the course of building during the past winter (between October and April), and no interruption has arisen in the continuous progress of erection, notwithstanding the unusual severity of a more than ordinarily protracted season of storm and frost. These cottages are built near London (Greenhithe), and can be readily examined.

The above description of the salient constructive advantages illustrates the usefulness of the system for ordinary and moderate purposes. But in the ornamental and advanced decorative direction, we have had during the recent Paris Exhibition an opportunity of proving conclusively its great adaptability in the production of advanced architectural design.

In a prominent position in the "Street of Nations," Mr. Lascelles erected a building represented by Fig. 43, the design of which was furnished by R. Norman Shaw, R.A.

CONSTRUCTIVE CONCRETE APPLICATIONS. 263

Fig. 43.

Queen Anne House, Street of Nations, Paris Exhibition, 1878.

The whole of the parts of the structure were fashioned and prepared in London, shipped from thence, and erected by foreign workmen, under the superintendence of Mr. Lascelles' foreman, in the park or grounds of the Paris Exhibition. The design is after the purest time of the Queen Anne period, and the happy success both of design and construction is a subject for congratulation.

The whole building was faced with thin tiles of a pleasing red colour, permanent in character and made of the best Portland cement, with which was used metallic oxides ground and carefully mixed. These tiles are of two sizes, viz. $4\frac{1}{2}$ inches by $2\frac{3}{4}$ inches, and 9 inches by $2\frac{3}{4}$ inches, being cemented to the concrete slab, 3 feet by 2 feet, which had been previously fastened to the wooden framework by screws. Owing to the colouring ingredient being mixed with the cement in a homogeneous manner, the tile is throughout its whole thickness uniform in shade.

Fig. 44 represents the centre panel of the frieze, of elaborate design and successful execution.

The character of this carefully designed and substantially built structure was altogether pleasing, and the effect of its graceful parts was considerably heightened by the brilliant colouring of the window frames and balcony, the former having been gilt and the latter painted of a cream colour.

It will be as well to describe the mode of fixing the thin red tiles. On the wooden structure are fixed concrete slabs 3 feet by 2 feet, 1 inch thick, with screws in the ordinary way. On these carefully fastened slabs the courses are accurately lined, and the final or ornamental covering fastened thereto with Portland cement, the result being an accurate external surface superior to any attainable through the agency of rubbed brick or terra-cotta. The reason of this perfection is easily understood when the processes employed are examined.

CONSTRUCTIVE CONCRETE APPLICATIONS. 265

True and unchangeable form and colour are secured by the

Fig. 44. Centre Panel of Frieze of House

use of Portland cement of the best quality coloured with

metallic oxides; terra-cotta and coloured brick being usually liable to deterioration from frost and atmospheric chemical destructants in consequence of the impossibility of securing through the ordinary fire treatment perfectly true and enduring forms.

The merits and success of this building commended themselves to the Paris jurors who were called upon to adjudicate on its architectural and structural qualities, and they awarded to Mr. Lascelles a gold medal. In addition to such high recognition, he was further rewarded by having conferred upon him the honourable distinction of a Chevalier of the Legion of Honour. This latter distinguishing mark was the more remarkable owing to the small number selected for such honour.

The building has been presented by Mr. Lascelles to the French Republic, and the care with which it is to be preserved indicates that Frenchmen fully appreciate the ingenuity and liberality of the enterprising English donor.

We offer no excuse for thus detailing matters not strictly within the province or scope of this book, as they illustrate, in the most pleasing manner, the happy realization of what may be described as the most advanced outcome of novel construction in which Portland cement concrete plays a conspicuous part.

To enter into any lengthened detail of this building would almost be superfluous, seeing that our illustration (although in a most inadequate degree) gives a fair idea of the character and extent of its architectural excellence. A building of the exceptionally costly and elaborate character of that we have described, placed in such a prominent position and subject to the scrutiny and criticism of architects and constructors of all nations, must have been perfect in character to have passed such an ordeal so successfully and honourably.

A system of construction capable of prompt application in cases of emergency, as we have already stated, and also suitable for the most ornate architectural embellishment, places at our disposal resources hitherto unobtainable in any other direction. Its facile adaptation for any constructive purpose is well illustrated by a circumstance arising out of the exigencies of "Metropolitan Board of Works" government. Fig. 45 represents one of three projecting windows,

STUDIO WINDOW, KENSINGTON.

designed by Mr. Norman Shaw for the studio of a well-known artist, which was specially required to secure the most suitable light for artistic purposes. When some progress had been made in its construction, it was found that the wooden frame and its corbelling protruded beyond the charmed line of the "Building Act," thereby involving a modification, which was readily accomplished by Mr. Lascelles in the substitution of concrete window frames, as shown by the illustration. The substitution of concrete for timber met the difficulties of this case with a result highly creditable to all concerned.

In a beautifully illustrated book published by Mr. Lascelles are exhibited twenty-eight designs by R. Norman Shaw, R.A., of buildings suitable to a variety of purposes, from the entrance gate or gamekeeper's lodge to the mansion.

Fig. 46.

Country House.

Fig. 46 is a representation of a design for a country house, highly picturesque in character, and suitable for the occupation of a moderate family.

A circular dome, 20 feet in diameter, is now being designed of Portland cement concrete for India, to be erected for a native prince. There is also to be a lantern on the top for lighting the building. This novel structure is to be made in segments, and arrangements provided by grooves so that when it reaches its intended destination the parts may be readily put together. The thickness of the wall is $4\frac{1}{2}$ inches at the springing, tapering to $2\frac{1}{2}$ inches at the crown.

There can be no better evidence of the confidence acquired in the use of Portland cement concrete under the Lascelles system than this daring novelty in construction. The whole process, from the initiatory designing of the necessary moulds to the packing of the completed parts and their shipment to a far distant point in the East, will entail an amount of care and anxiety of no ordinary kind and character. This dome is going to a district in India where labour is both abundant and cheap, and stone of the best quality exists, from a combination of which a structure of the required character could readily be produced. All the more credit is due therefore to the designer and the constructor of this dome, who can, by ingenuity and adaptation of home resources, compete successfully under such apparently adverse circumstances.

Other buildings prepared under the same system are also in progress of construction for various places in the East.

Many advantages are commanded by the "Lascelles system of construction" for the erection of houses for hot climates generally. The more important advantages are—1st, the readiness with which a building arriving at an out settlement or distant station can be built without the necessity of any very high quality of skilled labour being employed; 2nd, the security of having the structure in all its parts free from the dangers and accidents attending the use of unseasoned, and therefore unsuitable, materials in tropical countries; 3rd, the desirable advantage of portability, securing the removal of the building from one point to another when desired; 4th, the possibility of rendering such a building even fireproof by the special treatment of the timber.

This last advantage may not reach the perfect desideratum of a non-inflammable structure, for practically under existing circumstances there is no such thing as a fireproof building. The contents of a house is the most frequent cause of fire, and when once the match, as it were, applies the initial start, no walls are proof against the action of an intensely heated chamber. The granite and marble with which the major part of the buildings of Chicago were built, became useless under the high temperature to which they were subjected during the great fires, causing the one to be reduced to sand and the other to lime, which a subsequent storm of wind literally blew away.

"But," says the artistic architect who pins his faith in true old-fashioned style to existing usage, " there are no bricks used in such a building as that produced under the 'Lascelles system of construction,' and how can you build a house without bricks?" Undoubtedly such is the fact, but the very reason why is because it has been found that a good and substantial dwelling can be built in their absence, and advantages secured which bricks, by the utmost stretch of ingenuity, fail to command.

The idea which ultimately led to the design of the house built in the "Street of Nations" at Paris, originated from an experience acquired in the erection of perhaps one of the most successful examples of cut-brick building in London. It was designed by Mr. Shaw, and built, in the most careful manner, at a great cost. The style and character of the structure involved the employment of no less than three distinct qualities of bricklayers, viz. one to lay the ordinary stock-bricks, another to cut and rub the facing brick to the required form, while the third, the most skilled of all, had to lay with the utmost care the carefully cut red ornamental bricks. Besides the disadvantage of employing such varied qualities of skilled workmen, structural difficulties arose to increase the anxiety of the builder. The stock-bricks and the finely rubbed facing-bricks were unavoidably different in their sizes, causing the exercise of the greatest care to maintain uniform lines of jointing. By incessant attention a satisfactory result was achieved, but at a cost greatly increased from the difficulties inseparable to such a style of building.

The same brains which planned and carried out the above building were at work in realizing the house represented by Fig. 43, and at least two of the skilled bricklayers were eliminated from the scheme of construction, the only special workman being the one who laid the moulded bricks, and all the assistance which he received was from foreign labour of divers nationalities in Paris: far distant from the point where all the parts of the structure had been fashioned.

Mr. Lascelles would in certain cases, instead of facing with the thin coloured tiles, use bricks of the ordinary size and thickness (made of his concrete), having the desired colour faced to the proper depth. In such application the large slab might have a roughened surface, and the brick also which it was to receive similarly treated, thereby ensuring a more than usually perfect bond.

Portland Cement Concrete Tiles.

We are generally familiar with the many beautiful productions in mosaic and other encaustic tiles, manufactured during the last half century by the famous firms of Minton, Maw, and other potters. The ingenuity displayed in the designs for these tiles of world-wide reputation, is highly creditable to the artistic skill brought to bear in the various processes of their manufacture. Tiles for floors, walls, and many other purposes show the adaptability of clay for such ornamental application. Beginning with nearly the simplest type of minerals for the raw materials, which by highly ingenious treatment become what is so much admired when the products are placed in their ultimate position, there is hardly a limit to the embellishment or strictly useful direction which the mosaic and encaustic tiles are incapable of reaching. Chemistry and mechanical science have alike aided in bringing about the present highly satisfactory condition of the tile decorative art and its increasing applicability to many new purposes of domestic utility hitherto disregarded.

In the varied processes attending the production of clay tiles, there is much that interests the inquirer who takes the trouble to examine the divers ingenious appliances brought to bear in the mosaic tile potter's art. From the initial stage of excavating the clays, their accurate and scientific blending before they are suited for conversion into the requisite form, their subsequent careful treatment to avoid distortion, or damage while drying, and the anxiety attending the kiln or baking operation—indeed at every stage of the process of manufacture—much delicacy of manipulation is essential to secure profitable results. The most experienced manufacturers, commanding the highest class machinery and skill, do not, however, secure the desired accuracy, because the circumstances inseparable from their art preclude its possibility. The chemical ingenuity of the potter may successfully apportion or blend the various plastic ingredients, and mechanism of a special character secure their conversion into mathematical accuracy of form in their raw condition. The final and all-important process of burning, however, introduces uncontrollable dangers, unless surrounded with an amount of care

CONSTRUCTIVE CONCRETE APPLICATIONS. 273

and precaution at a cost which the value of a comparatively cheap mosaic tile of the ordinary kind cannot stand. Hence in the great majority of kiln-burnt tiles, imperfect forms are produced which add materially to their ultimate cost when laid, and prevent their general adoption owing to the difficulty of fitting them so as to harmonize in decorative designs of a highly ornate character.

The great efforts made in modern times to satisfy a craving for highly coloured floor and wall embellishment have almost become exhausted, and the artistic eye now longs for that repose which it could not obtain from the gaudy and unnatural colouring of the most advanced growth of tile-making—that, together with the great difficulty surrounding the obtainment of accuracy of form, has diverted public taste to the original mosaic tile of the Roman, or more modern Italian period. The recent introduction of the mosaic pavement or tiling, composed of marble "tesseræ" imbedded in cement, has led to great changes in the treatment of halls, lobbies, and corridors in public and private buildings. The costly and tedious character of the Italian mode of accomplishing the best examples and designs of this system of floor decoration interferes with its more general adoption, and substitutes of an equally effective and doubtlessly more enduring character are now receiving much favour and support.

Among many important works executed on the old lines of "Italian mosaic," perhaps that of the Manchester Town Hall may be considered one of the largest, the extent of corridors amounting to a total surface of some hundreds of square yards, at a high cost which indicates to a certain extent its superiority over the best mosaic tiles which could have been laid at half the price. That work, executed under Mr. Waterhouse's direction, shows the advantages which the system secures—good foothold to the

T

walker, and repose to the eye from the quiet tone and harmony of its colouring. Such a pavement or flooring laid in the monolithic form as usually adopted, involves great delay in execution, owing to the necessity of putting each small piece in its place by hand while the operator is on his knees. The cement used requires to be quick in its initial set, and owing to such necessity is composed of such ingredients as will prevent its reaching any very high point of ultimate induration.

In the cement tiles produced by Messrs. J. and H. Patteson, of Manchester, all the advantages of the mosaic paving are obtained at a much less cost, and in addition to that important advantage they are also more durable. Tiles of any size can be made by their method, and accurate colouring and form secured with the utmost certainty. We will endeavour to show how the qualities we ascribe to these cement concrete tiles are secured.

To manufacture tiles of the character represented on the accompanying plate, it is necessary that the materials for their fabrication should be of the best kind. Portland cement of unexceptionable quality is of course, as in all good and true concretes, a *sine quâ non*, and rough, clean, gritty sand and pieces of marbles of various shades. The latter is readily secured by Messrs. Patteson from their extensive marble works, and the cement is obtained from the best manufacturing sources. The tiles are of two kinds, one being uniform throughout its thickness, and the other merely faced with the marble chippings and backed up with a less expensive aggregate. The proportions used are one of the best cement to two of aggregates. After being thoroughly mixed in a dry state, the mixture is wetted and put in moulds of the required size and form. While the paste is moist, the pieces of marble are laid according to the design intended. Afterwards, and when the tile has become

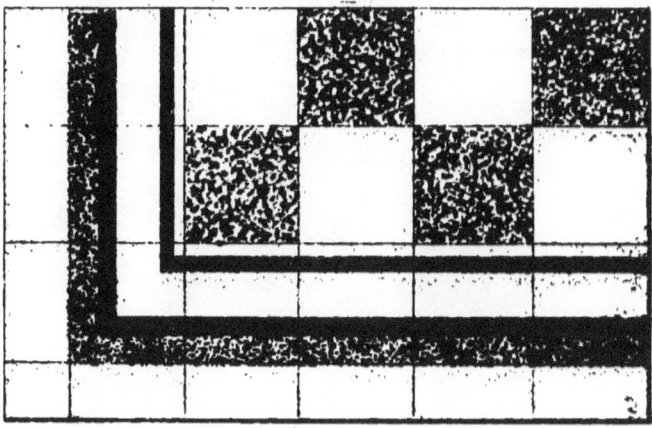

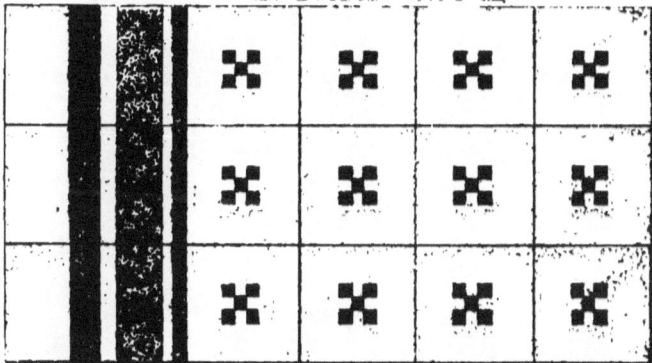

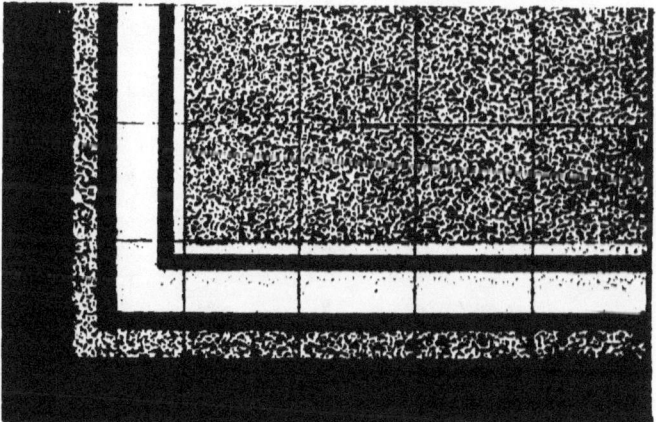

Scale ¾ Inch to 1 Foot.

hard enough, its surface is smoothed by a revolving polishing table of the usual kind. By assimilating the size of the tiles, a large amount of varied adaptation of colour and design is by this means secured, the limit being of course measured by the extent and character of the coloured marbles at command; a wide field of selection in this and other countries, and one difficult of exhaustion. .

On the plate is delineated some of the designs, in natural marble colours, of paving tiles and bordering, and the same system is also applied in walls and other internal decorations; but for this purpose larger slabs are generally employed, which may be polished to any desired extent.

The mosaic tiling of the Italian kind, while securing the object of agreeable tone and easy foothold, does so at high cost and with doubtful permanency for the reason we have named. The concrete tile of Messrs. Patteson does quite as much at infinitely less cost, with the additional advantage of greater durability. The use of Portland cement as the matrix of concretion not only secures the early and successful binding of the aggregates and surface colouring tesseræ, but at the same time provides for the continuous and eventually high strength and density of the carefully concreted mass.

Form in its truest condition is also secured, as no changing of the lines of the tile is possible, except from fracture caused by violence or carelessness. Such an advantage secures the utmost accuracy in joining the various patterns with the greatest facility.

The numerous examples of halls and other works executed in various parts of the country testify to the excellence of this system of Messrs. Patteson.

Many attempts of late years have been made to introduce quasi-ornamental pavements in ordinary monolithic concrete, but not generally with any degree of success, sometimes the failures being due to an ignorant treatment of the Portland cement, or too ready acceptance of pseudo-cements used without undergoing any preliminary test. Such pavements when laid internally, and more especially if a quick-setting cement has been used, are incapable of becoming thoroughly hard, because the conditions to secure such a result are wanting. One of the commonest dangers surrounding such application of concrete is the absence of the necessary moisture to secure eventual crystallization of the mass. From such dangers the concrete tiles we have described are free; because they are not laid until a considerable degree of induration has arisen, and therefore beyond the risk of any abnormal accidents by which the monolith of common concrete is encompassed. The numerous examples of defective internal as well as external paving of this class with which we are familiar, lead us to caution in the strongest manner against such an application for paving purposes.

Frost, so destructive to all kinds and characters of building materials, in times of great winter severity, plays sad havoc with some qualities of tiles, but in no case has it, even in this or other countries, injured Portland cement concrete when properly made with the right materials.

The introduction of tiles of this character, capable of being produced without difficulty and free from the dangers surrounding the production of similar forms made of clay, is likely to lead to their extensive use for general purposes of internal paving and wall decoration. We thus hurriedly refer to these excellent tiles for the purpose of showing still another branch of industry based on Portland cement.

The tiles are usually made one inch in thickness, but of course that limit may be readily extended to suit the requirements of increased size and other special demands on the resources of this successful and artistic concrete development.

The introduction of this unique industry must lead eventually to a great expansion of house decoration in floors and walls.

Hornblower's Patent Fireproof Girders, Floors, &c.

In the various substantial and ornamental adaptations of constructive concrete we have referred to, the all-important question of the fireproof quality of the material has not been entered upon. While, however, claiming for well-combined Portland cement concrete, generally, a certain amount of fire-resisting capacity, we feel that without special provision that great desideratum is not fully attainable. The numerous kinds and varieties of concrete floors prepared with plaster of Paris and other cement matrices, in conjunction with aggregates of every sort, have to a considerable extent the advantage of strength at a comparatively low cost.

The prime objects in the greater number of the floors we refer to, are the obtainment of a substitute for the ordinary joists and flooring boards with the advantage of greater strength and less liability to destruction by fire. For moderate spans, monolithic concrete blocks or slabs have been resorted to, but the comparatively limited resources realized from such an application led to the adoption of iron in a variety of designs in conjunction with the concrete. The heedless use of iron has resulted in floor combinations of a highly questionable character, which, although occasionally securing additional strength, import a new and dangerous element of liability to destruction when under the influence of high temperature.

The great fires of modern times have illustrated in a most painful manner the dangerous character of iron when left exposed to the heat action resulting from raging and uncontrollable conflagrations. In the case of wrought iron, the great heat causes its distortion, while cast iron is liable to fracture, in both cases utterly destroying their influence as constructive aids when so assailed. So much has this danger from iron thus applied been estimated from experience, that those engaged in the hazardous duties of fire extinction, dread entering buildings in which iron, in an exposed form, has been largely employed in their construction. Those who have witnessed such a burning as that of the warehouses in Tooley Street during the great fire in 1862, and more recently the destructive fires at Chicago, can readily understand why unprotected iron floors or walls failed to check, but rather, on the contrary, intensified the loss occasioned by those several disasters.

Iron, an all-important constructive agent when judiciously applied and protected from the dangers which high temperatures occasion, is now capable of much more extensive adaptation through the system patented by Mr. Hornblower.

The primary object of this invention is the combination of iron, terra-cotta, or fireclay tiles, with cement concrete, so as to secure strength and complete immunity from the dangers of fire. A reference to Fig. 47 shows the floor adaptation of this system of fireproof construction. These are two transverse sections showing the relative positions of the various materials, and the judicious disposal of the iron ingredient, so as to ensure its protection from the damaging influence of the highest temperatures, being enclosed in a fireclay tube or tile, and bedded in concrete.

In the floor represented by the upper section, the rolled iron joists are placed 2 feet apart, the intermediate spaces being filled up with hollow fireclay tubes of a wedge shape, the upper or floor portion being covered by Portland cement concrete. By this arrangement there are three combined elements of strength, namely, the iron, the fireclay tube, and the concrete, either of which in proper proportions would

CONSTRUCTIVE CONCRETE APPLICATIONS. 279

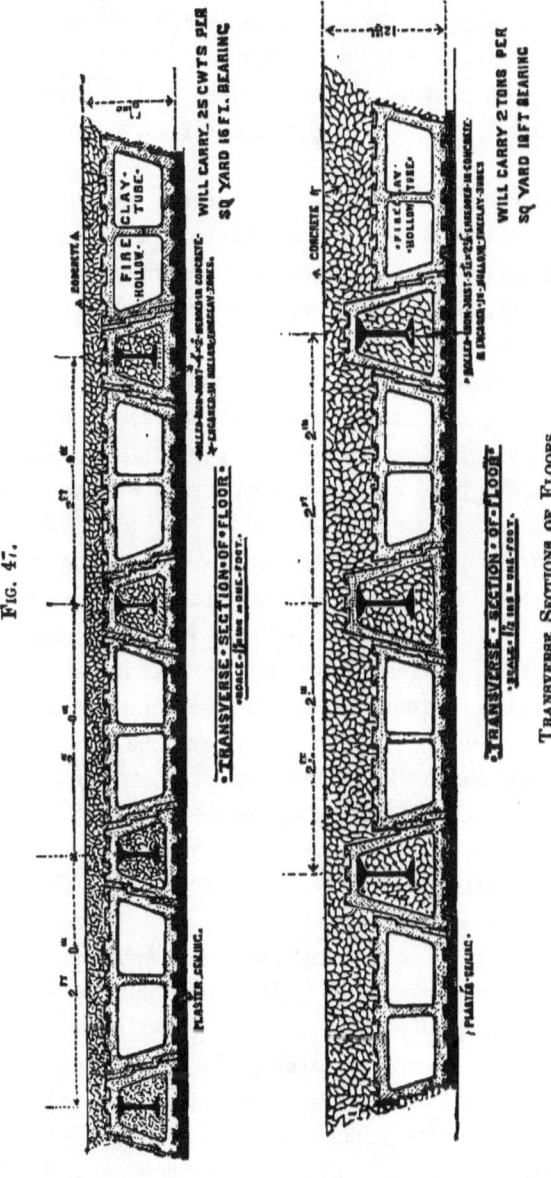

Fig. 47.

Transverse Sections of Floors.

provide a floor more or less suited for ordinary purposes. The floor, according to this design, is 9 inches deep, a large proportion of which is hollow. The weight of this floor is 5½ cwt. per square yard of floor surface, and is capable of sustaining a weight of 25 cwt. per square yard, with a 16-feet bearing.

The upper and under faces of the tubes are corrugated so as to bond with the concrete of the floor, or the plaster of the ceiling. The lower illustration shows transverse section of a floor capable of sustaining greater weight than that shown in the upper one.

The rolled iron joists are 2 feet 2 inches apart, the tiles being of a slightly varied form and 12 inches in depth. Such a floor weighs 7 cwt. per square yard, and is capable of sustaining a weight of 2 tons per square yard with an 18-feet bearing.

A careful examination of the sketch will show that a floor so constructed, while securing the maximum quality of strength, at the same time protects its most important member from all dangers of fire, and provides in the readiest manner for ornamentation and other desiderata of a house to be used as a dwelling. The floor proper may be made of wood parquette, or mosaic tiles of the most elaborate kind. The ceiling also admits of being highly decorated, and at the same time beyond the influence of damage incidental to a warping or twisting floor made solely of timber in the shape of joists, floor boards, and lath.

Fig. 48 shows an isometrical view of a floor constructed with rolled iron girders 16½ inches apart, encased in tiles, having the space filled in around the girder with Portland cement concrete or ground fireclay, the spaces between the girder tiles being filled in with similar tiles inverted. The sides of the tiles are splayed and rebated, and when put together the joint is made with Portland cement or fireclay;

the floor being finished in the usual way as a Portland cement concrete floor and the under side rendered for plastering.

Fig. 49 shows a view of a floor in every respect similar to

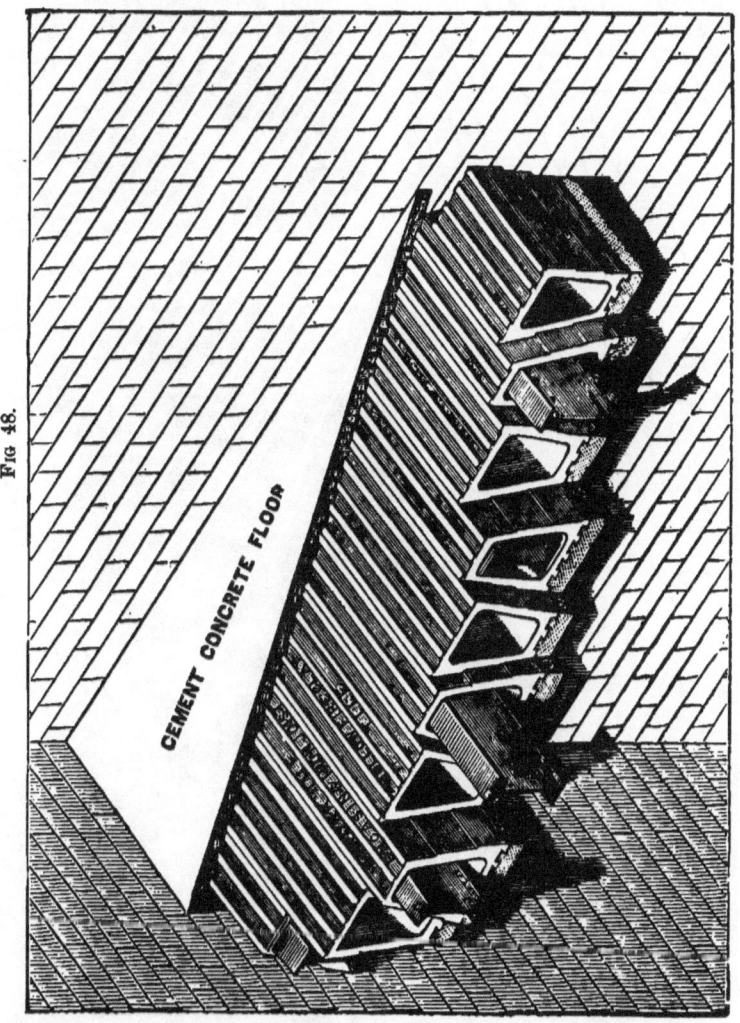

that shown by Fig. 48, the girders being 2 feet apart; the tiles in which they are contained are larger, involving the filling in of a greater quantity of concrete between the crown of the intermediate tiles and the floor.

FIG. 49.

CONSTRUCTIVE CONCRETE APPLICATIONS 283

Fig. 50 is another isometrical sketch of a floor having iron girders similar in form and strength to those shown in

Fig. 50.

Figs. 48 and 49, but one foot apart. The tiles in this design are of uniform pattern throughout, and differ from those in Figs. 48 and 49 in having a cell or hollow in its under side in the one on which the iron girder is placed. This is intended to serve as a conduit for water, and when so supplied from the main is at all times available in case of fire. Between each tile facilities are afforded for laying or placing iron or other pipes intended for warming or any required purpose. An arrangement of this kind possesses many advantages, not the least being the ease with which, during a fire, water may be intercepted and controlled, thus limiting the damage to the floor in which it occurs.

Fig. 51 illustrates another adaptation in a cheaper form of floor, having smaller girders and the tiles or tubes with curved top surfaces. A floor of this simple character would be well suited for ordinary dwellings where the rooms were not required to be very large.

We have, we hope, said sufficient to show, in conjunction with the illustrations, that this principle of Mr. Hornblower's fireproof construction, admits of wide and general application. Although primarily intended for floors, it is equally suited for walls, roofs, and numerous other constructive purposes. In extensive buildings where the necessity of using continuous floors arises, such as dock or similar warehouses, it is important that the supporting cast-iron or other pillars should be protected from the damaging influence of high temperatures. Such an arrangement of protection is easily reached by the aid of fireclay tubes, which can be adapted to any form or design.

A pavement constructed on this system might be designed to meet all the difficulties inseparable from the existing plan, with its inconvenient and dangerous surroundings. All the pipes for water and other purposes, or telegraph tubes, could be so arranged as to dispense with at least nine-tenths

CONSTRUCTIVE CONCRETE APPLICATIONS. 285

of the inconvenience of the present method of footway construction. Not only would such a plan secure more

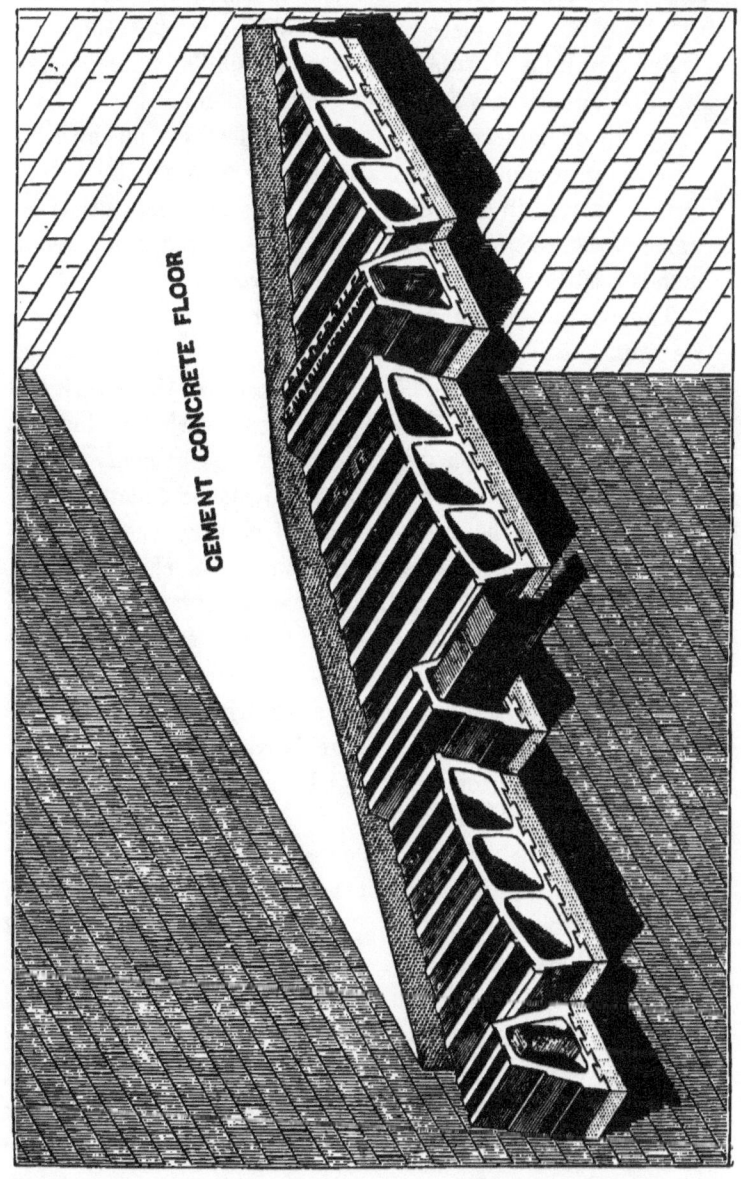

Fig. 51.

comfort to the passer by, but it would also utilize in the most economical manner the spaces below the pavement, which are now completely wasted.

For breweries, distilleries, and other similar industries, or for abattoirs, such a system of construction as that we have described offers peculiar and beneficial features not hitherto recognized or appreciated. The facility with which gratings can be inserted in such floors, and the convenient outlets for waste water secured by the tube or tile arrangement, commends itself for many divers useful purposes.

The system thus hurriedly detailed in its main and leading features, is the profitable combination in the manner defined of iron, clay, and concrete, all of which are of the easiest and cheapest obtainment. While claiming for such combinations the various advantages described, there may be many modifications of the principle applied. For instance, concrete pipes or tubes might be used instead of fireclay ones. These admit of more accurate and in many cases cheaper forms, but it would not be expedient for the present at least to depart from Mr. Hornblower's well-considered and beneficial adaptations of his fireproof system of construction.

This fireproof construction is specially applicable to roofs of factories, cavalry and other barracks, or indeed any building where substantiality and freedom from the danger and loss by fire are desirable.

In the various patterns of floors we have illustrated, a fair idea can be realized of their adaptability to all classes of buildings, from the modest cottage or dwelling to the most substantial warehouse.

This system of fireproof construction cannot be regarded as in its experimental stage only, for there are already many applications of the principle. Amongst others, the following important buildings:—The Cornbrook Pantechnicon at Manchester, in which there is 3400 square yards of Mr. Hornblower's fireproof floor. For such buildings this construction is specially applicable, seeing that their destined object is the safe custody of property of great value and every kind. The comparatively recent total destruction of the London Pantechnicon illustrated in a remarkable degree the utter unfitness of that building to protect property from the most ordinary dangers of its surroundings. In the Liverpool Corn Exchange, these floors were largely used, and at the Bradford Station (Manchester Road Station of the Great Northern Railway) it was also adopted; in other parts of England, notably Wolverhampton and Birkenhead; while in Scotland, at Glasgow, Greenock, Paisley, and Aberdeen, its merits have been appreciated, and in every case of its application the utmost satisfaction has been experienced in its adoption.

The bearing of these fireproof floors can be increased to 30 or 35 feet span without supporting columns, simply by using wrought-iron box girders fireproofed. The Liverpool Corn Exchange floor is 27 feet span. The Cornbrook Pantechnicon, 24 feet span without supporting columns.

The disregard which many modern architects display in refusing to make use of such preservative systems of constructions as those we have here and elsewhere discussed, is not creditable to their intelligence. Had the architect of the late Birmingham Free Library been conscious of the high trust reposed in him, the lamentable destruction of the great Shakespeare collection, and other valuable books, would not now be deplored. The building can be replaced, but where are we to find the literary treasures which no effort or pecuniary means can reinstate? A simple fire insurance frequently leads to a too confiding reliance on the mere money value of a building, and the means for its reconstruction in the event of loss. No policy of insurance, however large or comprehensive in its scope, can restore life or resuscitate from their ashes the immortal fruits of the departed painter, sculptor, and author's skill and genius.

The fire at Clumber the other day, destroyed works of art of national importance. Indeed, wherever we look we find that our art treasures, the national property, are at the mercy of the most trifling act of carelessness on the part of some subordinate. There are plenty of appliances to contend with the fire when kindled, but they seldom succeed in averting the dangers, the protection from which primarily was the business of the architect.

If fireproof construction interfered with, or rendered impossible, the fullest and most ornate display of external embellishments, its non-adoption in so many cases might be to some minds excusable, but it does not prevent in the smallest degree the highest efforts of the most ambitious artistic constructor. For all these and other mercies we must rest contented, until the good time we hear so much of really does come.

Before the full reliance on the quality of the concrete, so essential an ingredient in the construction of floors, walls, &c., was established, it was perhaps difficult, if not impossible, to realize the best and most convenient results in this direction. Increased knowledge, both in the chemical and mechanical direction, of iron, pottery, and cement, reduces the whole question to a level within the reach of the most ordinary constructive intelligence, and free from the risks inseparable from the ignorant practices of the past.

Besides the above-mentioned tests of the adoption of this system of fireproof construction in buildings subject to more than ordinary risks and strains, the following experiment was made to prove its resistive capacity of an actual fire.

In 1874, at the request of the Liverpool Fire Insurance Office, Mr. Hornblower erected two rooms of the following description: ono 18 feet by 16 feet, having floors, as shown by Fig. 49, capable of sustaining a load of 2 tons per square yard, and another 16 feet by 14 feet, capable of carrying 25 cwt. per square yard, as per Fig. 51. The carrying power of the floors had been accurately calculated, according to the Government formula, to sustain a working load to break which the floors would have to be loaded to an extent three times greater than the ordinary safe-working load. These floors were supported by 18-inch brick walls, being covered with timber roofs. Flues were so constructed as to introduce a sufficiency of air; the floors were loaded with bricks, and the most inflammable materials introduced so as to encourage combustion, and a raging fire was maintained for a continuous period of forty-eight hours. The Liverpool Salvage Corps had the management and control of this extemporized conflagration. The fire police were engaged in the suppression of this "fire," and after playing on it with an unlimited quantity of water for six hours, it was extinguished, when on examination it was found that, although the supporting brick walls had been red hot, the fireproof structures had received no damage, and were in reality "as good as new." Mr. Hornblower was disappointed however, for such a successful result did not secure the "certificate of merit or reward" which it was tacitly understood was to follow in the event of success. Fire Insurance Companies have too high a vested or prescriptive interest in fires to encourage the erection of buildings which would largely curtail their lucrative business, and sensibly reduce its profits.

Lish's Patent Z Blocks.

There have been a great many schemes for economizing material, and reducing the cost of labour in the block direction for house-building purposes. Some of the systems similar in character to those we have already referred to, as practised in the United States of America, have had for their objects the lessening of the weight and facilitating divers plans for ventilating and warming. Mr. Lish's plan has in view the combination of many of the various advantages derivable by the block system, and as it may be regarded as the latest outcome of this principle, we shall shortly refer to it.

Fig. 52 represents the simple application of this principle for ordinary or external walls, and consists of two parts, viz. the block and slab, from the combination of which it appears that considerable advantages are secured. Fig. 1 shows the plan, Fig. 2 the facility its form commands for packing, when it may be necessary to send it a considerable distance by rail or road. Fig. 3 a view of wall built and capped in the ordinary way. Fig. 4 shows the plan of building with every alternate block reversed and the mode of fixing the slabs in the rebates. By breaking joint, considerable advantage is obtained, and if desired the hollow spaces may be filled with ordinary concrete, or retained in a hollow form as shown in the sketch. These blocks and slabs may be made of any size, although the drawing illustrates a wall of 10 inches in thickness, and the horizontal joints 20 inches apart.

Fig. 53 illustrates in isometrical perspective, Fig. 5, the corner of a building composed entirely of Z blocks, showing their application to the purposes of house building. Fig. 6 gives the plan, showing the mode of jointing, and the adaptation of the angle blocks in forming quoins, &c.

It will be seen that while for block and slab walling these blocks build 20 inches by 10 inches, when used for the most solid work they build only 10 inches by 10 inches by 6 inches deep.

These blocks can be made of any thickness, according to the concrete process by which they are prepared. If made by impact and of the best matrix and aggregate, they need

CONSTRUCTIVE CONCRETE APPLICATIONS. 291

not be more than 1½ inch thick, and walls so constructed would be of extremely moderate weight, rendering their use of great advantage for internal walls.

FIG. 52.

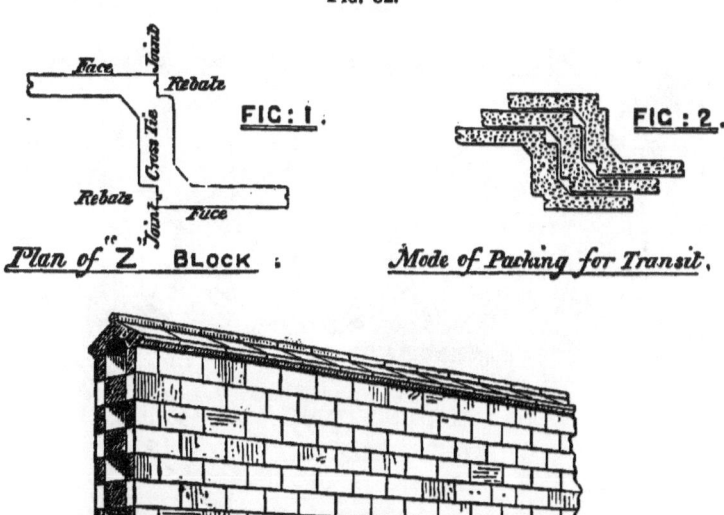

Plan of "Z" BLOCK.   *Mode of Packing for Transit.*

Wall built of "Z BLOCK and SLABS."

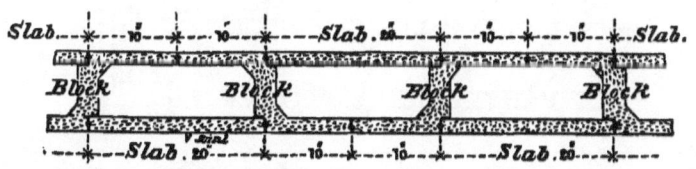

FIG: 4. *Plan of Wall* "Z BLOCKS *and* SLABS".

u 2

There may be some objection made to the form of these blocks, owing to the numerous angles and a liability to fracture in handling. Blocks of a similar form made from

FIG. 53.

FIG : 5 .

Perspective View of "Z Block" *Building.*

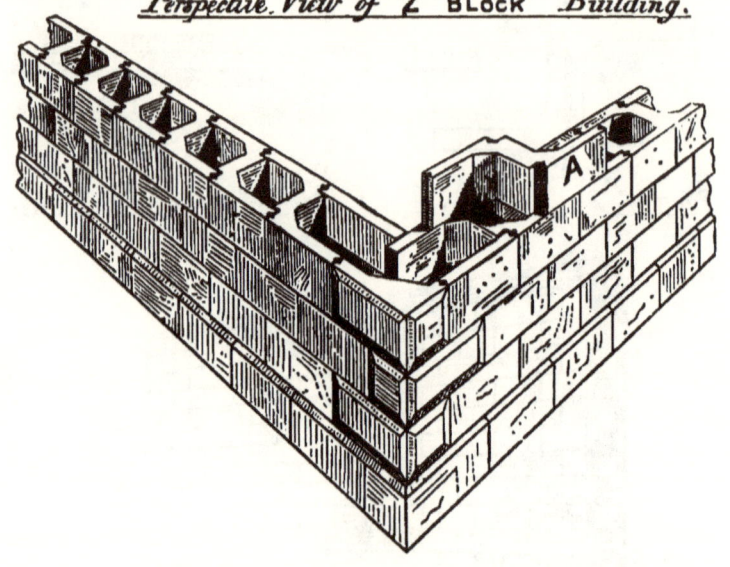

FIG : 6 .

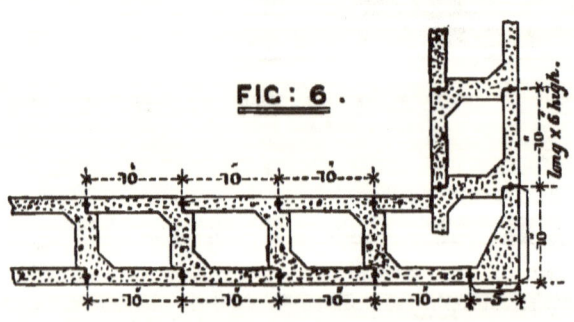

Plan of "Z Block" *Building.*

clay, and hardened by heat influence, would doubtless suffer considerably in the various processes involved in their production, but if made of a good concrete mixture, no such danger need be apprehended.

These illustrations are merely suggestive, for it will be apparent that this system admits of much and varied application to ornamental as well as solid construction.

It would be tedious to give a full detail of all the various plans and systems more or less in vogue for building in concrete, and we have therefore limited our remarks and observations to those which we regard as most useful.

The attention directed, however, to the subject generally will, we trust, awaken the desire for further inquiry on the part of those about to enter on the study or practice of concrete operations.

CHAPTER XI.

IMPORTANT ENGINEERING CONCRETE WORKS.

IN the various industries and systems of concrete construction which we have referred to, sufficient has been said to show with what celerity the various adaptations have proceeded, more especially in those examples where a Portland cement matrix is used. Generally speaking, we have limited our observations to what may be considered domestic outcomes of concrete, which are capable of familiar illustration. In the more elaborate and important works of an engineering character, we do not consider any lengthened reference necessary, as these undertakings have been more or less referred to in the various journals published by scientific societies.

We shall only refer shortly to some of the more recent works of engineering novelty and importance, beginning with

The Tay Bridge.

Mr. Thomas Bouch was the engineer of this undertaking, spanning the Firth of Tay (which at the line of crossing is about two miles broad) on eighty-five piers, supporting girder arches of varying widths. The piers on which the girders rest are lined internally with brickwork and concrete composed largely of Portland cement. In an interesting article by Mr. A. Grothe, C.E., manager of the Tay Bridge Contract, which appeared in the February and March (1878) numbers of 'Good Words,' describing the character and magnitude of this great structure, it is said :—" Without this (Portland cement) or some substance of similar properties, the building of the Tay Bridge would in all probability have been impossible;" and "It plays also a most conspicuous part in the construction of the piers, the first fourteen from the south side being entirely built with it up to the very top, and all the others up to 5 feet above high water, where the iron begins." All the work of this important building was put together on shore and floated out to the points of submergence or erection, without the usual expensive scaffold appliances. Even under such arrangements, the difficulties were great in contending with the natural influences of storms, and it is a matter of congratulation to know that such a work was begun, continued, and finished, with a comparatively limited amount of life sacrifice. The Tay Bridge is the longest in the world, being 2 miles and 50 feet in length. There are a variety of gradients (the steepest 1 in 73) in its course, and it resembles the figure S in plan, having at the land approaches two curves of a quarter of a mile radius each.

The illustration, Fig. 54, taken from Mr. Grothe's "narrative," shows the manner of floating out the piers to their destined sites. On the foreshore a level space was made, on which were built up the piers bit by bit, and when the required height was reached, they were floated off as shown. Some of these piers weighed 200 tons, and no difficulty, comparatively speaking, was experienced in floating them out to the prescribed position, as their weight successfully withstood the disturbing action of the most violent storm. The floatage of the smaller pieces, however, weighing from 40 to 80 tons, was more difficult, owing to these weights failing to possess the required amount of resistive capacity against storm action.

Apparently, on looking at the simple means by which this great work was prosecuted, the whole operation of preparing such huge compound masses of iron, brick, and concrete, was an easy matter. Such a conclusion, however, would be an erroneous one, for if we examine only the

296   A PRACTICAL TREATISE ON CONCRETE.

details of pier building, a fair estimate may be formed of the anxiety and care incidental to such a work. From the preliminary operation of making the foreshore bed sufficiently sound and suitable for the accurate reception of such considerable weights, to the final launching between the twin barges, continuous and ever-watchful care must have been displayed on the part of the several controlling authorities.

Fig. 54.

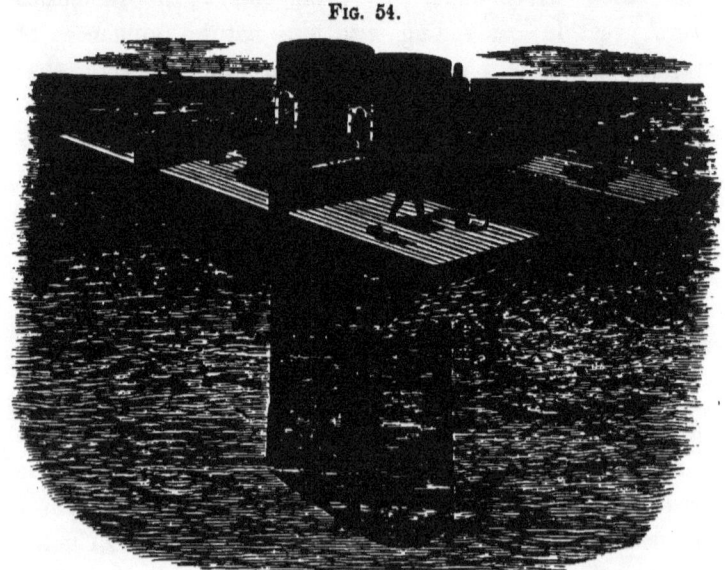

Floating Tay Bridge Piers.

When these huge piers had been safely deposited on their foundations, the anxiety regarding them was, comparatively speaking, at an end. A variety of geological substratum involved in some cases the application of ingenious means to secure the deposition of each pier on a solid base, and whether that was the trap rock of the country, or the dense diluvial gravel, a secure footing had to be obtained. The subsequent filling in of the hollow space of the iron and brick cylinders with the best Portland cement concrete, still further added to the solidity and permanency of the carefully built up piers. Mr. Grothe says:—

"The very first foundation, floated out on August 27th, 1875, was severely tested by the weather. When it left the shore in tow of two steamers a gentle breeze was blowing, but by the time it reached the place where it had to form part of the bridge, the breeze had increased to a strong gale, and the waves washed over the barges so that the hatchways had to be caulked to prevent filling and sinking. It was left in that position nearly three days, till the gale moderated sufficiently to allow the operation to proceed. With the smaller pieces, this could not have been done. They weighed only from 40 to 80 tons, and were entirely at the mercy of waves, which on the larger ones produced little or no effect."

From this short reference to the Tay Bridge operations it is shown that huge masses accurately put together are readily subordinated to the exigencies of the boldest engineering schemes. Until, comparatively recent times European engineers did not venture on the use of large monolithic masses for marine works, and the system has sprung from the novel and daring use of great blocks by Mr. Stoney in the works of the River Liffey improvement at Dublin, he having shown the capacity of Portland cement concrete for producing blocks of great size and weight.

During a recent visit paid by Mr. Gladstone to the works at North Wall, Dublin, he was shown by Mr. Stoney the mechanical appliances used for lifting concrete blocks weighing 350 tons. The engine power applied in that operation was only 16 horses nominal, but by the successful adaptation of pulleys, &c., that apparently limited amount of energy was rendered competent to hoist the above-named weights on to the pontoons by which they were ultimately floated to the desired points of submergence.

It will be observed that in these examples of advanced engineering practised at the Firth of Tay and the River Liffey, two systems were adopted. In that of the Tay bridge the piers were only partially built up, and when fixed in their sea-bed subsequently completed by filling up the space between the inner brickwork rim, adding thereby very considerably to their weight and stability. The blocks at Dublin were *literally built*, and as far as possible completed, on shore, when, after a sufficient time had been allowed for their induration, they were hoisted on to the pontoon and ultimately submerged by the guiding agency of specially constructed diving-bells. The Tay Bridge piers were not so handled, for being built considerably under high-water level the buoyant action of the tide afforded the means of floating them, as shown by Fig. 54, to the points where they were required to be placed.

In the Scotch operations the hydraulicity of the concrete was proved at the recurrence of every tide, while in the Irish practice no test of its capacity of resisting the action of water was realized (except the proving of the cement) until the blocks were deposited in their beds. In neither one case nor the other, however, were the most searching tests wanting to ensure that no cement of a doubtful or dangerous character could by any possibility be used. The advanced intelligence controlling these important undertakings could not afford to risk the use of faulty cement, and we may be sure that the most elaborate tests were made to secure the quality being irreproachable.

It will be impossible for us, even if it were desirable, to give many more illustrations of the efficacy of the Portland cement concrete agent in other engineering works at home and abroad, and we will finish our argument in this direction by a short reference to Messrs. Milroy and Butler's system of concrete cylinder building.

The best example of building wharf or harbour walls is that executed by Messrs. Brassey and Co. in the works of the Clyde Trust at Glasgow. Mr. Milroy, representing the contractors, describes the mode of constructing these brick concrete cylinders thus. After describing the platform on which the cylinders were built, he says:—"On this platform the rings or annular sections of brick cylinder were moulded in frames of suitable size, constructed of wood, in four sections, bolted together. The woodwork consisted of two or three ribs each, of three thicknesses of $1\frac{1}{2}$-inch planking, formed to the circle, nailed together and lined with 1-inch boarding." Various other details of the moulds are given. The rings took about 2000 bricks of the ordinary shape, as time did not admit of the preparation of the proper radiated ones. The Portland cement used was of the best quality, being mixed with one of clean gritty sand, so as to hasten the setting of the mortar. Five days were allowed for induration, when the surroundings of the moulds were released and the ring removed by special mechanical means to its required position. Each ring was 2 feet 6 inches in depth, weighing between nine and ten tons. Altogether 1200 rings were made, being equal to 3000 lineal feet of cylinders, measuring 8000 cubic yards, and weighing about 11,000 tons. Fourteen moulds sufficed to accomplish this work, and although subject to considerable wear and tear, they maintained their accuracy of form to the last. The use of bricks for the cylinders was due to their being at the time of the execution of these works exceptionally cheap in price. In a subsequent extension of the same quay wall concrete was substituted, and with equal success.

Although we have fully discussed in Chapter III. the necessity of accurate and careful testing of the cement for all constructive purposes, we would here add a few observations to show what dangerous consequences would have arisen if faulty cement had been used in any of these important works we have just described. The piers of the Tay Bridge, if blowing or over-limed cement had been used, must have become distorted, and from its expansive power would eventually have burst the protecting iron cylinders. The stability of the whole structure would have become endangered under such circumstances.

The river walls at Dublin and Glasgow could not have been made permanent if bad cement had been used in their construction, and the whole expense incurred jeopardized by neglecting the necessary precautions. Bad or doubtful iron, or imperfectly made bricks, exert only a negative quality of danger, but *untrue* Portland cement possesses inherent capacity of destruction avoidable only by a vigilant attention to its accurate properties, which should *precede* and *not follow* its use.

Birkenhead Docks.

In these extensive operations, carried on under the direction of Mr. Lyster, the engineer to the Liverpool Dock Board, a considerable quantity of concrete has been used for the walls and other works. The aggregate used was principally broken stone and the best Portland cement in the proportion of ten to one, six to one, and three to one of the former to one of the latter, according to the nature of the work. The proportions are first mixed together in a dry state, and then put through a concrete mixer, shown in Fig. 55, the general arrangement of which is the invention of Mr. Le Mesurier, the resident engineer.

The machine as illustrated is ready for work, and may be driven by a 5 horse-power engine.

A A are hoppers of about 2 cubic feet capacity, into which are put the aggregate, either direct from the stone breaker or by hand, and as they revolve on the turntable on which

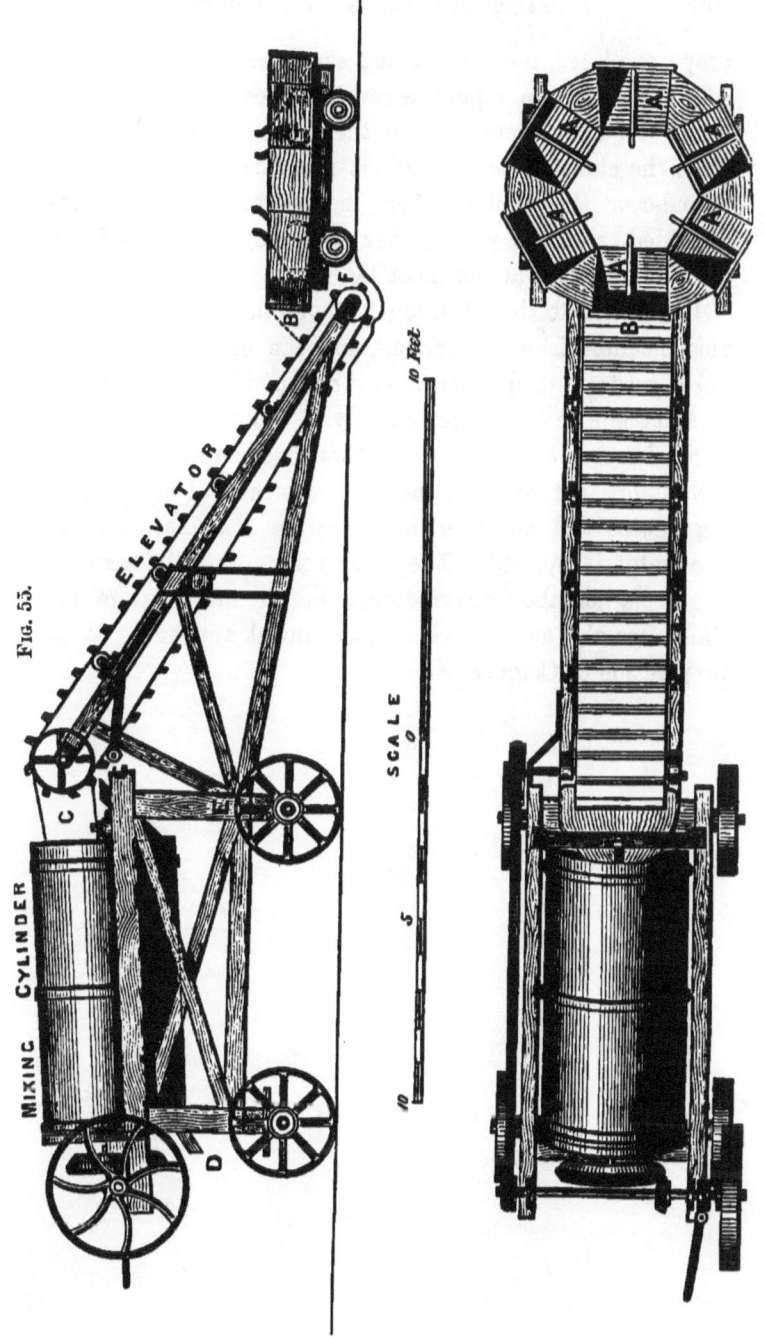

Fig. 55.

they are placed, the cement and sand (when necessary) are added. When the hopper in revolving reaches the points B, a valvular arrangement effects the deposition of its contents on to the elevator band, by which it is carried to the mixing cylinder, at the point C, where the desired amount of water is added, and the whole, after mixture, is delivered into barrows or trucks at the shoot D.

In addition to the advantages of simplicity and portability this machine is most efficient, and can easily produce 150 cubic yards of good concrete at a cost of $4d.$ per cube yard.

Although we have referred only to this concrete mixer, it is not because there are none other, but simply to show the advantages, derived from mechanical means, of bringing the aggregates and matrices into accurate combination before the water is applied. The reference to the Malaxator at page 165, and the Greyveldinger mortar mill at page 177, illustrates the advantages of mechanical appliances in the preparation of Coignet béton.

(303)

CHAPTER XII.

GERMAN PORTLAND CEMENT.

In the various remarks made in the preceding pages regarding the difficulty experienced by *concrete* makers and others interested in important works in obtaining the required quality of cement, it is hoped that sufficient reason has been shown to induce English manufacturers to improve their produce. That there is no difficulty in doing so is already apparent to those engineers who understand the subject, and we avail ourselves of an opportunity which offers of showing that all and even more than we ask or desire is now being supplied by the Portland cement factory, "Stern," Toepffer, Grawitz, and Co., Stettin, Germany.

During a considerable number of years the proprietors of these works have guaranteed their Portland cement to be equal to a tensile strain of about 500 lb. per square inch for the seven days' test, and to a degree of fineness equivalent to a rejection of 3 per cent. through a 2500-mesh sieve.

The adoption of this guarantee step on the part of that firm was quite voluntary, but before they could comply with the exigencies which it involved they determined to manufacture Portland cement of the *best* quality only. Indeed, to our minds such a condition was obligatory, for it encouraged a uniform and unvarying system of manufacture. In England there is, unfortunately, a too willing concession on the part of cement-makers to meet the whims of consumers, and in consequence much confusion prevails. A good Portland cement, prepared on accurate technical lines, is competent to perform any work required of it in an ordinary way. Unfortunately those least acquainted with the properties of a good cement insist on a quick-setting quality, which is usually incompatible with strength and those other advantages which invariably follow.

The reputation which Messrs. Toepffer, Grawitz, and Co. have acquired in Germany and other places proves that the principle they have adopted is the right one. They make a Portland cement regardless of the sacrifices involved in its honest and careful production, and thus they know its irreproachable quality and have no hesitation in guaranteeing it to be what they represent, and expect to be paid for accordingly. Under such circumstances the engineer or architect using such cement feels an amount of confidence and comfort in dealing with a commodity thus produced, and furnished on such assuring terms. It does not, of course, obviate the necessity of protective testing, which should be regarded as obligatory on all cement consumers, from whatever source the supply may be obtained.

It is somewhat remarkable that this lesson of "*guaranteed cement*" should be furnished from Germany, where they have only followed England in becoming manufacturers, for until comparatively recent times that country, as well as France, and indeed the whole of Europe, were solely dependent on England for supplies. The difficulties, however, of obtaining uniform and reliable qualities of the article elsewhere have compelled the introduction of the manufacture in Germany, and it will be well for English makers to seriously consider the probability of German cement not only being sold in our best foreign markets, but the home markets also. There is the startling fact before us that German cement has been sold in London and elsewhere in England, and of such a quality as to show indisputably that our English makers have still a great deal to learn about their business.

In the following particulars of the Portland cement

factory, "Stern," and its doings, it will be shown that German cement of this brand is competent to contend in the market against English makers, and eventually prove a formidable rival everywhere.

The Portland Cement Factory, " Stern," of Messrs. Toepffer, Grawitz, and Co.—These works are situated at Stettin, Pomerania, having been established in 1861, made during the first year of their operation 30,000 casks of Portland cement. Since that time a gradual increase of the output has been obtained, the annual produce now being increased to 240,000 casks. "Stern" means star, which is the trade mark of the firm.

This production is realized under certain disadvantages, for owing to the severity of the Pomeranian climate the works are always idle for two or three months in the winter season, so that practically the large make above named is obtained from about nine months' operation only. The net tonnage made at the Portland cement factory, " Stern," after deducting weight of casks, amounts to about 40,000 tons a year. The exact weight of cement in each cask being 3 cwt. 1 qr. 11 lb., or in round numbers 375 lb., making with the cask a gross weight of 400 lb., the usual and now generally acknowledged standard for shipping and export purposes.

The works, from having the command of railway and river facilities for transit, possess great advantages in despatching cement by both inland and ocean outlets; the system of German railways securing in the one case access to whatever inland points the cost of carriage permits, and the river on the other hand securing a convenient and ready outlet by the Baltic Sea, at the port of Stettin, to the most distant countries.

We are accustomed in England (and the cement manufacturers of London loudly proclaim it) to think that high-class Portland cement of unexceptionable quality is only possible where chalk and clay exist, such materials being alone capable of conversion by the highly vaunted wet process. The Portland cement factory, "Stern," was established before the dry process had been much, if at all, discussed, but the surroundings of the works themselves necessitated the use of the local materials, and indeed their original establishment must have been due to their favourable character.

There is abundance of the best quality of chalk and clay on the premises, the analyses of which, and the resulting cement, are as follows:—

CHALK MARL.

Carbonate of lime	87·29
Sulphate of lime	0·37
Lime in combination with silica	1·12
Magnesia	1·38
Silica	7·43
Alumina	0·83
Oxide of iron	1·48
Alkalies	trace.

CLAY.

Silica	49·52
Alumina	23·85
Oxide of iron	7·02
Bisulphide of iron	3·54
Lime	3·96
Lime in combination with silica	1·45
Magnesia	0·82
Alkalies	2·88
Organic matter and moisture	5·14

The analysis of the cement produced from the careful and accurate combination of these materials being as follows:—

"STERN" PORTLAND CEMENT.

Silica	22·850
Alumina	5·511
Oxide of iron	2·760
Lime	64·409
Magnesia	1·235
Potash and soda	0·923
Sulphate of lime	2·865

The "Stern" cement has obtained a high reputation in Germany, and at the Philadelphia Exhibition of 1876 received first-class testimonials as to its great constructive value. Indeed, General Gillmore, President of the Jury, characterized it as "*the best cement sent to the Exhibition.*"

The testimonial from the Jury of the Philadelphia Exhibition was as follows:—

"International Exhibition,
"Philadelphia, 1876. No. 235.

"The United States Centennial Commission has examined the report of the Judges, and accepted the following reasons, and decreed an award in conformity therewith.

"Philadelphia, 2nd Dec., 1876.
"Report on Awards.
"Product.—Portland Cement.

"Name and address of Exhibitor: Toepffer, Grawitz, and Co., Portland Cement Factory, 'Stern,' Stettin, Germany.

"The undersigned having examined the product herein described, respectfully recommend the same to the United States Centennial Commission for Awards, for the following reasons, viz. :—

"The Portland cement is of excellent quality, is finely ground, and produces a superior mortar of great strength, hardness, and density. No Portland cement superior to it is exhibited.

"Q. A. GILLMORE, Judge."
"Approval of Group Judges.

E. T. COX.   C. DE BUSSY.
Dr. G. SEELHORST.   R. H. SODEN SMITH.
HECTOR TYNDALE.   ARTHUR BECKWITH.

"A true copy of the Record.

"FRANCIS A. WAEHN,
"Chief of the Bureau of Awards.

"F. L. CAMPBELL, Secretary. B. 1488."

The view of the Portland cement factory, "Stern," shown by woodcut will give some idea of the extent and character of Messrs. Toepffer, Grawitz, and Co.'s establishment. The raw materials are brought by wire rope, and other railways, to the several points of the factory, and washed in a somewhat similar manner to that prevailing in England. The washed materials flow into a series of substantially constructed backs or tanks, from which in due course they are taken in the condition of partially dried "slurry" to the drying plates and hacks. The drying arrangements are of an extensive character, the heat being for this purpose supplied from the coke ovens (where the coal is carbonized), of which there are sixty in number. From the plates and hacks, after being thoroughly dried, it is placed in the kilns. These kilns are of the English type, 80 feet high, on the open intermittent system, and of unusually large size; so much so, indeed, as to require but fourteen to produce the yearly quantity of cement we have named.

There is not much difference in the treatment of the clinker, from that prevailing in England and other countries, except that, after leaving the millstones, the powder is passed through a series of sieves securing a quality of fineness much beyond anything obtained in England.

The total horse-power realized from the various steam engines is 382. The number of men employed is somewhere about 800.

The author has had recent opportunities for a thorough examination of this "Stern" Portland cement, from having been employed by Mr. Gustav Grawitz to test the quality of a shipment received in London during the winter of 1878–79. From the tests made and the results obtained, there can be no question about the correctness of the encomium passed upon it by General Gillmore and the Jury of the Philadelphia Exhibition.

"STERN" PORTLAND CEMENT WORKS OF TOEPFFER, GRAWITZ, AND CO., STETTIN.

To face page 308.

Before recording his own tests, the author will place before the reader a series of experiments undertaken by the Pomeranian branch society of German engineers and others. The first series were conducted under the personal supervision of a Commission appointed by the Society in December, 1876. The "Stern" cement necessary for these experiments was obtained, and all the details of the tests were conducted by the Commissioners themselves; the breaking of the briquettes having been witnessed by a large number of the members of the Society. The whole results were published in No. 3 (dated January 20th, 1877) of the weekly Journal of the Society of German Engineers.

The briquettes were moulded in metal moulds placed on a plaster absorptive bed. The following are the results:—

TESTS SEVEN DAYS IN WATER.

AVERAGE OF TEN TESTS.

Neat cement; one cement and one sand; one cement and three sand.

Lb. per sq. in.	Lb. per sq. in.	Lb. per sq. in.
764·49	347·04	212·63

TESTS TWENTY-EIGHT DAYS IN WATER.

AVERAGE OF TEN TESTS.

Neat cement; one cement and one sand; one cement and three sand.

Lb. per sq. in.	Lb. per. sq. in.	Lb. per sq. in.
833·46	419·58	259·57

It should be observed that these tests having been made from briquettes moulded on a plaster of Paris bed, give higher results than those made on a non-absorptive bed, such as marble or glass. This difference in value is equal to about 20 per cent. for neat cement.

Another series of experiments made by Dr. Wilhelm Michaelis, of Berlin, on a non-absorptive plate of glass or marble, produced the following results. The samples were taken on June 1st, 1878, from the bulk at the Government arsenal buildings, Berlin, for the purpose of experiments.

Tests Seven Days in Water.
Breakings of Ten Tests.
Neat Cement.

No.		lb. per square inch
1		512·03 lb. per square inch.
2		611·59 ,, ,,
3		586·70 ,, ,,
4		576·03 ,, ,,
5		593·81 ,, ,,
6		604·48 ,, ,,
7		643·60 ,, ,,
8		547·58 ,, ,,
9		611·59 ,, ,,
10		604·48 ,, ,,
Or an average of		589·18 lb. per square inch.

Tests Twenty-eight Days in Water.
Neat Cement.

No.		lb. per square inch
1		721·82 lb. per square inch.
2		789·38 ,, ,,
3		807·16 ,, ,,
4		704·04 ,, ,,
5		768·04 ,, ,,
6		704·04 ,, ,,
7		832·04 ,, ,,
8		778·71 ,, ,,
9		810·71 ,, ,,
10		746·71 ,, ,,
Or an average of		766·26 lb. per square inch.

In some very interesting experiments made by Mr. Kuhn, civil engineer, at Dresden (published in 'The Journal of the Society of Architects and Engineers at Hanover,' vol. xxiv., year 1878, No. 4) it was found that "Stern" cement eclipsed all other brands, both German and English, that had been examined. For further information on this subject, see chapter on Bastard Mortars, page 344.

The manufacturers of the "Stern" Portland cement, (Messrs. Toepffer, Grawitz, and Co.) have made in their own testing house upwards of 80,000 tests, which have been duly registered, and give even higher results than those made by Dr. Michaelis and others.

The actual details of the tests at the works are as under, and are duly registered only from 1864 inclusive.

Year	No. of Tests	Year	No. of Tests
1864	500	1872	1,816
1865	600	1873	1,761
1866	1,000	1874	1,941
1867	1,500	1875	5,604
1868	2,247	1876	16,600
1869	2,076	1877	16,650
1870	1,726	1878	27,060
1871	1,880		82,961

The strict control exercised over the details of the manufacture of cement produced at these works warrant Messrs. Toepffer, Grawitz, and Co. in *guaranteeing* their produce, and, until the end of 1878, that guarantee was as follows:—

NEAT CEMENT.

7 days in water, 35 kilos. per sq. centimetre, or 497·81 lb. per square inch.
28 ,, ,, 45 ,, ,, ,, 640·04 ,, ,,

ONE OF CEMENT WITH THREE OF SAND.

7 days in water, 10 kilos. per sq. centimetre, or 142·23 lb. per square inch
28 ,, ,, 14 ,, ,, ,, 199·12 ,, ,,

ONE OF CEMENT WITH SIX OF SAND.

7 days in water, 6 kilos. per sq. centimetre, or 85·34 lb. per square inch.
28 ,, ,, 8 ,, ,, ,, 113·78 ,, ,,

The briquettes to be moulded in metal moulds placed on non-absorptive beds, and put in water after setting.

These guaranteed figures were from 18 per cent. to 20 per cent. below the actual records of the factory tests. That fact, and the increased confidence which the proprietors of the "Stern" Portland cement factory have in their own capacity to still further improve the quality of their manufacture, has induced them to increase the guarantee according to the following scale:—

NEAT CEMENT.

7 days in water, 40 kilos. per sq. centimetre, or 568·92 lb. per square inch.
28 ,, ,, 50 ,, ,, ,, 711·15 ,, ,,

ONE OF CEMENT AND THREE OF SAND.

7 days in water, 15 kilos. per sq. centimetre, or 213·35 lb. per square inch.
28 ,, ,, 20 ,, ,, ,, 284·46 ,, ,,

ONE OF CEMENT AND SIX OF SAND.

7 days in water, 6 kilos. per sq. centimetre, or 85·34 lb. per square inch.
28 ,, ,, 10 ,, ,, ,, 142·23 ,, ,,

The cause which induced these bold and enterprising cement makers to increase their guarantee, and therefore their liability, was owing to the fact of their latest factory tests having often given the following results :—

NEAT CEMENT.

7 days in water, 50 kilos. per sq. centimetre, or 711·15 lb. per square inch.
28 ,, ,, 60 ,, ,, ,, 853·38 ,, ,,

ONE CEMENT AND THREE OF SAND.

7 days in water, 18 kilos. per sq. centimetre, or 256·01 lb. per square inch.
28 ,, ,, 25 ,, ,, ,, 355·58 ,, ,,

ONE CEMENT AND SIX OF SAND.

7 days in water, 7 kilos. per sq. centimetre, or 99·56 lb. per square inch.
28 ,, ,, 11 ,, ,, ,, 156·45 ,, ,,

The great importance of being able to rely with confidence on the quality of a cement, cannot but be fully appreciated by all architects and engineers, whilst the security and comfort which such reliance must engender, should eventually add greatly to the consumption of the article. Here we find in the "Stern" guarantee, a lesson which the English cement manufacturers ought to take seriously to heart and ponder over. The result of the determination to institute such a safeguard as that we have described, could only have been possible where a thorough technical knowledge of the question existed, for the great responsibilities which so novel and unprecedented a course involves, are of no ordinary kind. A rule-of-thumb system of manufacture could not produce such confidence and the processes of manufacture, and their careful technical direction must be performed with an accuracy approaching that of almost mathematical precision. It is to be hoped that a similar course of treatment may soon take place in England, when much of the prevailing bickerings would cease, and consumers of cement be relieved from the necessity of such incessant and disagreeable precautions to protect themselves against the shortcomings and supineness of careless and over-confident manufacturers.

It has been said in times past by an English cement maker, that it was, in his opinion, unwise to make cement stronger than the materials it was intended to bind together. But from a concrete or mortar point of view, the quality of Portland cement is of the utmost consequence, for these tests which we here and elsewhere record, indicate beyond question the importance of a high tensile value in an economical direction. *Good* cement must be paid for at a higher price, and yet its increased cost is practically attended with advantage, as the following table prepared by Dr. Michaelis proves. This table was produced by this eminent cement authority at a meeting of engineers, architects, and others interested in the question, held at the "Stern" factory on December 10, 1876.

DR. MICHAELIS'S TABLE, SHOWING RELATIVE VALUES OF DIFFERENT QUALITIES OF CEMENT.

	NUMBERS.					
	1.	2.	3.	4.	5.	6.
	Lb. per sq. in.	Lb. per sq. in.	Lb. per sq. in.	Lb. per sq. in.	Lb. per sq. in.	Lb. per sq. in.
Tensile strength of six different cement briquettes, having been immersed in water 7 days ..	213·35	284·46	355·58	426·70	497·81	568·92
After one year's immersion ..	398·24	540·47	625·81	768·04	853·38	896·05
Cost of 1 cubic mètre having a tensile strength equal to 284·46 per sq. in.	36/5	26/9	24/4	21/8	18/5	16/0
A cask of cement being worth, according to these tensile values	9/3	12/6	13/9	15/5	18/2	20/9

314   A PRACTICAL TREATISE ON CONCRETE.

To show that this guarantee is no mere alluring trade device, we partially copy as under a translation of the advertised guarantee.

"The Stern Portland cement manufactured by Messrs. Toepffer, Grawitz, and Co., of the 'Stern' (star) Portland cement works in Stettin, is now *guaranteed* to be of the following *average minimum strength* when tested according to the German official and acknowledged *rules*.

	Age.	Breaking per sq. centimetre.	Strain per square inch.
Pure Stern cement	7 days	40 kilos.	568·92
„ „	28 „	50 „	711·15
1 Stern cement and 3 of sand	7 „	15 „	213·35
„ „	28 „	20 „	284·46
1 Stern cement and 6 of sand	7 „	6 „	85·34
„ „	28 „	10 „	142·23

The Stern works manufacture this *one superior quality* of cement only.

The high value of Stern cement is best proved by the fact that the Stern works guarantee that a *mixture of one of Stern cement* and *six of sand* is equal to a minimum strength prescribed by the official German *rules* for a mortar of one cement with *three of sand*.

From one barrel of Stern cement when used for mortar or concrete as much work can be obtained as from two barrels of cement of the minimum strength prescribed by the German official rules.

Portland cement therefore should be valued according to its tensile strength and the quality of mortar which it can produce.

Then follows the testimony of General Gillmore and the extent and capacity of the Stern works which we have elsewhere referred to. This cement has obtained a high reputation in Germany, and all the technical journals speak in praise of its excellent qualities.

The salient and more valuable points of the cement

pressed on the public notice are to show that a high tensile quality in Portland cement is the test of its value, for although a lower priced cement may appear preferable to some minds, it will be found that, having regard to the work it performs, it is in reality the dearest. The degree of strength permits of a larger admixture of sand, and thus with a good cement the cost of mortar is less, as is well shown by the table prepared by Dr. Michaelis at page 313.

When our English makers have the wisdom and courage to publish such a table of guarantees as those just named, it will indeed be a satisfactory solution of the long-existing cement difficulty. An examination of the view of the Stern works will show that there is nothing in the shape or character of the kilns which brings about results of so satisfactory a character, and the analyses of the raw materials are not better in quality than those obtained from a valuation of English chalks and clays.

Although a cursory examination of the woodcut does not discover much difference in appearance between the Stern works and those on the rivers Thames and Medway, yet a close inspection discloses the existence of a "Technical Director" residence. The whole secret of success lies here, for the constant and vigilant attentions of a scientific manager ensures the carrying out of all necessary details with the utmost accuracy and carefulness.

The cement samples examined by the author were taken from a shipment which came to London during the winter, and had been for a considerable time exposed on a river-side wharf. Notwithstanding such disadvantages, the following tests and experiments will show that no very appreciable decrease in value has taken place.

Weight.

One hundred and twelve pounds per imperial bushel.

Fineness.

1st. Rejection of 5 per cent. when passed through a sieve having 2500 meshes to the square inch.

2nd. Rejection of 10 per cent. by a 5000-mesh sieve.

Hydraulicity, good.—*Colour*, good.

Tensile Strength.

Briquettes in all cases moulded in brass moulds on glass. Briquettes placed in water twenty-four hours after moulding.

NEAT CEMENT.

	Lb. per square inch.
7 days an average from 10 tests of	517
14 ,, ,, ,,	562
21 ,, ,, ,,	592
28 ,, ,, ,,	629

ONE CEMENT AND ONE OF THAMES SAND.

7 days average of 10 tests	237
14 ,, ,,	245
21 ,, ,,	370
28 ,, ,,	420

ONE CEMENT AND TWO OF THAMES SAND.

7 days average of 10 tests	194
14 ,, ,,	203
21 ,, ,,	225
28 ,, ,,	270

ONE OF CEMENT AND FOUR OF BOURNEMOUTH SAND.

7 days average of 10 tests	201
14 ,, ,,	242
21 ,, ,,	270
28 ,, ,,	304

ONE CEMENT AND ONE OF SCOTCH GRANITE.

28 days average of 10 tests	478

ONE CEMENT AND TWO OF SCOTCH GRANITE.

28 days average of 10 tests	316

ONE CEMENT AND THREE OF BROKEN POTTERY.

28 days average of 10 tests	308

Owing to the difference between the English and German mode of testing, these tests are about 20 per cent. less in value, as explained in page 371.

The whole of the above tests being 190 in number.

There is another advantage realized from the use of

"Stern" cement, and that is the possibility of using it *direct* from the mill without the necessity of waiting as in the case of inferior cements, which sometimes require lengthened periods of seasoning or purification. And a greater advantage still consists in the knowledge that only one kind and quality of cement is produced, thereby avoiding the risk and danger consequent on a supply of inferior or quick-setting quality. In this manufactory of Stern, unanimity reigns, and the consumer is not bewildered by a display of various bins or parcels of different values prepared under *special instructions* from this and that engineer or architect. It would indeed be amusing to see the analysis of each parcel of cement so prepared to order, a record of its fineness, and the test required, and the methods of their estimation. Also what preliminary treatment, by turning over and mixing, was resorted to in preparing such cement for its ultimate destination. Still further interest would arise if the different shipments of cement could be followed to their final field of *usefulness*, and their treatment by the constructor examined. The German system of a uniform and well-considered test overcomes all the difficulties which bewilder the English consumer and manufacturer of Portland cement, and until some such method of valuation is adopted in England differences of opinion must prevail. We have in this country divers methods of estimating cements outside the chemical and mechanical tests, such as *smell*, *taste*, and *touch*, still practised by the essentially *practical* man who prides himself on being beyond the guidance of technical knowledge or its instruments of estimation.

The prominence here given to the German cement will not, the author fears, speedily induce the necessary stimulus to improvement, for if all the "Stern" produce was sold in England it could only supply a mere bagatelle of the cement required by English engineers at home and abroad.

In addition to the above experiments the author has had another opportunity of obtaining a portion of "Stern" cement from a recent shipment to London on its way southward. It is new cement, and has not had the disadvantage of being exposed on a river-side wharf during inclement weather.

The results obtained from this second series of tests are as follows:—

NEAT CEMENTS.
AVERAGE OF TEN TESTS.

	Lb. per square inch.
7 days old	523
14 „	574

ONE CEMENT AND ONE THAMES SAND, SIFTED THROUGH A 400-MESH SIEVE.
AVERAGE OF TEN TESTS.

	Lb. per square inch.
7 days old	287
14 „	316

ONE CEMENT AND THREE OF UNSIFTED THAMES SAND.
AVERAGE OF TEN TESTS.

	Lb. per square inch.
7 days old	223
14 „	294

ONE CEMENT AND THREE OF FINELY GROUND ARGYLESHIRE GRANITE.
AVERAGE OF TEN TESTS.

	Lb. per square inch.
7 days old	247
14 „	295

Since the above experiments were made and the remarks on the "Stern" cement recorded in these pages, the author has visited the manufactory at Stettin, and after a thorough examination of the whole operations carried on by Messrs. Toepffer, Grawitz, and Co., is not surprised at the high quality of the cement produced. Every part of the process of manufacture is designed to produce one undeviating and true quality of Portland cement.

The accompanying table for the conversion of kilos. per square centimetre into English lb. per square inch, has been kindly prepared for this work by Mr. Gustav Grawitz.

GERMAN PORTLAND CEMENT. 319

The formula on which the table is based is as under :—

1 kilogramme equal to 2·2046 lb. English.
6·4514 sq. centimetres equal 1 sq. in. English.

lb. English. sq. Centimetres. lb. English.
Therefore 2·2046 × 6·4514 = 14·22275.

A tensile strain of 1 kilo. per square centimetre is consequently equal to 14·223 lb. per square inch.

As many of the experiments recorded in this book are given in kilos. per square centimetre, the following table, for their conversion into lb. English per square inch, may be found convenient.

TABLE OF COMPARISONS BETWEEN GERMAN AND ENGLISH MEASURES AND WEIGHTS.

Per sq. cm.	Per sq. inch.	Per sq. cm.	Per sq. inch.	Per sq. cm.	Per sq. inch.	Per sq. cm.	Per sq. inch.
0·25	3·56	26	369·80	54	768·04	82	1166·29
0·50	7·11	27	384·02	55	782·27	83	1180·51
0·75	10·67	28	398·24	56	796·49	84	1194·73
1·00	14·223	29	412·47	57	810·71	85	1208·96
2	28·45	30	426·70	58	824·93	86	1223·18
3	42·67	31	440·91	59	839·16	87	1237·40
4	56·89	32	455·14	60	853·38	88	1251·62
5	71·12	33	469·36	61	867·60	89	1265·85
6	85·34	34	483·58	62	881·83	90	1280·07
7	99·56	35	497·81	63	896·05	91	1294·29
8	113·78	36	512·03	64	910·27	92	1308·52
9	128·01	37	526·25	65	924·50	93	1322·74
10	142·23	38	540·47	66	938·72	94	1336·96
11	156·45	39	554·70	67	952·94	95	1351·19
12	170·68	40	568·92	68	967·16	96	1365·41
13	184·90	41	583·14	69	981·39	97	1379·63
14	199·12	42	597·37	70	995·61	98	1393·85
15	213·35	43	611·59	71	1009·83	99	1408·08
16	227·57	44	625·81	72	1024·06	100	1422·28
17	241·79	45	640·04	73	1038·28	200	2844·55
18	256·01	46	654·26	74	1052·50	300	4266·83
19	270·24	47	668·48	75	1066·73	400	5689·10
20	284·46	48	682·70	76	1080·95	500	7111·38
21	298·68	49	696·93	77	1095·17	600	8533·65
22	312·91	50	711·15	78	1109·39	700	9955·93
23	327·13	51	725·37	79	1123·62	800	11378·20
24	341·35	52	739·60	80	1137·84	900	12800·48
25	355·58	53	753·82	81	1152·06	1000	14222·76
Kilos.	English lb.	Kilos.	English lb.	Kilos.	English lb.	Kilos.	English lb.

CHAPTER XIII.

BUILDING FRAMES.

A WORK on concrete would be regarded by many as defective, if no allusion was made to the comparatively modern system of monolithic construction through the agency of building frames. In all the varied systems of concrete adaptations to which we have alluded, more particularly in the various American methods, it is shown that building by blocks is preferred, while by the "Coignet béton" plan, temporary aids by timber frames are in use. The frame system adopted in this country partakes of neither of these plans, for in the former scaffolding must be resorted to, while by the latter stiffened timber guides rigid enough to resist the impacting blows of the rammer, are used in making "béton aggloméré" for building purposes.

For many years a large number of inventions have been patented for the use of building frames, and experience in their application has developed defects in their principle of construction and application which have more or less led to their partial if not total abandonment. While we have, in former works, cautioned against the ignorant use of unsuitable frames which were *pushed* into notice regardless of the more vital question of the quality of the concrete which they were intended to fashion or shape, we were not altogether opposed to the frame system when carefully applied.

The great object in resorting to the use of frames at all is to dispense as much as possible with skilled labour. The early frame inventors were too ambitious to accomplish work involving considerable complication in the scheme of frames, and their want of profitable success may in a great measure be traced to that cause. When, however, the saving of skilled labour in one direction only leads to an increase of the same difficulty in another, it cannot be said that any great advantage is realized. That was pretty much the effect produced during the early introduction of building frames, and their inventors did not sufficiently recognize the necessity for attention to the quality of Portland cement or trouble themselves about its properties. Hence much of the loss and dissatisfaction resulting from the pioneer frame-building operations, was due to the character of the cement, which could not by any possibility be controlled by the most elaborately devised framework.

We here illustrate a simple and ingenious method of building frames, called Henley's revolving and self-fastening apparatus, which offers considerable facilities for house-building purposes.

The character of the several parts of this apparatus is shown in the illustration, Fig. 56.

This illustration is a view of buildings in course of construction, an examination of which will indicate the methods practised by this system. First, at each corner are fixed the angle standards, which in fact are somewhat similar to the pegs usually driven in the ground to define the outline of the house. On these standards holes are made at the desired points to secure the boards or plates to which are fastened the bolts or pivots, and on these the plates or boards turn. Beginning at the ground level, the plate is fastened outside and inside to the required thickness of the wall, and after the concrete has been carefully filled and arranged round the whole extent of the building, the work in that part ceases, and after an interval of twenty-four hours (to allow for the setting of the concrete), the plates are turned up on their hinges or pivots. This process is continued until the desired

Fig. 56.

BUILDING FRAMES. 323

height has been reached, and after, of course, due provision has been made for window or door openings, and other necessary apertures or ornamentations in the walls. By this arrangement it will be seen that during the deposition of each course of concrete, the whole structure, so far as it

FIG. 57.

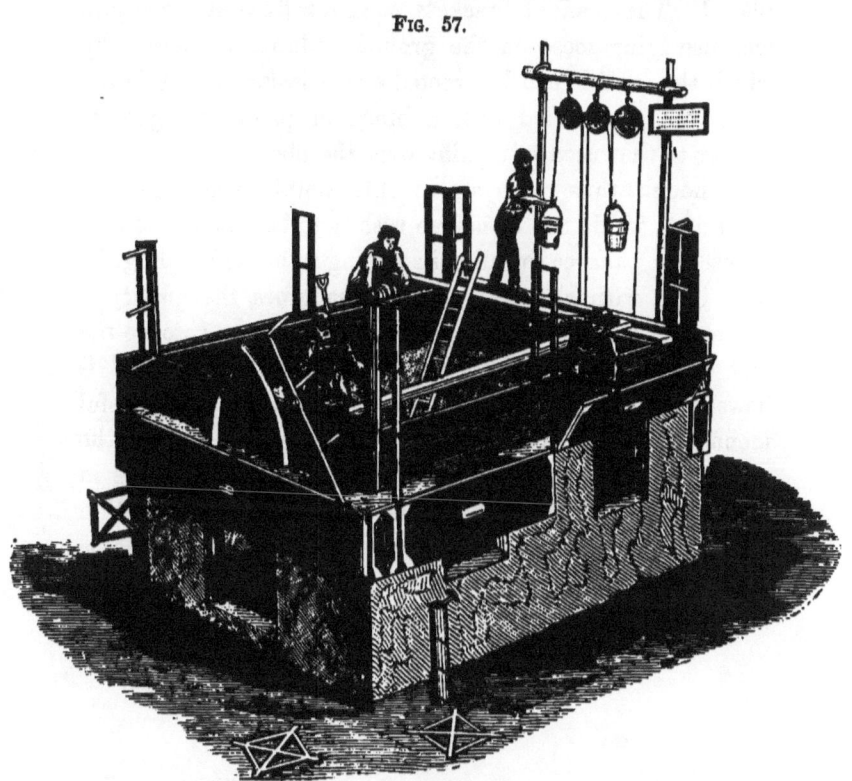

is done, is compactly bound together by the standards and plates, thereby preventing any possibility of bulging or distortion of the walls.

Fig. 57 shows in a more detailed and special manner the method of using the apparatus.

There are two kinds of standards, one being used for angles, and the other in an intermediate position to accommodate the varying lengths of the plates. Lying loose in front of the apparatus are flue cores, provided with an arrangement by which their diameter can be reduced or enlarged according to the desired extent of the flues, and also to facilitate their withdrawal after the concrete has been placed. The scaffold brackets, as shown fixed at the angles and also lying loose on the ground, indicate the means by which the scaffold can be erected either inside or outside the walls, being provided with a hinge or pivot arrangement similar to that used in turning over the plates.

Although thus referring to this simple and ingenious apparatus, we do not of course wish it to be understood that its use is a panacea for all evils arising in house building, for with it, as with all other cognate machines, the quality of the matrices and aggregate used must in every case be the first consideration. The "Henley concrete apparatus" is however a useful one, as it secures in a very successful manner the continuous building, without the necessity of the removal or disturbance of any of its leading parts. It is in fact a sort of telescopic and self-regulating machine within the control of ordinary labour, and therefore very suitable for the construction of moderate buildings in localities where labour is scarce.

CHAPTER XIV.

CONSIDERATIONS AS TO TESTING GENERALLY.

AFTER what has been said in these pages it will be asked, and very fairly asked, Why this necessity for such watchful vigilance in precautionary measures for detecting bad and unsafe building materials? The question is not difficult to answer. During the last forty years, and more especially since the introduction of the railway system, works of an extensive character have been carried out in nearly every part of the world. The heavy masonry, brickwork, and ironwork, besides the tunnelling and earthworks which the development of such a system involved, necessitated the introduction of a new or contractor element in the carrying out of the more substantial portions of the work. At the early stage of this change, the work to be executed was of such a comparatively limited character as to be within the personal control of the engineer and his well-trained and experienced staff. But expanding trade and industrial development hastened the conception and execution of railway works especially, so much so as to outstrip and render impossible the accurate supervision of their construction. Notwithstanding the creation of an almost hot-house character of immature subordinate engineering staff, much of this hurried work was ignorantly or carelessly performed, for the chief or leading engineers were too busily engaged in the concoction of other schemes in far distant and widely separated districts. Although the chief engineers had been entrusted with the carrying out of extensive works, their engagements prevented the possibility of devoting the necessary time to the personal control of those undertakings with which their names were associated. The natural consequences followed, and although more than one valuable life was prematurely shortened in the struggle to cope with almost superhuman tasks, the great works during nearly a quarter of a century had only at occasional and sometimes long intervals the care of the master-eye of the chief engineer. The result was that which is too familiar to modern constructive experience, namely, the controlling contractor element to whom all or nearly all details were entrusted, and in many instances (his increasing influence under such circumstances being paramount) the appointment of the engineer was vested in or influenced by himself. Such a state of things now happily no longer prevails, for the emergencies which created and fostered such eccentric management of important works are found to be inconsistent with sound finance or effective and honest engineering. The quality of work executed under such circumstances could not possibly be of a high-class character, and the future generation will have to meet the account with compound interest for its too obvious shortcomings.

Public works executed under such gambling conditions created an amount of recklessness, the calm consideration of which in these more prosaic times strikes the observer with a feeling of wonder at the possibility even of such misdirection of engineering talent and ability. The absorption into the schemes of public works of the contractor, relieved the engineer of the practical duty of seeing into the detail of the quality and character of the materials to be used. The contractor and his staff were in fact the operative branch of the business, and the engineer the theoretical or official element. The abstraction under such circumstances of that vital supervising department which had been so sedulously looked after by civil engineers, until the advent of the Stephensons, has very naturally drifted away from the proper executive, and, indeed, is almost regarded by the modern practitioner as outside of his province. Hence the prevailing misconception of civil engineers and architects generally as to the character of the materials used in works of construction under their charge. The duty of supervising their selection is therefore necessarily deputed to subordinates or specialists by whom they are generally led and guided. It is not possible, neither can it be desirable, to distract the attention of the controlling professional from his artistic or mathematical duties to criticise the quality of a brick, or the tensile strength of a bit of iron, for it would be indeed ludicrous to insist upon a Hawkshaw or a Waterhouse taking the time or trouble to prove the constructive value of either. Nevertheless these duties must be attended to, for the safety and stability of all structures, even if conceived by the highest intelligence, are alike subject to disaster and decay, in the absence of the required vigilance in challenging and testing the materials to be used. It is true, that under the supervision of the chief, certain *practical* men are entrusted with the duty of examining materials, and from their position receive the designation of "Clerks of Works." In the increasing complication of modern buildings, however, the security to be derived from the vigilance of such an officer can only be limited, and in the generality of cases all but illusory. How can it be otherwise when you are confined in the selection of your practical authority to the three trades of mason, bricklayer, and carpenter? Doubtless during a busy practice through a long series of years the clerk of the works acquires a certain amount of knowledge of those materials outside his own trade, but at best it can only be superficial and empirical in character, and therefore to a great extent unreliable.

Ordinary practitioners being therefore in a great measure dependent on external aid for the honest and successful realization of their best and most matured plans, unhappily disregard in too many instances the guiding support of technical skill, relying rather on the character and reputation of the firms from whom the various materials are obtained. The security and confidence thus obtained is supplemented by the responsibility of the builder or contractor, who conveniently intervenes in case of accident or disaster as a safety valve or elastic cushion through or on which all the energy of blame becomes comfortably eliminated.

Recent disasters, such as those of the concrete floors at Cambridge, and the downfall of a house in the Haymarket, London, by which accidents loss of life arose, were entirely due to the use of improper materials, such as bad cement and honeycombed cast iron. These accidents occurred in different localities and received the attention due to the lamentable results with which they were attended, according to the value of the controlling authority by which they were examined. Little or no public attention was directed to that at Cambridge, but in London the coroner when performing the duties connected with his office required such evidence of a technical character as would guide the jury in arriving at a verdict on the case. A perusal of that testimony discloses an amount of complication rendering it difficult to understand to whose office or position was delegated the responsible duty of examining the qualities of the materials used in building up the houses, which without warning became wrecks of resulting rubbish.

The legislature has from time to time, in the interest of public safety, instituted such laws and regulations as in its wisdom seemed calculated to avert accidents of the kind named. While, however, clearly defining the rules for external and official guidance, the central government authority fails to state, with the necessary precision, the relative position and duties of the several subordinate bodies to whom such vital and important functions are delegated. The machinery by which the practical details of such various and important works is carried out, must be defective in character to admit of any doubt as to whom the blame is attached when accidents occur. Practically it may be said that the public look to the district surveyor of the parish or vestry division for protection under such circumstances, but a calm consideration of the whole surroundings of that officer's position, will prove that it is impossible for him to exert the required vigilance. And yet but comparatively little addition to the existing rules or bye-laws would render every district surveyor competent to grapple with the difficulties with which he is supposed to contend, and over which he should have sole and arbitrary control.

The facility with which the quality of all classes of building materials can be estimated and their variety of constructive values ascertained renders the neglect of their examination short-sighted, if not culpable; but to impose such duty on the already overtasked district officer is clearly out of the question, and its performance must be deputed to some properly qualified central authority, having the confidence and acting under the guidance of the combined interests, who shall be possessed of and competent to use the required machinery of precision. Such assistance would not necessarily release the district surveyor from the responsibilities of his official position, but would enable him confidently to use materials which had been tested and proved by a trustworthy coadjutor, and therefore practically, under the proper conditions, acts of his own. By such an arrangement you would constitute the district surveyor an effective officer, responsible to the public not only for the external lines of a building and details of its walls and their thicknesses, but also for the character and suitability of the materials used in their construction. Make him in reality what he is now only in name, the conscientious and reliable guardian of that part of the public safety connected with his office.

Ancient civilizations, and more especially the great Roman people, exercised the utmost vigilance in the control of all matters relating to building operations, and the comparatively unimpaired character of some of the more remarkable structures of ancient Rome and other districts in Italy testify to the advantages derivable from such surveillance. We look around in vain in this country for examples of buildings of such a character as those remains of ancient architecture, but we are familiarly acquainted with the quality of the mortar used in London about a century ago, from the amount of dust it occasions when the houses in which it was incorporated are pulled to pieces. In fact there was no *mortar* used at the beginning of this century or even since until quite recently in this country, and much of what is now called, by a more than ordinary elastic stretch of the imagination, lime mortar, is but a delusion and a snare. The adverse conditions under which the majority of our purely domestic structures are run up preclude the chance of either sound mortar or anything else strictly entitled to that designation. Their very conception is based on unsound and dangerous principles, for the ground on which they stand is so tied and fettered as to preclude any healthy development in the building direction. The speculating builder, bound hand and foot by the land, money, and material jobbers, has only of necessity the alternative of doing the best for himself, regardless of the ulterior consequences which must inevitably spring from his operations. How long, it may be seriously asked, are such doings to continue?

During the last twenty years, and about the time of the inauguration of the Metropolitan Board of Works, the subject of testing Portland cement received serious attention, so much so as to induce Sir Joseph Bazalgette to institute the requisite conditions by which it had to be performed. In connection with that system the author was consulted, and afforded the necessary advice for its primary organization. The progress of that system, however, under the misleading guidance of the engineering staff of that Board, has unhappily led to an amount of chaotic misunderstanding and confusion, which has prevented the establishment of a proper or reliable standard which can be safely followed. Ignorance of the chemistry of cement has had much to do with this state of things, and in consequence an ambitious desire exists amongst engineers to imitate the eccentricities of such leaders, with the result of leaving the cement-testing question in a most unsatisfactory and undesirable condition. Unlike other countries, as France, Germany, and America, England has no fixed and reliable standard, and much difference of opinion still prevails as to the best instrument for the examination of Portland cement. Until such a consummation takes place, engineers and manufacturers will be alike harassed in the performance of their duties and responsibilities, and the public generally unable to agree in the absence of that unanimity which should long since have been established. Although no fixed rules prevail under the Metropolitan Board for the accurate assessment of the values of materials of construction, the practice of even a fluctuating and uncertain test for Portland cement has been attended with a moderate amount of advantage. The basis on which the first tests were laid has had the merit at least of directing engineering practitioners to the qualities of a matrix which now enters so largely into concrete compounds prepared for various purposes, and some of which in these pages we have endeavoured to describe.

These remarks, it is hoped, may lead to a serious consideration of the question of cement testing, and induce users and makers to agree upon a common standard by which all may be guided and controlled.

Unless some defined system of testing cement is established, the results obtained from a variety of machines and divers methods by which they are used can only be unsatisfactory. In Germany a general agreement has been arrived at, whereby the makers and consumers of cement bind themselves to be controlled and guided; the character of that test is fully described at page 58.

Such a machine as that used in Germany might with much advantage be adopted in England, although the character of the test itself does not seem sufficiently reliable for ordinary practical purposes. The selection of the sand and its preparation for mixture with the cement involve too much risk and trouble for adoption; unless the size of its grains and general physical conditions were identical, the result would inevitably be unsatisfactory and fluctuating. There might be a standard cement, because its chemical and mechanical values are capable of accurate estimation, but to impose on the cement maker a new ingredient foreign to his industry would be unfairly harassing. Let the standard therefore by which the producer is to be measured be free from conditions which under the most favourable circumstances would be difficult of performance. It is now pretty well established that neat cements truly and reasonably examined are capable of being valued in relation to their mortar or concrete value; Dr. Michaelis's table at page 313 satisfactorily establishes this important point. Do not throw all the duties of testing on the manufacturers, for practically when the cement leaves the factory where it was made all control on their part ceases, and it enters on the ultimate stage which ought to be controlled by the engineer, architect, or general consumer. Divers qualities of sands give very variable results, as shown by Lieut. Innes's tables at pages 67 and 68, and therefore no experiments or tests can be reliable unless due allowance is made for their different physical values.

For laboratory experiments, where it is required to examine cements and their behaviour with extreme accuracy, any extent of refinement is permissible; but for the ordinary exigencies of practical construction the test should be of such a character as to permit its being performed by a workman of ordinary intelligence.

CHAPTER XV.

CHARACTER OF BUILDING MATERIALS.

IN the selection of building materials the important question of their heat-conveying or heat-resisting capacities is generally disregarded, and in many cases entirely ignored. Stability and endurance are all-important considerations, but these essential qualities are sometimes insisted upon without reference to their conductivity or heat-absorbing and heat-resisting capacity.

The specific gravities of wood, stone, and metal vary considerably, and so also does their value of specific heat, which means the varying amount of fuel required to warm them to any definite temperature. All bodies are not equally susceptible of heat influence, and some, such as open-grained woods, are differently affected according to the manner in which they are presented to the heating agent. Thus, for instance, the capacity of oak for conducting heat in the line of its fibres is represented by 5, while when applied across the grain it is $2 \cdot 83$. Fir has been found to be in the proportions of 5 to $2 \cdot 05$.

Next to the question of ventilation stands that of heat absorption of the walls, and according to the attention bestowed on this subject of domestic sanitary provision will be the amount of comfort realized by the dweller in the house. The consumption of fuel required to maintain the rooms of a dwelling house at a temperature sufficiently high to ensure comfort and health during cold and inclement weather is, owing to English indifference of this vital question, something enormous. It is sufficient to be burdened with the loss occasioned through the outlets our ignorance and persistent indifference continue to maintain, in the shape of a well-stoked furnace, the major part of whose costly heat is ingeniously withdrawn by the draught of a chimney specially devised for that purpose. But if the experiments referred to at page 123, conducted by so eminent an authority as Professor Pettenkofer, are to be regarded as trustworthy, we shall in future have to consider the desirability of not only giving more attention to the heat-conducting capacity of our building materials, but to their permeability also. If an hospital constructed of slag bricks possesses such admirable advantages as to maintain the degree of efficiency recorded, it will be well for us to consider whether or not our treatment of the internal walls of our dwellings is calculated to promote our comfort or health.

Architectural ingenuity has elaborated, it is true, some highly decorative designs for wall and ceiling ornamentation, and the recent revival of leather and kindred applications for these purposes indicate too truly that the question of the permeability of the wall is utterly overlooked. Layers of paper, coat upon coat, stuck together with vegetable paste, has been justly blamed as the cause of more than one outbreak of malignant fever, and although doubtless the poisonous colours and other surroundings of the paper-hanging system must have contributed their deadly quota, the air-tight room with its own absorbent and choked-up pores must also have been an important factor in producing the fatal result.

The slag bricks so favourably placed in the experiments of Professor Pettenkofer are simply concrete in one of its many forms, and if the sanitary benefits under such circumstances as those described as existing at the George-Marien-Hutte Hospital were realized, we ought seriously to consider the lesson which such a circumstance teaches.

The vital question from the above point of view should be examined not only with regard to the permeability of building materials, but also in reference to their capacity of specific heat and conductivity. For the purpose of assisting in the discussion of this important subject we give the following tables, the first being that of specific heats of various materials used in building, with their respective specific gravities:—

	Specific Gravity.	Specific heats, equal weights.	Specific heats, equal bulks.
Oak	·569	·4042	·2302
Beech	·744	·4431	·3297
Fir	·426	·5174	·2205
Firebrick	2·201	·1917	·4219
Stock brick	1·831	·1860	·3405
Malm brick	1·602	·1720	·2755
Asphalt	2·572	·2150	·5529
Hair and lime	1·691	·0905	·1530
Lath and plaster	1·542	·2065	·3184
Roman cement	1·560	·2099	·3274
Plaster and sand	1·308	·2109	·2758
Plaster of Paris	1·176	·2163	·2544
Keene's cement	1·230	·1855	·2281
Slate	2·788	·1924	·5364
Yorkshire flag	2·360	·1930	·4554
Portland cement	2·157	·1928	·4158
Bath stone	1·858	·1891	·3512
Chalk	1·549	·1827	·2830
Lead	10·560	·0292	·3082

It is not advisable to give the values of Portland cement or the concrete of which it forms the matrix, for they depend too much on the character of the materials used and the mode of their combination, and would therefore be under existing circumstances approximate, to a certain extent misleading. With the progress now being made, however, in concrete industry, we may look forward at an

CHARACTER OF BUILDING MATERIALS. 337

early period to accurate values from such products as the Victoria stone, silicated stone, and rock concrete, for they are produced on definitive lines, and therefore capable of comparatively accurate estimation.

The conducting power of these different materials and their rate of cooling in the same order is shown by the following tables, the two standards being fir at 100 and slate at 100:—

CONDUCTING POWER.

	Fir at 100.	Slate at 100.
Plaster and sand	67·72	18·70
Keene's cement	68·85	19·01
Plaster of Paris	73·36	20·26
Roman cement	75·62	20·88
Beech wood	81·26	22·24
Lath and plaster	92·55	25·55
Fir wood	100·00	27·61
Oak wood	121·90	33·66
Asphalt	163·66	45·19
Chalk	203·37	56·38
Stock brick	217·83	60·14
Bath stone	221·22	61·08
Firebrick	223·48	61·70
Malm brick	264·11	72·90
Portland stone	272·01	75·10
Slate	326·30	100·00
Hair and lime	396·16	109·38
Yorkshire flag	401·81	110·94
Lead	1888·30	521·35

COOLING POWER.

	Fir at 100.	Slate at 100.
Hair and lime	54·60	37·93
Keene's cement	79·81	55·36
Oak wood	80·31	55·79
Plaster of Paris	87·52	60·81
Plaster and sand	90·65	63·31
Fir wood	100·00	69·44
Roman cement	104·58	72·63
Lath and plaster	107·48	74·66
Chalk	107·52	74·58
Malm brick	112·19	77·96
Bath stone	115·82	83·96
Beech wood	122·26	84·71

COOLING POWER—*continued.*

	Fir at 100.	Slate at 100.
Portland stone	134·19	95·07
Lead	137·73	95·67
Stock brick	139·63	96·97
Slate	143·94	100·00
Yorkshire flag	146·48	102·29
Firebrick	149·07	103·13
Asphalt	151·95	105·57

These three tables will give some idea of the relative values of the materials named for building purposes.

In connection with this subject the very important question arises as to the capacity of walls to absorb and retain the impurities of their surroundings. In Professor Pettenkofer's experiments it will be seen that the conditions of permeability to air were altered when the materials under examination were fully charged with moisture, showing that so soon as the pores of the natural or artificial stone were closed externally the passage of air under pressure was checked. In some of our modern houses built of ordinary stock bricks, and even of some kinds of stone, it is quite a common thing to find that under pressure during violent gales the rain water is forced through, and the internal surface of the walls drip with water. An intervening hollow space in a wall so circumstanced would have intercepted the passage of the water and prevented the danger or damage to the interior. Water thus introduced through the walls of any building involves not only the loss occasioned to their decoration, but necessitates the eventual elimination of the contained moisture by an outlay of fuel, otherwise disease will be engendered to the danger of the inmates of a house so circumstanced.

From these observations and experiments it will be apparent that a non-porous or non-absorptive wall is practically impossible of realization from ordinary building material sources. The following tables of absorption of moisture will show the difference of value in that direction of some well-known natural and artificial constructive agents:—

Description.	Specific Gravity.	Absorption by Weight.	Absorption by Bulk.
Aberdeen granite	2·708	2·00	5·42
Carrara marble (white)	2·717	3·10	8·42
Shetland flag stone	2·691	3·25	8·74
Caithness „	2·638	3·27	8·62
Slate	2·788	3·50	9·76
Asphalt	2·572	5·00	12·86
Arbroath flag stone	2·477	20·50	50·77
Firebrick	2·201	32·00	70·43
Portland stone	2·157	34·25	73·87
Yorkshire flag	2·360	40·00	94·40
Bolsover stone	2·164	40·10	86·77
Bath stone	1·858	78·00	144·12
Teak	0·750	82·50	61·85
Stock brick	1·831	109·00	199·57
Hair and lime	1·691	109·12	184·52
Malm brick	1·602	116·50	186·53
Keene's cement	1·230	126·50	155·59
Chalk	1·549	133·50	206·79
Roman cement	1·560	133·56	208·35
Plaster and sand	1·308	147·00	192·27
Beech	0·744	185·50	138·04
Plaster of Paris	1·176	187·50	220·50
Oak	0·570	224·75	128·04
Fir	0·426	622·75	265·41

Professor Doremus, of the Buffalo Medical School, U.S.A., made very instructive and interesting experiments in reference to the permeability of sandstone. A block of sandstone such as is usually employed for window caps and sills, and about 12 inches square and 4 or 5 inches thick, had a panel half an inch deep sunk in each side. In each panel was fitted a block which was perforated by a piece of common gas-pipe, and this was cemented about the edges. The whole was then coated with an impervious varnish. Air now entering the pipe on either side had access to the clean surface of the stone beneath the panel; and it was found that if the mouth be applied to the protruding pipe on one side and a candle be placed in front of the opposite one, it could very readily be blown out by the air which with very little effort was forced through the stone. When a rubber tube was connected with the house gas-pipe on one side of the stone and a burner was attached on the opposite side, the simple pressure from the gas mains was sufficient to force the gas through the stone till it was lighted at the burner on the opposite side. When by any means the pressure was increased a very large flame was then produced. This shows the permeability of building stones. Brick walls and the plastering of rooms are much more porous; and it is readily seen that unglazed tiles, or stone or brick sewers, afford but little security against the escape of sewer gas.

In old buildings, and more especially in those used as prisons and hospitals, in which ventilation and warming had received but little consideration, or where their proper application was difficult or impossible, the walls became literally diseased. In illustration of this vital question we extract from Devey's 'Hygiène de Famille' the following particulars:—

"And more especially to the stones of the walls, especially those of a porous nature; there they are condensed when fresh with the humidity of transpiration. The ventilators which renew the air are almost without effect with regard to the mephitism of the walls. Disinfecting vapours cleanse the atmosphere of the impurities with which it is impregnated; but in general the infection of the walls escapes their action.

"The mephitism of the walls results from the accumulation and infiltration in the walls of the miasms of the surface in the interstices of the stones, the mortar, the plaster, &c., &c. There are formed, so to speak, nests of miasmas, which are hotbeds of infection, acting as a poison upon the blood."

From the time of Moses until now there has been "leper of houses," and numerous instances are recorded where prisons and hospitals, owing to their uncleanness, have been the means of propagating and perpetuating disease and death. In the gradual march of sanitary science more attention is now directed to the cleansing of the body and its clothing than formerly, but until we also recognize the necessity of cleansing our houses, and so constructing them as to facilitate that desirable operation, no satisfactory or substantial progress is possible. The external atmosphere is not the most dangerous of the difficulties to be contended with, for the internal gases, generated by the most complex means through the agency of bad ventilation and over-crowding, is in reality the source from which the greatest danger is to be apprehended. The high temperatures in which the votaries of fashion and victims of necessity spend so much of their time are favourable to the accumulation of concentrated poisons, while unsuitable walls or their dangerous clothings too readily conserve the future disease-spreading germs.

From recent fatal occurrences in connection with the colouring of wall papers, we are reminded that many of the colours used and their bases are of a poisonous character, either in the shape of lead or arsenic. In the richer class of paper or other wall-hanging specialities, owing to their flossy character, convenient means are readily afforded for the absorption and condensation of the vitiated germs generated in the heated atmospheres of the rooms, and also the impurities of the external air which is sucked in by the draughts in an unfiltered state.

In the 'Dangers to Health,' recently published, Mr. Teale, from careful and exhaustive examinations of the sanitary state of houses, very clearly shows that much of the evil is remediable, but holds out but little hope of amendment until "*the education of the public in the details of domestic sanitary matters*" is more advanced.

In 'Chambers's Journal' of January last is an excellent article on "Scamping," in which Mr. Teale's book is favourably reviewed; the writer (Dr. William Chambers), in the introductory matter describing his own experience of a house he had purchased, says (speaking of the "expert's" doings whom he had employed to assist him):—" It was interesting to observe the way in which the expert made his diagnosis of the ailments under which the house seemingly laboured. Like a hound trying to get on the scent of game, he sniffed about in all directions, and applied his nose along the walls and skirting boards, until he fixed on the spots whence issued the malarious odours. These spots were opened up, skirting boards were removed, and floors lifted. What hideous circumstances were revealed! The principal soil-pipes running underneath a passage were broken, from having been laid on soft earth, that had sunk; with the result that the sewage, instead of getting away, had poured into the foundations of the house, causing a filthy quagmire. A metal soil-pipe coming down an interior wall was cracked, in consequence of a bend having been roughly made, fumes from the crack escaping into a bedroom."

This house was bought for 2400*l*., and the cost of rendering it habitable by an overhaul of its faulty parts was 300*l*. Luckily the dangers were discovered before occupation, so that the chances of safe and comfortable occupancy were secured. There are many houses, however, which their owners never suspect, and confidently occupy without a suspicion of the danger until sickness and perhaps death arouse them to a knowledge of their credulous or mistaken faith in the builder. Of whom Mr. Chambers remarks, "It appeared as if cheapness had been alone consulted, and that the builder—honest man!—wished only to get the house off his hands, no matter what might be the result."

Mr. Teale, in his introductory remarks, says, "When disease

arises which we call 'preventable,' depend upon it some one ought to have prevented it." And again,

"This book will show work defective from *ignorance*, and work defective from *dishonesty*."

And assuredly the fifty-six illustrations in Mr. Teale's valuable contribution to sanitary science reveal a condition of things which must make the most careless pause and consider, and ask how much longer *must* these dangers be allowed to continue?

We hope that in these pages it has been shown that a correct knowledge of concrete has much to do with substantial and healthy homes, and at all events if the benefits which it may command are taken advantage of, a considerable amelioration of existing dangers will be secured.

CHAPTER XVI.

SPURIOUS OR BASTARD MORTARS.

EVER since the desire arose for a mortar to set fast, or one competent to resist the degrading action of water, the compound character of the mixture was unavoidable, because the knowledge of the properties of natural hydraulic limes was unknown until quite recent times. The Romans were the first who gave this question the required consideration, and they originated the mixture of fat limes with natural or artificial puzzolanas. It may be said that until the experiment of our famous countryman Smeaton, the subject of the hydraulicity of mortars had received but little attention, for it was the investigations made by him to solve the Eddystone Lighthouse problem that cleared away the mist surrounding the question of the properties of water limes, as they were at that time named. The condition and comparatively limited character of the harbours and piers on the Mediterranean, of Roman engineering, sufficed to meet the wants of trade until modern times, when the construction of larger vessels was commenced, for the reception and protection of which these ancient structures were unsuited. Hence the desire and necessity for more comprehensive shelters on exposed coasts wherein both vessels of war and peace could remain with safety, when trade or tactical necessity rendered such protection desirable.

Belidor and other modern writers on hydraulic engineering have informed us of the means adopted in the construction of such refuges, and we are struck with wonder almost at the result realized in the absence of those scientific

SPURIOUS OR BASTARD MORTARS.   345

guidances so familiar to the engineers of our own times. The necessity of a mortar competent to resist the degrading action of water must have been at an early date apparent to the Roman architect, and the abundance of the lava deposits contiguous to the sites of the early examples of marine engineering fortunately resulted in the introduction of the puzzolana, or hydraulic agent. Such compound mortars received unchallenged acceptance by engineers, until the labours of Smeaton, Vicat, Pasley, and Treussart illustrated the necessity and possibility of obtaining first-rate hydraulic cements from mineral deposits abundantly existing in almost every country. The ultimate practical outcome of these investigations was the invention of Portland cement by Aspdin, who, without chemical knowledge and guided by his own experience as a working bricklayer, arrived at the effective result by a simple combination of the raw materials which secured an artificial hydraulic cement of surpassing value. Notwithstanding the advantages which this discovery, or rather its practical application secured, much reluctance has been displayed even now (fifty years since its introduction) in accepting Portland cement with that confidence which such a material, possessing so many constructive merits, deserves. Although its quality since the time when Aspdin obtained his patent has much improved, owing in a great measure to the assistance rendered by the science of chemistry and the more rational treatment which it receives at the hands of engineers and architects, yet the information afforded by a study of these pages will show, however, that there is still a great amount of indifference displayed by both makers and consumers of Portland cement, which prevents its reaching the point of perfection. This may be due in a great measure, to the desire of overcoming the difficulties undoubtedly still surrounding this important question by an erratic, we might almost say a quixotic, desire to assist the fancied weaknesses of this cement. Hence the wish to impart quack remedies to remove ills of an abnormal and preventable character. From such mistaken assistance, has sprung the very reprehensive practice of spurious or bastard mortar making. The existence of this class of mortar may be said to be as old as scientific construction itself, and until our own time the necessity for its continuance was unavoidable, hence the various compounds in vogue to meet the desire for an hydraulic mortar, composed of fat limes and natural or artificial puzzolanas. A better knowledge of impure limestones and their property, after being decarbonized, of setting under water when used with sand, paved the way for the introduction of Portland cement, and secured the information which has ultimately resulted in a full appreciation of its valuable properties.

There are a variety of acknowledged systems of what may be termed adulteration in producing bastard mortars. First and the most frequent being the addition of lime mortar to Portland cement. Second, the introduction of gypsum or plaster of Paris to expedite the setting of laggard cements. Third, the combination of limes and cements as practised in the preparation of Coignet béton and such like concretes. Fourth, the introduction of some chemical agent, both fluid and in powder, for the purpose of neutralizing the tendency of mechanical disturbance in mortars made with impure Portland cement. In addition to these more general practices, there are numerous other eccentric interferences with the natural action of Portland cement. In the United States of America, owing to the necessity of importing foreign Portland cement, it was and is still in some districts the fashion, for the purpose of cheapening the cost of mortar or concrete, to mix it with the well-known Rosendale and other natural cements of the country. From the experience gained, however, during a lengthened practice of that method, it is now generally conceded that no compound of the artificial with the natural cement is attended with beneficial results, and General Gillmore, in remarking on the system, says:—

"All attempts to cheapen a matrix of Portland cement by the *substitution* of common lime for a portion of the cement, result in a sacrifice of strength in proportion to the extent of the adulteration, and the ratio of loss is not materially changed by the increased induration due to age. This is specially true in thick walls or other large masses of masonry, of which the portion which hardens by desiccation and the absorption of carbonic acid at the surface forms but a small portion of the entire mass."

In experiments made by General Gillmore to illustrate this subject, he found that neat Portland cement seven days old gave a tensile strength of 481 lb. per square inch, and when three-fourths of common lime powder was added to one of Portland cement the tensile strength was only 132 lb. per square inch. In reference to the above breaking of neat Portland cement the experimenter remarked, "Portland cement rarely attains this strength in seven days." Some of the experiments, now for the first time recorded in these pages, show, however, that the maximum and exceptionally high breaking, which General Gillmore characterized as a rarity, is now much exceeded, more especially in the case of the "Stern" Stettin cement. And it should be remembered that these high breakings are obtained without resorting to any "tricks of trade," so as to evade or pass by strategy the various members of the stipulated test. The weight is legitimate and not increased by spurious mixtures, proving that the chemical combinations of the raw materials from which the cement was produced were perfect, realizing a double silicate of lime and alumina, and therefore a true

Portland cement without any *"free lime"* in its constitution.

In some elaborate experiments made by M. F. Kuhn, the engineer engaged in the construction of the bridge over the river Elba at Dresden, the following results were obtained. We extract from No. 4, vol. xxiv., for 1878, of the 'Journal of the Society of Architects and Engineers of Hanover.'

There were a large number of interesting experiments with stones and other building materials, but we will limit our extracts to the matter referring especially to the question of "Bastard Mortars."

There is an old custom, still held in some favour, of mixing lime with cement, a proceeding sanctioned by ancient rules and instructions for "Improving a lime mortar by adding cement." In some cases this compound realizes higher compressive results, and in all other directions it fails to command any beneficial influence whatever, but on the contrary, as the following table will show, tends to depreciate the value of all mortars where such a combination is made.

TABLE OF BASTARD MORTARS PREPARED AND EXPERIMENTED UPON BY THE BOARD OF THE ALBERT BRIDGE, DRESDEN.

Nos.	Proportions of the Mortar Mixture.			Strength in kilogrammes per square centimetre and lb. per square inch, English.			
	Cement.	Lime.	Sand.	Compressive value.		Tensile value.	
				kilos.	lb.	kilos.	lb.
1	Two	One	Two	100	1422·3	13·8	196·2
2	,,	,,	Three	90	1280·0	9·0	128·0
3	One	,,	One	70	995·6	8·2	116·6
4	,,	,,	Two	90	1280·0	8·7	123·7
5	,,	,,	Three	80	1137·8	9·8	139·4
6	,,	Two	Two	15	213·4	6·1	86·8
7	,,	,,	Three	15	213·4	6·9	98·1
8	,,	None	One	240	3413·5	19·6	278·8
9	,,	,,	Two	190	2703·4	18·2	258·9
10	,,	,,	Three	180	2579·1	16·8	238·9
11	,,	,,	Four	155	2204·6	17·4	242·4

It is not recorded what the age of the above briquettes was, but we may assume them to be twenty-eight days old, the general time agreed upon in Germany. Whatever their age may be we have reliable information to guide us as to the value of mortars so concocted, and indisputable evidence as to their value when compared with legitimate mortars. A reference to the table will show that the increase of lime reduced the value, while an addition of sand when the lime was omitted pushed up the breakings to a higher point.

CHAPTER XVII.

ORIGINAL EXPERIMENTS ON CONCRETES, ETC.

ALL the tensile experiments on briquettes were made with a double lever machine and on inch sections.

In the various tests and experiments referred to in the preceding pages such references were intended to illustrate the subjects under immediate discussion, and their source and authority are generally stated. The following experiments were undertaken and conducted by the author for the purpose of showing the quality and character of those artificial concretes now in favour and extensively used for paving, pipe-making, &c. The chromo-lithographs on the frontispiece give favourable illustrations of the three leading artificial stone products, viz. "Victoria," "Rock concrete," and "Silicated stone." These sections faithfully represent the three concretes, being the natural size and colour of the originals. The artist in producing these coloured drawings was guided solely by the examples of each of the pieces placed in his hands for the purpose of illustration. The author's original intention was to have given extended and varied examples of natural as well as artificial concretes, and he had numerous sections cut for that purpose, but found that their accurate delineation would have occupied too much time and still further retarded the publication of this book, which has been already too much delayed. At some future time, however, he hopes to follow out his original scheme, for the study of concretes by this means is both interesting and instructive. The advantage of being able to examine *scientifically* prepared concretes by means of these reliable illustrations will tend, it is hoped, to awaken a desire for further information on this important subject.

The following experimental briquettes and cubes of artificial stone have been made with much care from the mixtures used in the manufacture of the several paving slabs, pipes, &c., and any apparent discrepancies in the recorded results are due to the ordinary or abnormal causes which we have elsewhere in this book referred to. In any view of the case, however, these tests and experiments mark emphatically the great progress of the concrete art, and show that with such increasing knowledge we may hope to find future experiments excel as much those we now record, as they themselves do most certainly eclipse those of the past.

We shall take the tests in the order we have discussed the various English concrete industries, beginning with the Victoria stone.

Victoria Stone Tests.
Compression, cubes $3'' \times 2''$.
Made in February, 1879, broken 27th May, 1879.

No. 1	..	..	..	..	..	3773 lb. per square inch.
" 2	..	..	..	..	..	5179 " "
" 3	..	..	..	..	..	5179 " "
" 4	..	..	..	..	..	3294 " "
" 5	..	..	..	..	..	5179 " "
" 6	..	..	..	..	..	3925 " "

26,529

Average, $4421\frac{1}{2}$ lb. per square inch.

Cubes same Size.
Tested 27th May, 1879.

No. 1	..	..	..	..	..	3766 lb. per square inch.
" 2	..	..	..	..	..	3766 " "
" 3	..	..	..	..	..	3766 " "
" 4	..	..	..	..	..	3578 " "
" 5	..	..	..	..	..	3766 " "
" 6	..	..	..	..	..	4237 " "
" 7	..	..	..	..	..	4237 " "
" 8	..	..	..	..	..	3955 " "
" 9	..	..	..	..	..	3766 " "
" 10	..	..	..	..	..	4708 " "

39,545

Average, $3954 \cdot 5$ lb. per square inch.

The above cubes had been gauged with water and remained in a silicate bath for about a fortnight, and since then they have remained in the air until fractured. Had they been kept in water or in a moist situation, there is no doubt that much higher results would have been realized. The character of the fractures was generally clean and defined, producing but little dust, and an examination showed that a profitable combination of aggregate and matrix had been realized.

One of the chromo illustrations was taken from a polished or rubbed cube afterwards broken. The colour is not so deep as that of the briquettes, owing probably to a less successful application of the silicate bath.

1st *Series, Tensile Tests.*

Victoria stone briquettes. Moulded 30th January, 1879. Broken 22nd May, 1879.

Leicestershire granite 3 parts (2 coarse and 1 fine), and 1 of Portland cement, gauged with silicate solution and placed for 11 days in the silicate bath.

No. 1	672 lb. per square inch.
,, 2	625	,, ,,
,, 3	630	,, ,,
,, 4	685	,, ,,
,, 5	710	,, ,,
,, 6	520	,, ,,
,, 7	710	,, ,,
,, 8	570	,, ,,
,, 9	610	,, ,,
,, 10	590	,, ,,
	6322	

Average, 632·2 lb. per square inch.

Fractures clean and angular, with an almost inappreciable quantity of dust. The mass thoroughly brecciated, with but little evidence of porosity.

ORIGINAL EXPERIMENTS ON CONCRETES. 353

2nd Series, Tensile Tests.

Victoria stone briquettes. Moulded 1st February, 1879. Broken 22nd May, 1879.

Leicestershire granite 3 parts finely ground mixed with 1 of Portland cement, gauged with silicate solution, and placed for 11 days in the silicate of soda bath.

No. 1	503 lb. per square inch.	
„ 2	585	„ „
„ 3	575	„ „
„ 4	630	„ „
„ 5	625	„ „
„ 6	640	„ „
„ 7	600	„ „
„ 8	625	„ „
„ 9	635	„ „
„ 10	655	„ „
	6073	

Average, 607·3 lb. per square inch.

Fractures clean and free from dust showing a small percentage of vacuities in section.

3rd Series, Tensile Tests.

Victoria stone briquettes. Moulded 4th February, 1879. Broken 24th May, 1879.

Guernsey granite 3 parts fine and 1 of Portland cement, gauged with silicate solution, and placed in silicate bath for 11 days.

No. 1	585 lb. per square inch.	
„ 2	577	„ „
„ 3	600	„ „
„ 4	520	„ „
„ 5	595	„ „
„ 6	495	„ „
„ 7	650	„ „
„ 8	525	„ „
„ 9	550	„ „
„ 10	685	„ „
	5782	

Average, 578·2 lb. per square inch.

Dark coloured concrete, with a clean fracture, but owing

to the absence of sufficiently fine aggregate the amount of porosity was slightly in excess of the previous briquettes (1 and 2 series).

4th Series, Tensile Tests.

Victoria stone briquette. Moulded 7th February, 1879. Broken 24th May, 1879.

Guernsey granite 3 parts (2 coarse and 1 fine) and 1 of Portland cement, gauged with water, and placed in silicate bath for 11 days.

No. 1	642 lb. per square inch.
,, 2	685	,, ,,
,, 3	650	,, ,,
,, 4	650	,, ,,
,, 5	700	,, ,,
,, 6	685	,, ,,
,, 7	680	,, ,,
,, 8	683	,, ,,
,, 9	678	,, ,,
,, 10	680	,, ,,
	6733	

Average, 673·3 lb. per square inch.

A strong compact concrete having a clean fracture with small amount of dust.

5th Series, Tensile Tests.

Victoria stone briquettes. Moulded 12th March, 1879. Broken 24th May, 1879.

Guernsey granite 3 parts (2 coarse and 1 fine), gauged with water, and placed in silicate bath for 11 days.

No. 1	680 lb. per square inch.
,, 2	705	,, ,,
,, 3	685	,, ,,
,, 4	618	,, ,,
,, 5	770	,, ,,
,, 6	675	,, ,,
,, 7	840	,, ,,
,, 8	595	,, ,,
,, 9	640	,, ,,
,, 10	772	,, ,,
	6980	

Average, 698 lb. per square inch.

Clean brecciated concrete, dense and free from fractural dust.

6th Series, Tensile Tests.

Victoria stone briquettes. Moulded 25th February, 1879. Broken 31st May, 1879.

Guernsey granite 3 parts (2 coarse and 1 fine) and 1 of Portland cement, gauged with water, and placed in silicate bath for 11 days.

No.	lb. per square inch
1	720 lb. per square inch.
2	700 ,, ,,
3	955 ,, ,,
4	800 ,, ,,
5	790 ,, ,,
6	815 ,, ,,
7	855 ,, ,,
8	732 ,, ,,
9	810 ,, ,,
10	765 ,, ,,
	7942

Average, 794·2 lb. per square inch.

The best of the granite series, exhibiting in all its points the characteristics of a well-balanced and accurately proportioned concrete.

7th Series, Tensile Tests.

Victoria stone briquettes. Moulded 3rd February, 1879. Broken 31st May, 1879.

Thames ballast 3 parts and 1 of Portland cement, gauged with water, and placed in silicate bath for 11 days.

No.	lb. per square inch
1	520 lb. per square inch.
2	635 ,, ,,
3	510 ,, ,,
4	480 ,, ,,
5	595 ,, ,,
6	600 ,, ,,
	3340

Average, 556·6 lb. per square inch.

This concrete exhibits defects due to its constitution, the fracture being imperfect, owing to the spherical character of the aggregate. In some instances the aggregate has been actually withdrawn from the cementing agent in a clean and distinct manner, showing that the matrix had an unfavourable surface on which to exert its best influence.

8th Series, Tensile Tests.

Victoria stone briquettes. Moulded 1st February, 1879. Broken 26th May, 1879.

Leicestershire granite 2 parts fine and 1 of coarse furnace clinker, with 1 of Portland cement, gauged with silicate solution and placed in a silicate bath for 11 days.

No. 1	.. ..	550 lb. per square inch.
„ 2	.. ..	500 „ „
„ 3	.. ..	610 „ „
		1660

Average, 553·3 lb. per square inch.

The object of this series of experiments was to ascertain if any advantageous use could be made of the furnace clinkers of the manufactory.

Compressive tests of Leicestershire granite used in the manufacture of Victoria stone.

Tested to the point of fracture only.

Cubes 4" × 4" very imperfectly squared, and the results therefore against the stone.

No. 1	.. ..	7200 lb. per square inch.
„ 2	.. ..	7416 „ „
„ 3	.. ..	6709 „ „
		21,325

Average, 7108·3 lb. per square inch.

Guernsey granite, also used in making Victoria stone.

Cubes 2" × 2" 11,300 per square inch to the point of fracture only. This cube was carefully dressed and accurately shaped.

Rock Concrete.

Rock concrete briquettes. Tensile experiments. Broken pottery and a small portion of fine siliceous sand 3 parts and 1 of Portland cement, gauged with water, afterwards placed in a water tank for one month. Broken 2nd April, 1879.

Moulded
17th April, 1878. No. 1 .. 590 lb. per square inch.
" 2 .. 635 " "
" 3 .. 495 " "
―――
1720

Average, 573·3 lb. per square inch.

Moulded
21st May, 1878. No. 1 .. 485 lb. per square inch.
" 2 .. 485 " "
" 3 .. 560 " "
―――
1530

Average, 506 lb. per square inch.

Moulded
21st June, 1878. No. 1 .. 455 lb. per square inch.
" 2 .. 500 " "
" 3 .. 415 " "
―――
1370

Average, 456·6 lb. per square inch.

Moulded
August 25, 1878. No. 1 .. 470 lb. per square inch.
" 2 .. 450 " "
" 3 .. 415 " "
―――
1335

Average, 445 lb. per square inch.

Moulded
September 23, 1878. No. 1 .. 595 lb. per square inch.
" 2 .. 475 " "
" 3 .. 490 " "
―――
1560

Average, 553·3 lb. per square inch.

Moulded
October 23, 1878. No. 1 .. 430 lb. per square inch.
„ 2 .. 450 „ „
„ 3 .. 520 „ „
―――
1400
Average, 466·6 lb. per square inch.

Rock concrete, also made by Messrs. Henry Sharp, Jones, and Co., silicated under the Victoria stone patent process, gauged with water, 3 of broken pottery with a small portion of fine siliceous sand and 1 of Portland cement, placed in the silicate bath for 7 days. Broken 26th May, 1879.

Moulded
January 25, 1879. No. 1 .. 605 lb. per square inch.
„ 2 .. 510 „ „
„ 3 .. 485 „ „
―――
1600
Average, 533·3 lb. per square inch.

February 24, 1879. No. 1 .. 680 lb. per square inch.
„ 2 .. 705 „ „
„ 3 .. 700 „ „
―――
2085
Average, 695 lb. per square inch.

The briquettes made in January suffered from frost while setting and were therefore imperfect.

These results show a marked increase in value when the concrete is silicated. All the fractures were clean, the section indicating dense and sound concrete.

Experiments with briquettes and cubes made of the pottery-ware from the refuse or waste of which the rock concrete is prepared.

BRIQUETTES ONE INCH SECTION.
No. 1 .. .. 780 lb. per square inch.
„ 2 .. .. 685 „ „
„ 3 .. .. 675 „ „
„ 4 .. .. 875 „ „
„ 5 .. .. 750 „ „
„ 6 .. .. 680 „ „
―――
4445
Average, 740·8 lb. per square inch.

These briquettes were broken shortly after they were taken from the kiln in which they were burnt, and it is superfluous to state their age, as no increased value could arise from being kept however long. The fractures were clean, and indicated that in their centre a more intense heat had semi-vitrified them.

Cubes made from pottery-ware, as in last experiments.

Cubes 2" × 2" submitted to the point of fracture by hydraulic pressure.

```
No. 1  .. .. .. .. .. 4240 lb. per square inch.
 „  2  .. .. .. .. .. 4802   „       „
 „  3  .. .. .. .. .. 3401   „       „
 „  4  .. .. .. .. .. 4802   „       „
 „  5  .. .. .. .. .. 3672   „       „
 „  6  .. .. .. .. .. 5650   „       „
                      ──────
                      26,567
         Average, 4427·8 lb. per square inch.
```

The fracture of these cubes was very complete, and a considerable amount of dust produced. They did not show as if they had been so perfectly burnt as the preceding briquettes, but the general appearance of the broken pieces indicated their suitability for a concrete aggregate.

Experiments on rock concrete cubes 2" × 2" made with 3 parts of broken pottery and a small portion of siliceous sand, with 1 part of Portland cement, gauged with water, and immersed in a water bath for 30 days, afterwards kept in the air until fractured. Age six months.

```
No. 1  .. .. .. .. .. 3107 lb. per square inch.
 „  2  .. .. .. .. .. 5650   „       „
 „  3  .. .. .. .. .. 3531   „       „
 „  4  .. .. .. .. .. 2119   „       „
 „  5  .. .. .. .. .. 4376   „       „
 „  6  .. .. .. .. .. 3107   „       „
 „  7  .. .. .. .. .. 3401   „       „
 „  8  .. .. .. .. .. 3107   „       „
 „  9  .. .. .. .. .. 3107   „       „
 „ 10  .. .. .. .. .. 5650   „       „
                      ──────
                      37,155
         Average, 3715·5 lb. per square inch.
```

These cubes would doubtless have given much higher results if they had been kept in water until fractured.

Experiments on silicated stone briquettes, Messrs. Hodges and Butler, manufacturers.

Silicated Stone.

1st Series.

Three parts of Lincolnshire slag (finely ground) and one of Portland cement. Gauged with water, and after being exposed to the air for 24 hours, placed in a silicate bath for 11 days. Moulded 11th December, 1878. Broken 3rd February, 1879.

No.		
1	410 lb. per square inch.	
2	520	,, ,,
3	540	,, ,,
4	530	,, ,,
5	478	,, ,,
6	760	,, ,,
7	643	,, ,,
8	725	,, ,,
9	855	,, ,,
10	650	,, ,,
11	815	,, ,,
12	823	,, ,,
13	520	,, ,,
14	658	,, ,,
15	768	,, ,,
	9695	

Average, 646·3 lb. per square inch.

The fractures of the above briquettes were exceedingly clean, and a very small amount of dust was produced. Texture close and tolerably free from porosity.

2nd Series.

Three parts of Belgian granite and one part of Portland cement, gauged with water, kept in the air 24 hours, and afterwards placed in a silicate bath for 12 days.

ORIGINAL EXPERIMENTS ON CONCRETES. 361

Moulded 30th November, 1878. Broken 3rd February, 1879.

No. 1	580 lb. per square inch.
„ 2	710 „ „
„ 3	675 „ „
„ 4	825 „ „
„ 5	665 „ „
„ 6	460 „ „
„ 7	680 „ „
„ 8	652 „ „
„ 9	910 „ „
„ 10	610 „ „
„ 11	745 „ „
„ 12	787 „ „

8299

Average, 691·6 lb. per square inch.

An exceedingly compact, dense concrete making but little dust in fracture. Dark in colour, and represented in the frontispiece, Fig. 6.

3rd Series.

Three parts of river ballast (crushed by a Blake's crusher) and one of Portland cement. Gauged with water, kept in the air for 24 hours, and afterwards placed in a silicate bath for 10 days.

Moulded 8th December, 1878. Broken 3rd February, 1879.

No. 1	540 lb. per square inch.
„ 2	414 „ „
„ 3	435 „ „
„ 4	650 „ „
„ 5	595 „ „
„ 6	720 „ „
„ 7	753 „ „
„ 8	410 „ „
„ 9	565 „ „
„ 10	625 „ „
„ 11	290 „ „
„ 12	393 „ „
„ 13	600 „ „
„ 14	290 „ „
„ 15	410 „ „

7690

Average, 512·6 lb. per square inch.

An excellent concrete, and in the case of the low breakings of the briquettes, Nos. 11, 12, and 14, it was apparent their weakness was due to the presence of an unusual quantity of partially rounded aggregates at the point of fracture. This concrete is exceedingly dense, and can be cut easily into sections of any thickness, which when polished exhibited a beautiful tortoiseshell appearance.

Two inch cubes of silicated stone made with 3 parts of Thames ballast and 1 of Portland cement, gauged with water, and put in the silicate bath for 11 days; about 12 months old fractured as follows :—

No. 1	4237 lb. per square inch.
„ 2	5650 „ „
	9887

Average, 4943·5 lb. per square inch.

Experiments with magnesite (Sorrel patent) and emery briquettes, made by Messrs. Hodges and Butler. Proportions of magnesite and of emery unsilicated.

Made, 1878. Broke, April 9th, 1879.

No. 1	1025 lb. per square inch.
„ 2	890 „ „
„ 3	685 „ „
„ 4	743 „ „
„ 5	675 „ „
	4018

Average, 803·6 lb. per square inch.

A close-grained concrete, used in the manufacture of union wheels; dark grey, almost black in colour.

Experiments with briquettes of Ransome's siliceous stone, prepared by Messrs. Williams and Son, Hale Cliff, near Liverpool.

ORIGINAL EXPERIMENTS ON CONCRETES. 363

1st Series.

No. 1	..	..	..	..	..	565 lb. per square inch.
,, 2	..	..	..	..	..	675 ,, ,,
,, 3	..	..	..	..	..	365 ,, ,,
,, 4	..	..	..	..	..	715 ,, ,,
,, 5	..	..	..	..	..	790 ,, ,,
,, 6	..	..	..	..	..	680 ,, ,,
,, 7	..	..	..	..	..	680 ,, ,,
,, 8	..	..	..	..	..	880 ,, ,,
,, 9	..	..	..	..	..	625 ,, ,,
,, 10	..	..	..	..	..	820 ,, ,,

6795

Average, 679·5 lb. per square inch.

2nd Series.

No. 1	..	..	..	..	..	440 lb. per square inch.
,, 2	..	..	..	..	..	585 ,, ,,
,, 3	..	..	..	..	..	445 ,, ,,
,, 4	..	..	..	..	..	680 ,, ,,
,, 5	..	..	..	..	..	620 ,, ,,
,, 6	..	..	..	..	..	680 ,, ,,
,, 7	..	..	..	..	..	590 ,, ,,
,, 8	..	..	..	..	..	515 ,, ,,
,, 9	..	..	..	..	..	760 ,, ,,
,, 10	..	..	..	..	..	765 ,, ,,

6080

Average, 608 lb. per square inch.

A beautifully compact and regularly coloured artificial stone, capable of being moulded into the most ornamental forms. These briquettes were specially made for the author's experiments, but as they attain when finished their highest value, it was not necessary to detail their age. They were, however, about three months old when broken.

Experiments with a slab of "granitic breccia" stone, made by Mr. Buckwell by his impact process, about twenty-five years ago. The slab was obtained by accident, having been found in the yard of an extensive pavior in London. Its history, however, is something as follows. Made by Mr. Buckwell, and laid down by him in the Poultry, London, where it was used from eight to ten years, or probably more. When a new pavement was necessary, owing to the uneven and irregular character of the York landing, this piece of "granitic breccia" was removed with the others. Its original dimensions were apparently 6 feet × 3 feet × 3 inches in thickness. The aggregate used was Anston stone of large and irregular size, with a heavy dark Portland cement. This slab was placed at the author's disposal, of which he has made use in the following tests.

The colour of the aggregate is a pale yellow, and the cement a dark (almost blackish) blue. Judging from the appearance of the matrix, it must have weighed 125 or 130 lb. per imperial bushel. A close examination of the concrete shows it to be remarkably compact and free from porosity, weighing 143 lb. per cube foot. When cut in sections it may be fairly polished, considering the character of the aggregate.

1st Experiment.

The slab was cut to the following dimension, 3 feet × 2 feet 4 inches, and the thickness 3 inches unaltered. This piece of slab was supported on two timbers, on which it had a 3-inch bearing. Upon it was placed a weight of 3 tons 5 cwt. 3 qrs. 17 lb., which it sustained for about five hours when it broke with a fracture about midway through its length. This was equal to 1054 lb. per square foot, and the resistance of the concrete to such pressure indicated that it had not deteriorated in value during the long period that had elapsed since it was made. This is the more remarkable, because at the time of its manufacture Portland cement was not of a very reliable quality nor so good as at present.

ORIGINAL EXPERIMENTS ON CONCRETES. 365

2nd Experiment.

"Granitic breccia" cubes 3" × 3" were fractured under pressure with the following results:—

No. 1	8161 lb. per square inch.
„ 2	8886 „ „
„ 3	7533 „ „
„ 4	7533 „ „
	32,113

Average, 8028·2 lb. per square inch.

The fracture was exceedingly clean, and although there was in the concrete some pieces of coke, the whole appearance was sound, indicating that the stone and cement had been profitably put together.

3rd Experiment.

Briquettes 1 inch section carefully cut.

No. 1	710 lb. per square inch.
„ 2	625 „ „
„ 3	575 „ „
„ 4	625 „ „
„ 5	710 „ „
„ 6	495 „ „
„ 7	920 „ „
„ 8	630 „ „
„ 9	705 „ „
„ 10	690 „ „
	6685

Average, 668·5 lb. per square inch.

A concrete so rough in character was not very suitable for experiment with inch briquettes, and where low breakings were realized an examination of the fracture indicated why a low result was obtained. In the case of No. 6 there was not in the section (one inch) more than 3 per cent. of cement, a large irregular piece of aggregate having occupied nearly the whole space.

Fig. 58 represents a briquette (No. 1, natural size), by which it will be seen that the aggregate of this piece of granitic breccia is very uneven and irregular in character.

Fig. 58.

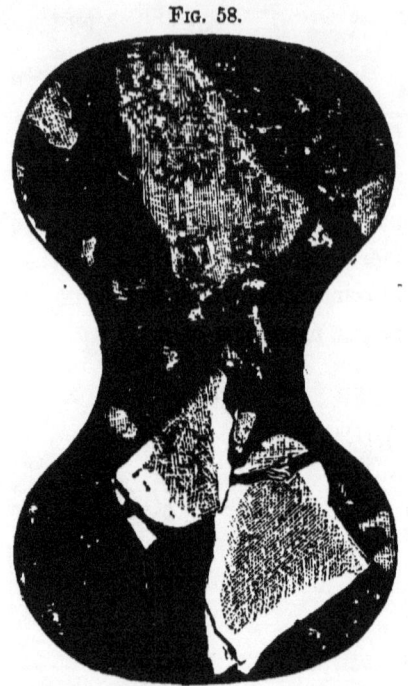

Fig. 59 represents a piece of Spinkwell sandstone enlarged to four times the natural size. This stone is of excellent quality, and has obtained a high reputation with architects. It has been used in many important buildings, among others the Manchester Town Hall. The stone is compact in character, and of a warm yellowish tint in colour.

The compressive value to the point of fracture is as follows: No. 1 cube, 4" × 4". No. 2 cube, 3" × 3".

No. 1 .. .. 5,650 lb. per square inch.
„ 2 .. .. 10,044 „ „

No. 2 specimen was obtained from the best bed, and No. 1 from a less compact one.

Aberdeen granite. Cubes 2" × 2", imperfectly faced. Compressive value to the point of fracture,
15,537 lb. per square inch.

Loch Etive granite, W. Sim and Co.'s quarries. Cubes 2" × 2", accurately faced. Compression to the point of fracture,

No. 1	19,775 lb. per square inch.	
„ 2	16,950	„ „
	36,725	

Average, 18362·5 lb. per square inch.

This is a fine, compact, close-grained, light-coloured granite, and was used by the author in the tests recorded at pages 316 and 318.

FIG. 59.

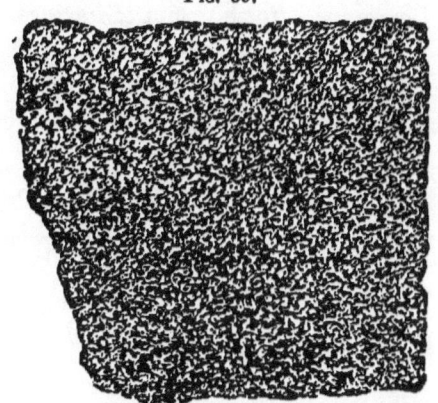

Mount Sorrel red granite. Cube 2" × 2", imperfectly faced on one side. Compression to point of fracture,
14,150 lb. per square inch.

Penmaen Mawr (Carnarvonshire) basalt. Used in paving the streets of Manchester and Liverpool. Cube 2" × 2". Compression to the point of fracture,

No. 1	28,250 lb. per square inch.	
„ 2	25,425	„ „
	53,675	

Average, 26837·5 lb. per square inch.

A close compact stone of great hardness, and susceptible of taking a high polish.

Carrara (white) marble. Cube 2″ × 2″. Compression to the point of fracture,

5650 lb. per square inch.

Slag Bricks.

No. 1. Middlesborough brick taken from a quantity being, used in a building at Redcar, Yorkshire. Cube $3\frac{1}{4}″ \times 4″$. Compressed to the point of fracture,

435 lb. per square inch.

A soft spongy material made apparently with a faulty matrix.

No. 2. Siegen slag brick forwarded by the makers. Cube $9\frac{3}{4}″ \times 4\frac{1}{2}″$. Compression to the point of fracture,

525 lb. per square inch.

Experiments in Ransome siliceous stone made by Messrs. A. H. Bateman and Co., East Greenwich. Inch briquettes.

No. 1 Tests.

Ransome's old calcium stone mixed with Maidstone sand:

No. 1	445 lb. per square inch.
,, 2	475 ,, ,,
,, 3	395 ,, ,,
,, 4	430 ,, ,,
,, 5	425 ,, ,,
	2170

Average, 434 lb. per square inch.

No. 2 Tests.

Calcium stone from Maidstone sand, different proportions from last:

No. 1	525 lb. per square inch.
,, 2	615 ,, ,,
,, 3	590 ,, ,,
,, 4	573 ,, ,,
,, 5	570 ,, ,,
	2873

Average, 574·6 lb. per square inch.

The colour of these stones is very good, the texture being firm and compact, showing great facility for accurate moulding.

Experiments on Apanite, made with emery as the base and the concrete submitted to a gentle heat of about 200° Fahr. for several hours:

Tensile Tests.

No. 1 .. .. 1105 lb. per square inch.
 „ 2 1070   „   „
                           ————
                           2175

Average, 1087·5 lb. per square inch.

An exceedingly hard, dense concrete, used especially for grindstones.

All the tests recorded in this chapter may be regarded as highly satisfactory, and they show what strides the concrete industry is making towards perfection; at all events it is clear that results are arrived at in other than haphazard or rule-of-thumb fashion.

The following tables of experiments made by Mr. Haslinger, Technical Director of the "Stern" Cement Works, illustrate the advantages derived from the use of finely ground cement when mixed with sand. These experiments also prove that fine cement when used neat does not reach the same value as the coarse; showing, therefore, the desirability of instituting a mortar test in this country similar to that for some time established in Germany. The degree of fineness of the cement used by Mr. Haslinger was of an unusually high standard, being 5000 meshes to the square centimetre, which is equivalent to 32,256 meshes per square inch.

The cement used in the experiments was taken from the winter stores of the Stern Cement Works; the briquettes were made during the months of October, November, and

December, 1878. It should be observed that the atmosphere of the testing room was maintained at a uniform temperature of about 60° Fahr. during the whole period in which the briquettes were under examination.

All the breakings were performed by the double lever German testing machine described and illustrated at page 59.

The decimal parts are omitted in the following tables.

The cement used in these experiments does not suffer in value however finely it may be sifted, for it is the product—fine or coarse—of one *uniform* and unvarying clinker. The weight scarcely ever exceeds or is less than 112 lb. per imperial bushel. In cements produced on careless or inexact lines a considerable proportion of imperfectly burnt clinker is produced as well as from wrong proportions of the raw materials. Such irregular products when sifted result in the elimination of the hardest or most valuable portions; hence the results with which we are familiar in England of fine cement weighing much less than coarse.

As considerable difference exists in this country as to the manipulation of the cement under examination for testing purposes, it should be observed that much of the discrepancy in results is due to the character of the bed or plate in which the briquette is moulded. In illustration of this subject we call attention to Mr. Haslinger's experiments on absorptive plaster of Paris beds, Table 2.

Owing to the careful manner adopted in making the briquettes, and the general management of cement-testing practised, better results are secured than those obtained by our careless and varying system in England. The briquettes are not put in .water until 24 hours after moulding, and remain in the cistern for the number of days required by the intended test; so that practically the German tests are all one day older than the English. That fact, coupled with the extra care bestowed on their manipulation, gives an advantage of at least 20 per cent. in their favour. This allowance should, in the writer's opinion, be made when comparing the recorded tests under both systems. Such allowance would help to reconcile the differences between the author's tests of "Stern" cement and those of other experimenters, more especially those recorded in Chapter XII.

TABLE No. 1.
Mr. Haslinger's Experiments with "Stern" Cement.

Cement of 5806 meshes per square inch fineness.

Proportions used, breakings per square inch in lb.

1878. Moulded in	Age. Days.	Neat.	1 sand.	2 sand.	3 sand.	4 sand.	5 sand.	6 sand.	7 sand.	8 sand.	9 sand.	10 sand.
Oct.	7	833	573	449	289	201	151	110	90	78	61	53
	28	844	602	525	349	265	191	157	137	114	98	91
	90	931	742	663	486	418	314	265	207	160	139	119
Nov.	7	771	532	421	262	174	129	101	83	72	58	50
	28	1004	650	532	398	272	206	162	142	117	100	97
	90	910	768	666	532	411	315	252	192	161	135	122
Dec.	7	859	529	388	256	188	149	106	90	78	60	53
	28	993	614	516	375	273	201	161	127	107	91	80
	90	905	708	603	465	356	277	206	166	137	127	114

Cement of 32,256 meshes per square inch fineness.

Proportions used, breakings per square inch in lb.

	Neat.	1 sand.	2 sand.	3 sand.	4 sand.	5 sand.	6 sand.	7 sand.	8 sand.	9 sand.	10 sand.
	617	543	548	432	377	262	215	169	141	132	111
	730	668	686	549	455	358	273	233	195	164	138
	748	732	745	641	509	420	343	311	233	215	181
	627	502	533	454	368	256	215	169	132	129	108
	833	737	741	599	465	363	290	247	202	171	143
	799	775	754	643	519	398	333	270	238	202	188
	644	553	538	431	347	251	207	166	143	120	107
	777	774	711	569	445	370	289	238	203	179	145
	707	760	779	651	516	371	323	267	229	201	159

Tests from an average sample of the above three casks.

Cement of 5806 meshes per square inch fineness.

Age. Moulded in	Days.	Neat.	1 sand.	2 sand.	3 sand.	4 sand.	5 sand.	6 sand.	7 sand.	8 sand.	9 sand.	10 sand.
Feb. 1879.	7	873	624	532	371	275	203	154	118	105	97	78
	28	1004	731	660	442	387	280	219	188	140	126	119
	90	893	760	646	540	424	331	262	219	171	137	119

Cement of 2322 meshes per square inch fineness.

	Neat.	1 sand.	2 sand.	3 sand.	4 sand.	5 sand.	6 sand.	7 sand.	8 sand.	9 sand.	10 sand.
	764	538	488	329	223	176	119	110	97	78	71
	919	650	586	415	294	230	181	138	107	100	92
	987	758	670	513	384	273	208	174	152	118	112

All of the above tests were made in brass moulds placed on plates of the same metal. The briquettes were kept in the air for 24 hours, and placed in water until tested.

ORIGINAL EXPERIMENTS ON CONCRETES. 373

TABLE No. 2.

Mr. Haslinger's Experiments with neat Portland Cement ("Stern") on plaster of Paris beds.

Age.	Fineness, 5806 meshes per square inch.				Fineness, 32,256 meshes per square inch.				Fineness, 2322 meshes per sq. in.
	1878. October Tests.	1878. November Tests.	1878. December Tests.	1879. February Tests.	1878. October Tests.	1878. November Tests.	1878. December Tests.	1879. February Tests.	1879. February Tests.
	lb. per sq. in.	lb. per sq. in.	lb. per sq. in.	lb. per sq. in.	lb. per sq. in.	lb. per sq. in.	lb. per sq. in.	lb. per sq. in.	lb. per sq. in.
7 days	1041	957	1051	1032	768	722	787	869
28 ,, ...	1117	1112	1120	1168	944	1080	1177	1041
90 ,, ...	1149	1211	1196	1118	942	1031	1011	1131
	3307	3280	3367	3318	2654	2833	2975	3041
Average ..	1102·3	1093·3	1122·3	1106	884·6	944·3	991·6	1013·6

The average of neat cement on metal beds being, as shown by Table 1, as under:—

| 902·6 | 895 | 919 | 923·3 | 698·3 | 753 | 709·3 | 890 |

These tests were made simultaneously with those in Table 1, and from the same cement sources, the amount of water used being according to the German official rules.

INDEX.

ABRADERS, millstones, &c., 143.
Absorptive capacity of granite, 103.
Absorption of walls and materials, 334, 339.
Adulteration of mortars, 346.
Aggregates, 101.
—— affected by Portland cement, 75.
——, machinery for reducing, 125.
——, their treatment, 150.
——, necessity of cleaning, 151.
——, their size and form, 152.
——, sizes used in silicated concrete, 243.
Air contained in concrete, 205.
—— influenced by percussion, 208.
Aluminate of potash, effect of, 160.
American cement enterprise, 176.
Analysis of a good Portland cement, 50.
—— when altered by exposure, 63.
—— of a good average cement, 65.
—— different limes, 87.
—— Thiel lime, 89.
—— Sheppey and Roman cements, 93.
—— Derbyshire limestone, 94.
—— felspar and mica, 104.
—— Portland stone, 112.
—— Oolitic limestones, 114.
—— Craigleith sandstone, 116.
—— Magnesian limestone, 114.
—— silicate of soda, 154.
—— Leicestershire (Groby) granite, 213.
—— Victoria stone, 223.
—— broken pottery, 228.
—— rock-concrete pipe, 232.
—— Lincolnshire slag, 243.
—— silicated stone, 245.
—— Kentish ragstone, 243.
—— of silicated pipes, 250.
—— chalk marl, Stettin, 306.

Analysis of clay (Stettin), 306.
—— cement, do., 306.
Andrews' experiments with granitic breccia stone, 201.
Ansted, Professor, on granite, 103.
—— experiments, 197.
Artificial stone, its advantages from a chemical point of view, 4.
—— products referred to, 26.
—— concrete, 28.
—— —— industries, 29.
—— ——, its superiority, 29.
—— stone, Ranger's, 97.
—— aggregates, 101, 119.
—— stone, Frear's, 179.
—— ——, American blocks, 180.
—— ——, Sorel's patent, 183.
—— ——, Union Co.'s, 183.
—— ——, its cost, 191.
—— —— compared with York, 192.
—— ——, Ransome's, 195.
Ashes (smithy) and cement, 75.
Aspdin, investigations of, 345.
Asphalte on concrete, 37.
Atmosphere, its influence on natural stones, 17.
Augustus, Emperor, marble used during his reign, 4.
Auvergne, granite of, 17.

BABEL, burnt bricks used in, 25.
Bailey's cement-tester, 56.
Basaltic rocks, 116.
Bastard mortars, 344.
—— experiments on ditto, 348.
Bath stone, 96.
——, its constructive strength, 198.
Bazalgette, Sir J., and main drainage of London, 84.

Bazalgette, Sir J., and establishment of testing, 331.
Beechey, Capt., coral rock formation, 30.
Belgian granite, 244.
Béton aggloméré, its accurate combination, 6.
——, description of, 163.
——, machinery used in its preparation, 164.
——, its crushing strength, 169.
——, experiments on, 189.
Billiard balls made from magnesite, 185.
Birkenhead docks, 300.
Blake's stone-breaking machine, 127.
Blocks of concrete, 204.
—— (building), Lish's patent, 290.
Bolsover magnesian limestone, 19.
Bonne cement, analysis of, 65.
Boston and Brooklyn (cities of), concrete pipes used in their drainage, 224.
Boulogne cement and Coignet béton,189.
——, analysis of, 65.
Bouch, Mr., engineer of Tay Bridge, 294.
Bournemouth drainage, rock-concrete pipes used in, 235.
Brick sewers, concrete foundation in, 42.
Bricks, patent concrete, 261.
——, ancient use of, 25.
——, slag, 121.
——, Osnabruck, 122.
Briquettes, care required in making, 56.
—— press, 58.
—— moulds, 61.
Broadbent's machine, 131.
Broken pottery, 120.
Brooklyn (U.S.A.) drainage works, 242.
Brunel, Roman cement used by, 95.
Buckwell's granitic breccia, 28, 199.
—— 90-ton blocks, 204.
—— introduced concrete slabs, 43.
—— original paving, 151.
—— manufacture discontinued, 209.
Buckland, Dean, on granitic breccia, 206.
Building frames and their inventors, 40.
——, description of, 320.
Buildings of concrete blocks, 41.

Builders (modern) can command scientific rules for their guidance, 2.
Building in concrete, 323.
Building materials, their character, 334.
—— ——, testing of, 325.
—— —— easily tested, 329.
—— ——, their specific gravity, 336.
—— ——, absorption and heat conducting characteristics, 334, 337.
—— ——, accidents from using bad, 328.
Building stones, difficulties surrounding, 32.
Burnt ballast, 120.
—— shales, 119.
Buxton limestone, 113.

CAEN stone, strength of, 198.
Calvert's, Professor, experiments with magnesian limestone, 183.
Calcium, silicate of, 158.
Carbonic acid, a powerful solvent, 16.
—— ——, dangerous if found in Portland cement, 62.
—— —— in magnesian limestone, 184.
Carboniferous limestone, 113.
Carbonate of magnesia, 184.
Carr's disintegrator, 147.
Carrara marble, 4.
—— compressive tests, 368.
Caustic soda, 211, 222.
Cements, new-fashioned, 49.
Cement-testing in Germany, 59.
Cements, mixture of, unsafe, 98.
Centimetre, its value in English inches, 319.
Central cement-testing authority, 330.
Chambers, Dr. William, on house-building, 342.
Chalk, clay, and cement analyses, 306.
Chemistry estimates, causes of decay, 4.
Chicago, Frear's blocks used in, 179.
——, granite buildings in, 270.
——, fire and its consequences, 278.
Chloride of calcium, use of, 195.
Chromo-lithographs, reference to, 350, 352.
Clay ballast, and cement, 75.
Clay, innumerable varieties of, 24.

INDEX. 377

Clark's, Ellice-, tests of Bournemouth pipes, 238.
Cleopatra's Needle, 3.
Clinker (cement) sent to New York, 85.
—— adulterated with limestone, 64.
—— cement, 51.
Clyde (River) Trust harbour walls, 299.
Cob (clay) dwellings, 10.
Coke in cement, 62.
Colour of good cement, 50.
—— in wall papers, 341.
Coignet's "béton aggloméré," 163.
——, its crushing strength, 169.
—— experiments, 189.
—— machinery used in making it, 164.
Concrete, reprehensible practices in using, 6.
——, historical observations, 7.
——, its meaning, 34.
——, natural formation of, 1.
—— making in Italy, 35.
——, classification of, 36.
—— treated with chemical solutions, 37.
——, reckless preparation of, as practised in London, 38.
——, its use in combination with iron, 39.
——, excessive use of water in its preparation, 45.
—— cottages, their downfall, 62.
—— from different sands, ashes, &c., 75.
—— seldom made during winter, 86.
——, limes that may be used for, 86.
——, lime and cement, relative values for, 90.
—— used in Douglas Harbour, 152.
—— silicated, 155.
——, its induration by age, 159.
—— hardened by aluminate of potash, 160.
—— established process of industry, 163.
—— used in Paris sewers, 169.
—— treated with gum-shellac, 179.
—— Ranger's, 182.
—— Sorel's, 183.
—— Union Company's, 183.
—— crushing strength, 189.
——, strength attained in course of time, 189.
——, comparative cost of American, 191.

Concrete, Ranger's, its value compared with York stone, 192.
——, ancient remains of, 210.
—— pipes, 224.
—— pipe-moulding machine, 231.
—— ——, strength of, 232, 237.
—— —— used at Bournemouth, 235.
—— pipe-testing machine, 237.
—— —— used at Brooklyn, 242.
—— constructive applications of, 257.
—— bricks (Lascelles), 261.
—— country house, 268.
—— tiles, 271.
——, fireproof quality of, 277.
——, important works in, 294.
—— building frames, 321.
—— floors at Cambridge, failure of, 328.
——, English table of crushing and tensile values of, 351, 352.
Construction, cost of, by frames, 40.
Consumers of Portland cement, their requirements, 69–71.
Conducting powers of building materials, 337.
Coode, Sir John, and purity of aggregates, 152.
Cornwall rock basins, 15.
Cornish granite, examples of, in bridge building, 13.
Coral reefs, their formation, 30.
——, character of, 32.
——, piece of, with a portion of submarine cable attached, 32.
Cottages of concrete, 261.
Country roads influenced by water, 44.
—— house in concrete, 268.
Cracks in large blocks, 7.
Craigleith stone, weight and analysis, 96–116.
Custom House, London, built of Portland stone, 17.
Cuzco, ruins of, 7.
Crushing weight of bricks, 122.
Crushers, rollers, &c., 126.
Crushing strength of granite, 218.

DAMP walls, 338.
Dangers to health (Teale), 341.
Dartmoor granites, their qualities, 16.

Dartmoor granite used in Nelson's monument, 16.
D'Aubuisson on French granite, 17.
Deacon's cement-testing apparatus, 54.
Decay of building stones, 22.
De la Beche's opinion of "Kettle and Pans," 17.
Devonshire granite used in London, 13.
Devey's 'Hygiène de Famille,' reference to, 340.
Dialysing process, 157.
Disintegrators, 147.
District surveyors and their responsibilities, 329.
Dolemieu on disease of granites, 17.
Dolomites described, 115.
Dome (concrete) for India, 269.
Doremus, Professor, experiments, 339.
Douglas Harbour, concrete used in, 152.
Dover Harbour, 194.
Drain pipes of concrete, 227.
Drainage scamping, 342.
Druidical influence attributed to natural results, 17.
Dust produced from fracture, 150.

EASTERN Pyrenees, granite of, 17.
Egyptian Pyramids, their condition, 3.
—— sphinx in Paris, 3.
—— granite used for a long time, 11.
—— —— analysis, 12.
Emery wheels of the Union Co., 185.
—— —— of Bateman and Co., 201.
English concrete industries, 203.
Engineering instincts of coral, polypi, and bee, 31.
Essex Bridge (Dublin), granite, 14.
Etheridge, Mr., on granites, 214.
Experiments by Lieut. Innes, R.E., 72.
—— on shrinkage, 76.
—— on heavy cement, 77.
—— on influences of temperature, 81.
—— on lime concrete, 90.
—— on Roman cement, 96.
——, Pasley's, 96.
—— with various stones, 96-111.
—— on slag bricks, 122.
—— on cements, Philadelphia Exhibition, 174.

Experiments on Van Derburgh blocks, 181.
—— on Sorel stone, 188.
—— Coignet's béton, 189.
—— Frear stone, 190.
—— Foster and Van Derburgh stone, 191.
——, Ranger's and York stone, 192.
—— by Dr. Frankland, 196.
—— by Professor Ansted, 197.
—— on granitic breccia, 200.
——, Ransome's stone, 196-201.
—— rock-concrete pipes, 233, 237, 238, 250.
—— on slag by Dr. Roscoe, 244.
—— on German cement, 309-312.
—— by Dr. Michaelis, 313.
—— by Professor Doremus, 339.
—— on adulterated cements, 347.
—— by M. Kuhn, 348.
—— on bastard mortars, 348.
——, authors on concretes, &c., 351.

FARNHAM silica, 43, 211.
Faraday, Professor, on concrete, 206.
Felspar, 104.
——, artificial, 161.
Fineness of Portland cement, 50, 67.
Fireproof concrete, 39, 277.
—— ——, 281-288.
Fire, destructive effects of, 287.
Fish-scale tile slabs, 260.
Flints, soluble, 195.
——, necessity of careful selection, 119.
Foster and Van Derburgh's block system, 190.
Footpaths made from tar and limestone, 113.
Foundations, lias lime concrete, 37.
Fountains in Victoria stone, 219.
Frankland's, Dr., experiments, 197.
Frear's artificial stone, 179, 190.
Freestone, 115.
Frieze on Paris Exhibition house 265.
Frost, influence of, 169.
Frost test, Dr. Jackson on, 185.
Frames for building, 320.
Frontispiece, description of, 350.

GAS forced through sandstone, 340.
Geological aspect of coral reefs, 30.
George-Marien-Hutte, hospital experiments, 123.
German testing-machines, 59, 60.
—— tests, 309-312.
—— Portland cement, 303.
—— and English measures, 319.
Gillmore, Gen., on cement-making, 172.
—— on German cement, 307.
—— on limes, 79.
—— on adulteration of cements, 347.
Gimson's duplex stone-breaker, 131.
Gold quartz crushing by elephant stamper, 139.
Goodman's stone-crusher, 132-134.
Graham's, Dr., experiments with silica, 157.
Granite of Auvergne and Eastern Pyrenees, 17.
——, effect of tool-dressing on, 15.
—— blocks, cause of fracture in, 7.
——, Cornish, weight, &c., of, 96.
——, its peculiarities explained, 103.
—— quarrying in U.S.A., 109.
Granites of Great Britain, 105.
——, wear and tear of, 107, 108.
Granitic breccia, transverse and tensile tests, 364-366.
—— —— stone, 199.
—— favourable example of concrete, 5.
Gravels unsuitable for concrete, 117.
Grecian Archipelago, discovery of a buried city, 9.
Grinding machines, varieties of, 143.
—— limes, 88.
——, horizontal, 149.
Greyveldinger mortar mill, 177.
Groby granite, analysis of, 213.
Grothe's, Mr., description of Tay Bridge, 294.
Greenwich Hospital built of Portland stone, 17.
Guaranteed German cements, 311.
Guarantee, translation of, 314.
Guernsey granites, crushing strength of, 214.
—— ——, compressive tests, 356.
Gum-shellac used in concrete, 179.

Gypsum, 99.
—— used to adulterate cements, 346.

HALL's stone-breaker, 127.
Haslinger's, Mr., experiments, 370, 372.
Haymarket, fall of house in, 328.
Heat and specific gravity of building materials, 336.
Heintzel's, Dr., experiments on cements, 74.
Henley's concrete building apparatus, 321.
Highton, H., inventor of Victoria stone, 43, 211.
Hodges and Butler, silicated stone works of, 242.
Hornblower's patent fireproof floors, 277.
Hospital (George-Marien-Hutte), slag bricks used in building, 336.
Houses of Parliament, blunder made in selection of building stone, 19.
Howard's, General, remarks on Frear stone, 190.
Humboldt on mortars of Peru, 8.
Hydraulicity of good cement, 50.
Hydraulic cements, 345.

IMPACT concrete (Buckwell's), 199, 204.
Impact process, necessity of using exact proportions of water in, 230.
Impact machine, 253.
Impingements, slabs made by, 43.
Induration of Victoria stone, 220.
Innes, Lieut., experiments of, 72.
Ireland, Portland cement industry in, 91.
Iron and other slags, 121.
Iron and concrete fireproof floors, 278.
Italian mosaic, 273.
Italy, practical knowledge of concrete-making in, 34.

JACKSON, Dr., on qualities and properties of Union Co.'s stone, 185.
Jerusalem, character of building and description of stone used at, 20.

380 INDEX.

KAOLIN deposits in Cornwall, 15.
——, its uses, 24.
Keene's cement, 99.
——, tests of, 100.
Kentish ragstone, 109.
——, analysis, 243.
——, weight and strength, 96.
Kiln tiles, disadvantages in using them, 261, 272.
"Kettle and Pans," St. Mary's, Scilly Isles, 17.
Kilogrammes converted into English lb., 319.
Kuhn's, Mr., experiments, 310.

LABOURER, his condition and shelter, 11.
Lascelles system of construction, 258.
—— concrete house at Paris Exhibition, 263.
Lava blocks used in building without mortar, 9.
Leicestershire granite, compressive tests, 356.
—— (Groby) granite analysis, 231.
Leper of houses, 340.
Lewisham, granitic breccia pavement at, 200.
Liffey River, large concrete blocks used in, 297.
Limestones, cementing agent of, 24.
——, analyses, 87.
Limes (building), 86.
——, necessity of fine reduction, 88.
Lime, celebrated, of Thiel, 88.
—— used successfully after being kept seven years, 89.
—— of Ardwick, Lancashire, 90.
Limes obtained in abundance in many localities, 87, 91.
Lincolnshire slag analysis, 243.
Lining slab (Lascelles' patent), 260.
Lish's patent building blocks, 289.
Liverpool streets, character of concrete used in paving, 37.
Loch Etive granite, compressive tests, 367.
London, its atmosphere prejudicial to building materials and metals, 4, 20.

London buildings generally injured by atmospheric influences, 13.
—— main drainage, 41.
—— Portland cement analysis, 65.
London Bridge, granite from Devonshire used in building, 13; granite of which it is built showing signs of decay, 15.
Lurman's machine for making slag bricks, 121.
Lyster, Mr., and Liverpool Docks, 300.

MACADAM, 44–46.
Machinery for reducing concrete materials, 125–143.
Madrepora, coral-forming, 30.
Maclay's, Mr., experiments in Portland cement, 79–81.
Magnesite grinding wheels, 201.
——, its high cost, 187.
Magnesian limestone, cement made from, 114, 183.
—— —— in Houses of Parliament, 19.
Malaxator mixing machine, 165.
Mansfield building stones, 114.
Manufacturers, cement, independence and shortsightedness of, 69.
—— —— and American testing, 83.
Martin's cement, 99.
Materials, building, testing, 325.
Matrix, derivation and meaning, 48.
Medallion in Victoria stone, 224.
Metallic oxides, their influence on cements, 210.
Michaelis, Dr., experiments of, 309.
——, table of cement values, 313.
Mica, 104.
Middlesborough slag bricks, 121.
Milroy and Butler's concrete cylinders, 298.
Minton, Maw, and mosaic potters, 271.
Mixing concrete, 245.
Mixing machine (American), 229.
—— (Le Mesurier's), 301.
Mixtures of different cements unsafe, 98.
Mortar, ignorance of its properties, 7.
—— mill (the Greyveldinger), 178.
—— of speculating builders, 330.

Mortar spurious, 344.
Mosaic tiles (Roman), 273.
—— (Patteson's), 274.
Motte's grinding machine, 144.
Moulding machine (pipes), 231, 246, 251.
Mount Sorrel granite (red) compressive tests, 367.

NATIONAL Portland Cement Co., New York, 173.
Nelson Monument, London, 16.
Nile mud, bricks made from, 25.
Norman strongholds, architecture of, 22.
Notre Dame, Paris (municipal buildings), 170.

OAK, its capacity of heat conductivity, 334.
Obsidian or volcanic glass, 9.
Ornamental articles from concrete, 43.
Osnabruck slag bricks, 122.
Oxland's, Dr., experiments with elephant stamper, 139.
Oxygen, injurious action of, 16.

PACIFIC Ocean, coral reefs of, 30.
Pallant's testing machine, 57.
Papers on walls, 335.
——, colours of, 341.
Papin's digester, 155.
Parian cement, 99.
Paris sewers, 169.
—— Exhibition, concrete house, 262.
Pasley's experiments on Roman cements, 95.
—— —— on Ranger's artificial and York stone, 192.
Patent Victoria stone process, 209.
—— concrete bricks, 261.
Patterson's elephant stamper, 136, 139.
Patteson's concrete tiles, 274.
Pavements, London, their imperfections, 25.
Paving stones (natural), laminæ clearly defined, 25.
—— slabs and pipes made from concrete, 36.

Paving slabs and pipes, machinery for making, 251.
—— slabs first introduced by Buckwell, 43.
—— —— of patent Victoria stone, 217.
Penmaen Mawr basalt compressive tests, 367.
Percussion, beneficial influence of, 208.
Peruvians ignorant of mortars, 7.
Peruvian climate favourable to preservation of buildings, 10.
Pettenkofer's, Dr., experiments, 335.
Philadelphia Exhibition, cement tests at, 174.
——, German cement at, 307.
Pickle Herring Wharf mooring blocks, 204.
Piers of the Tay Bridge, 295.
Pipes made by impact, 205.
—— of concrete, 227.
—— silicated by Victoria stone process, 232.
—— require age before being used, 241.
—— their thickness, 242.
——, moulds, and machinery for making them, 246.
——, concrete analyses, 250.
Plaster of Paris concrete floors, 36.
Pompeii, lava pavements of, 23.
Poole and Bournemouth clays, 228.
Porcelain, variety of kinds of, 24.
Porous concrete, its dangers, 41.
Port Said, Thiel lime used at, 88.
—— —— analysis of, 228.
Porphyry affected by moisture, 3.
Portland cement stronger than any old mortar, 5.
—— highly appreciated by engineers, 41.
—— the most important of all, 49.
—— a double silicate of lime and alumina, 49.
——, the properties of a first-class, 50.
——, ingredients from which it is made, 51.
——, dangers of manufacture, 52.
——, —— of carbonic acid, 62.
——, unreasonable desire for tests of, 55.
——, analysis of a faulty, 62.

INDEX.

Portland cement, its specific gravity, 63.
—, analysis of a sample at Dublin, 64.
—, analyses of six good sorts, 65.
—, eccentric breakings of, 66.
—, high degree of fineness important, 66.
—, English makers disregard fineness, 69.
—, example of improvement from do., 71.
—, dangers of coarse ground, 73.
— influence of light on, 74.
—, weight influences setting, 78.
— examinations in America, 78.
— at different temperatures, 81.
— freshly ground, 85.
— sometimes mixed with lime, 90.
— made in Ireland, 91.
— block three years old tested, 108.
—, suitable machine for testing, 138, 149.
—, effect of silication on, 162.
—, its cost in the United States, 172.
— of National Co., New York, 173.
— tests at Philadelphia Exhibition, 174.
—, strength of, reached in two years, 189.
— mixed with lime effects no saving, 191.
—, analyses of Thames and Medway, 215.
—, mode of testing by the Victoria Stone Co., 216.
— unaffected by frost, 276.
— made at Stettin, 303.
—, good quality commands higher prices, 313.
—, necessity of uniformity in testing, 331.
—— added to lime mortars is no saving; remarks by General Gillmore, 346.
Portland stone buildings in London, 17.
— analyses, 18.
— buildings affected injuriously by London atmosphere, 18.
—, suitability for concrete, 19.

Portland stone selected personally by Wren and Smeaton, 19.
— dust and cement, 75.
—, weight of, 96.
— described generally, 112, 113, 198.
Pottery (broken) as an aggregate, 120, 227, 228.
— tensile tests, 358.
— compressive do., 359.
Pumice stone quarried at Santorin, 9.
Pyramid, Great, description of, 11.

QUARTZ, 104.
Quartz (gold) crushed by stamper, 139.
Queen Anne House, Paris Exhibition, 263.
Queen Anne period, character of buildings, 27.

RANGER'S concrete and artificial stone, 97, 182.
— compared with York stone, 192.
Ransome's siliceous stone, 26, 155, 195.
— experiments, 363, 368.
— siliceous stone made by Williams and Son, tensile tests, 363.
— made by Bateman and Co., 368.
— calcium stone made by Bateman and Co., tensile tests, 368.
— apanite made by do., do., 369.
Red Mansfield stone, 114.
Road-making materials, effect of water on, 44–46.
Rock-concrete pipes, 223.
— analysis, 232.
— pipes, sizes and weights, 232.
— strength, 233, 237.
— tests, 239.
— (unsilicated), tensile tests, 357.
— silicated, do., 358.
— compressive tests, 359.
— pipes testing machine, 240.
Roman concretes, 7, 35.
— desire to decorate buildings, 3.
— cement, its manufacture, &c., 91.
— —, tensile strength, 94.
— buildings and materials, 330.
Rosendale cement, 172.

SANDSTONE, 115.
——, its permeability, 339.
Sand, Thames, experiments with, 77.
Sands, various, 75.
—— as aggregates, 118.
Santorin, entombed city at, 9.
"Scamping" drainage, 342.
Scotch granite and Cornish, 13.
Sedimentary deposits (rocks), 211.
Septaria, Roman cement from, 91.
"Setting" of Portland cement, 50.
Sewers, their primary object, 241.
——, egg-shaped, 254.
Shale (burnt), used as adulterant, 62.
Shaw, R. N., R.A., house designed by, 263.
Sharp, Jones, and Co's. rock-concrete pipes, 227.
Sheppey cement stone and analysis, 92.
Sholl's pneumatic stamps, 141.
Shrinkage of sand when wet, 76.
Siegen slag bricks, 122.
Sieve for testing fineness of cement, 50.
Silica (Farnham), 43, 211, 222.
Silicate process, 154.
Silicate of potash, 161.
Silicate of soda bath, 43.
——, manufacture of, 155.
——, Gossage's process, 156.
——, action of carbonic acid on, 158.
—— and chloride of calcium, 195.
Silicate used in Buckwell's process, 204.
—— —— Victoria process, 211.
Silicated stone manufacture, 28, 242.
—— —— pipes, 29.
—— ——, Hodges and Butler's tensile tests, 366.
—— slag mixture, 366.
—— Belgian granite, 361.
—— Thames ballast, 361.
—— compressive tests, 362.
Silication, its great importance, 161.
Siliceous stones (sweating of), 159.
Sills, sinks, &c., Victoria stone, 218.
Slabs used in building, 260.
Slag brick tests, 368.
——, Lincolnshire analysis, 244.
Slags, utilization of, 121.
Smeaton's search for limes, 86, 89.

Sorby, Mr., on granite, 103.
Spiller, Mr., on silication, 159.
Span in concrete, safe limit of, 40.
Speculating builders, 330.
Specification of a good cement, 50.
Specific gravities of building materials, 336.
Spinkwell stone compressive tests, 367.
Spurious mortars and experiments, 344, 348.
Squire, Mr., on the ruins of Cuzco, 8.
Stettin cement analysis, 65.
—— cement works, description of and analyses of materials, 303, 306.
"Stern" cement works, 303.
Stephenson, Mr., used Roman cement, 95.
Stone-breakers, 125.
——, Blake's, 127.
——, Hall's, 128.
——, Broadbent's, 131.
——, Gimson's, 131.
——, Goodman's, 132.
——, Patterson's stamper, 136.
——, Sholl's pneumatic stamps, 141.
——, Motte's, 144.
——, Hall's disintegrator, 147.
——, Carr's ditto, 147.
Stones, building, their weight and strength, 96.
—— improved by silication, 154.
——, porosity of, 197.
St. Paul's Portland stone, 17, 112.
St. Thomas's Hospital, béton aggloméré used near, 163.
Stoney corals, cementing agent of, 24.
Stoney, Mr., concrete blocks used at Dublin by, 297.
Studio window in concrete, 267.
Suez Canal, béton aggloméré used in constructing the, 163.
Syenite, its geological position, 3.
—— used for upwards of 4000 years, 11.
——, description and analysis, 12.

TABLE of German and English measures and weights, 319.
Tay Bridge, concrete used in, 294.

Tay Bridge, piers of, 296.
Teale, Dr., on defective drainage, 342.
Technical Director of Stern Works (Mr. Haslinger), 315.
Temperature influences cement, 81.
Templar Rocks, Lundy Island, 15.
Tensile strength of good cement, 50.
—— of building stones, 198.
—— of German cement, 303.
—— of Victoria stone, 352.
—— of rock-concrete, 357.
—— of silicated stone, 360.
Thames tunnel, Roman cement used in, 95.
Thiel lime used by French engineers, 88.
Terra cotta, its antiquity and value for building purposes, 28.
Testing cement without machine, 61.
——, English system, 53.
——, different methods, 53.
Testing machine, Deacon's, 54.
——, Bailey's, 56.
——, Pallant's, 57.
——, German, 59.
Testing considerations, 325.
Tests of cement uniform if good, 64, 331.
——, crushing and tensile, of English concretes, 351.
Tiles (kiln burnt) unreliable, 273.
——, Roman, 273.
Toepffer, Grawitz, and Co., Stettin, 305.
Tooley Street fire, 278.
Transverse strength of stone, 197.

Union Company's artificial stone, 183.
United States granite, 109.
——, importation of foreign cements, 81.
——, adulteration of cements in, 346.

Van Derburgh's stone-making process, 180.
Vanne Aqueduct (France), 168.
Vesinet Church (ditto), 169.

Victoria stone, 28.
—— process, 209.
——, its introduction, 43.
——, early mistakes, 151.
——, analysis of Portland cement used, 215.
—— paving described, 217.
—— silica tanks, 26, 221.
——, analysis, 223.
——, various products, 223.
—— crushing strain, 351.
—— tensile strain, 352.
—— ——, 353, 354, 355, 356.
——, compressive tests, 356.

Wallsend cement analysis, 65.
Walz, Dr. Isidor, experiments on béton agglomeré, 176.
Wall built of Lish's blocks, 291.
Waterloo Bridge, granite used, 13.
——, opinion of French engineers, 14.
Water beneficial to roads, 44.
Water, prejudicial effect when in excess, 45.
——, small quantity used in making granitic breccia stone—Faraday's remarks, 203.
——, effect of using different quantities, 72.
——, exact quantity necessary in making granitic breccia stone, 230.
Weights of various stones per cubic foot, 96.
Wind and water action on Cornish granite, 18.
Wollaston, Dr. (Bakerian lectures), 207.
Woolwich Arsenal (wharf wall), 193.
Wren's City churches, 18.

Yorkshire stone, weight of, 96.
—— compared with Ranger's artificial stone, 192.

Z Blocks and slabs, 291.
Zoophytes, coral-forming, 30.

1889.

BOOKS RELATING
TO
APPLIED SCIENCE,

PUBLISHED BY

E. & F. N. SPON,
LONDON: 125, STRAND.

NEW YORK: 12, CORTLANDT STREET.

———•———

The Engineers' Sketch-Book of Mechanical Movements, Devices, Appliances, Contrivances, Details employed in the Design and Construction of Machinery for every purpose. Collected from numerous Sources and from Actual Work. Classified and Arranged for Reference. *Nearly* 2000 *Illustrations.* By T. W. BARBER, Engineer. 8vo, cloth, 7s. 6d.

A Pocket-Book for Chemists, Chemical Manufacturers, Metallurgists, Dyers, Distillers, Brewers, Sugar Refiners, Photographers, Students, etc., etc. By THOMAS BAYLEY, Assoc. R.C. Sc. Ireland, Analytical and Consulting Chemist and Assayer. Fourth edition, with additions, 437 pp., royal 32mo, roan, gilt edges, 5s.

SYNOPSIS OF CONTENTS:

Atomic Weights and Factors—Useful Data—Chemical Calculations—Rules for Indirect Analysis—Weights and Measures—Thermometers and Barometers—Chemical Physics—Boiling Points, etc.—Solubility of Substances—Methods of Obtaining Specific Gravity—Conversion of Hydrometers—Strength of Solutions by Specific Gravity—Analysis—Gas Analysis—Water Analysis—Qualitative Analysis and Reactions—Volumetric Analysis—Manipulation—Mineralogy—Assaying—Alcohol—Beer—Sugar—Miscellaneous Technological matter relating to Potash, Soda, Sulphuric Acid, Chlorine, Tar Products, Petroleum, Milk, Tallow, Photography, Prices, Wages, Appendix, etc., etc.

The Mechanician : A Treatise on the Construction and Manipulation of Tools, for the use and instruction of Young Engineers and Scientific Amateurs, comprising the Arts of Blacksmithing and Forging ; the Construction and Manufacture of Hand Tools, and the various Methods of Using and Grinding them ; the Construction of Machine Tools, and how to work them ; Machine Fitting and Erection ; description of Hand and Machine Processes ; Turning and Screw Cutting ; principles of Constructing and details of Making and Erecting Steam Engines, and the various details of setting out work, etc., etc. By CAMERON KNIGHT, Engineer. *Containing* 1147 *illustrations,* and 397 pages of letter-press, Fourth edition, 4to, cloth, 18s.

B

CATALOGUE OF SCIENTIFIC BOOKS

Just Published, in Demy 8vo, cloth, containing 975 pages and 250 Illustrations, price 7s. 6d.

SPONS' HOUSEHOLD MANUAL:
A Treasury of Domestic Receipts and Guide for Home Management.

PRINCIPAL CONTENTS.

Hints for selecting a good House, pointing out the essential requirements for a good house as to the Site, Soil, Trees, Aspect, Construction, and General Arrangement; with instructions for Reducing Echoes, Waterproofing Damp Walls, Curing Damp Cellars.

Sanitation.—What should constitute a good Sanitary Arrangement; Examples (with illustrations) of Well- and Ill-drained Houses; How to Test Drains; Ventilating Pipes, etc.

Water Supply.—Care of Cisterns; Sources of Supply; Pipes; Pumps; Purification and Filtration of Water.

Ventilation and Warming.—Methods of Ventilating without causing cold draughts, by various means; Principles of Warming; Health Questions; Combustion; Open Grates; Open Stoves; Fuel Economisers; Varieties of Grates; Close-Fire Stoves; Hot-air Furnaces; Gas Heating; Oil Stoves; Steam Heating; Chemical Heaters; Management of Flues; and Cure of Smoky Chimneys.

Lighting.—The best methods of Lighting; Candles, Oil Lamps, Gas, Incandescent Gas, Electric Light; How to test Gas Pipes; Management of Gas.

Furniture and Decoration.—Hints on the Selection of Furniture; on the most approved methods of Modern Decoration; on the best methods of arranging Bells and Calls; How to Construct an Electric Bell.

Thieves and Fire.—Precautions against Thieves and Fire; Methods of Detection; Domestic Fire Escapes; Fireproofing Clothes, etc.

The Larder.—Keeping Food fresh for a limited time; Storing Food without change, such as Fruits, Vegetables, Eggs, Honey, etc.

Curing Foods for lengthened Preservation, as Smoking, Salting, Canning, Potting, Pickling, Bottling Fruits, etc.; Jams, Jellies, Marmalade, etc.

The Dairy.—The Building and Fitting of Dairies in the most approved modern style; Butter-making; Cheesemaking and Curing.

The Cellar.—Building and Fitting; Cleaning Casks and Bottles; Corks and Corking; Aërated Drinks; Syrups for Drinks; Beers; Bitters; Cordials and Liqueurs; Wines; Miscellaneous Drinks.

The Pantry.—Bread-making; Ovens and Pyrometers; Yeast; German Yeast; Biscuits; Cakes; Fancy Breads; Buns.

The Kitchen.—On Fitting Kitchens; a description of the best Cooking Ranges, close and open; the Management and Care of Hot Plates, Baking Ovens, Dampers, Flues, and Chimneys; Cooking by Gas; Cooking by Oil; the Arts of Roasting, Grilling, Boiling, Stewing, Braising, Frying.

Receipts for Dishes—Soups, Fish, Meat, Game, Poultry, Vegetables, Salads, Puddings, Pastry, Confectionery, Ices, etc., etc.; Foreign Dishes.

The Housewife's Room.—Testing Air, Water, and Foods; Cleaning and Renovating; Destroying Vermin.

Housekeeping, Marketing.

The Dining-Room.—Dietetics; Laying and Waiting at Table: Carving; Dinners, Breakfasts, Luncheons, Teas, Suppers, etc.

The Drawing-Room.—Etiquette; Dancing; Amateur Theatricals; Tricks and Illusions; Games (indoor).

The Bedroom and Dressing-Room; Sleep; the Toilet; Dress; Buying Clothes; Outfits; Fancy Dress.

The Nursery.—The Room; Clothing; Washing; Exercise; Sleep; Feeding; Teething; Illness; Home Training.

The Sick-Room.—The Room; the Nurse; the Bed; Sick Room Accessories; Feeding Patients; Invalid Dishes and Drinks; Administering Physic; Domestic Remedies; Accidents and Emergencies; Bandaging; Burns; Carrying Injured Persons; Wounds; Drowning; Fits; Frost-bites; Poisons and Antidotes; Sunstroke; Common Complaints; Disinfection, etc.

PUBLISHED BY E. & F. N. SPON. 3

The Bath-Room.—Bathing in General; Management of Hot-Water System.
The Laundry.—Small Domestic Washing Machines, and methods of getting up linen; Fitting up and Working a Steam Laundry.
The School-Room.—The Room and its Fittings; Teaching, etc.
The Playground.—Air and Exercise; Training; Outdoor Games and Sports.
The Workroom.—Darning, Patching, and Mending Garments.
The Library.—Care of Books.
The Garden.—Calendar of Operations for Lawn, Flower Garden, and Kitchen Garden.
The Farmyard.—Management of the Horse, Cow, Pig, Poultry, Bees, etc., etc.
Small Motors.—A description of the various small Engines useful for domestic purposes, from 1 man to 1 horse power, worked by various methods, such as Electric Engines, Gas Engines, Petroleum Engines, Steam Engines, Condensing Engines, Water Power, Wind Power, and the various methods of working and managing them.
Household Law.—The Law relating to Landlords and Tenants, Lodgers, Servants, Parochial Authorities, Juries, Insurance, Nuisance, etc.

On Designing Belt Gearing. By E. J. COWLING WELCH, Mem. Inst. Mech. Engineers, Author of 'Designing Valve Gearing.' Fcap. 8vo, sewed, 6*d*.

A Handbook of Formulæ, Tables, and Memoranda, for Architectural Surveyors and others engaged in Building. By J. T. HURST, C.E. Fourteenth edition, royal 32mo, roan, 5*s*.

"It is no disparagement to the many excellent publications we refer to, to say that in our opinion this little pocket-book of Hurst's is the very best of them all, without any exception. It would be useless to attempt a recapitulation of the contents, for it appears to contain almost *everything* that anyone connected with building could require, and, best of all, made up in a compact form for carrying in the pocket, measuring only 5 in. by 3 in., and about ¾ in. thick, in a limp cover. We congratulate the author on the success of his laborious and practically compiled little book, which has received unqualified and deserved praise from every professional person to whom we have shown it."—*The Dublin Builder.*

Tabulated Weights of Angle, Tee, Bulb, Round, Square, and Flat Iron and Steel, and other information for the use of Naval Architects and Shipbuilders. By C. H. JORDAN, M.I.N.A. Fourth edition, 32mo, cloth, 2*s*. 6*d*.

A Complete Set of Contract Documents for a Country Lodge, comprising Drawings, Specifications, Dimensions (for quantities), Abstracts, Bill of Quantities, Form of Tender and Contract, with Notes by J. LEANING, printed in facsimile of the original documents, on single sheets fcap., in paper case, 10*s*.

A Practical Treatise on Heat, as applied to the Useful Arts; for the Use of Engineers, Architects, &c. By THOMAS BOX. *With* 14 *plates.* Third edition, crown 8vo, cloth, 12*s*. 6*d*.

A Descriptive Treatise on Mathematical Drawing Instruments: their construction, uses, qualities, selection, preservation, and suggestions for improvements, with hints upon Drawing and Colouring. By W. F. STANLEY, M.R.I. Fifth edition, *with numerous illustrations,* crown 8vo, cloth, 5*s*.

B 2

Quantity Surveying. By J. LEANING. With 42 illustrations. Second edition, revised, crown 8vo, cloth, 9s.

CONTENTS:

A complete Explanation of the London Practice.
General Instructions.
Order of Taking Off.
Modes of Measurement of the various Trades.
Use and Waste.
Ventilation and Warming.
Credits, with various Examples of Treatment.
Abbreviations.
Squaring the Dimensions.
Abstracting, with Examples in illustration of each Trade.
Billing.
Examples of Preambles to each Trade.
Form for a Bill of Quantities.
    Do.   Bill of Credits.
    Do.   Bill for Alternative Estimate.
Restorations and Repairs, and Form of Bill.
Variations before Acceptance of Tender.
Errors in a Builder's Estimate.

Schedule of Prices.
Form of Schedule of Prices.
Analysis of Schedule of Prices.
Adjustment of Accounts.
Form of a Bill of Variations.
Remarks on Specifications.
Prices and Valuation of Work, with Examples and Remarks upon each Trade.
The Law as it affects Quantity Surveyors, with Law Reports.
Taking Off after the Old Method.
Northern Practice.
The General Statement of the Methods recommended by the Manchester Society of Architects for taking Quantities.
Examples of Collections.
Examples of "Taking Off" in each Trade.
Remarks on the Past and Present Methods of Estimating.

Spons' Architects' and Builders' Price Book, *with useful Memoranda.* Edited by W. YOUNG, Architect. Crown 8vo, cloth, red edges, 3s. 6d. Published annually. Sixteenth edition. *Now ready.*

Long-Span Railway Bridges, comprising Investigations of the Comparative Theoretical and Practical Advantages of the various adopted or proposed Type Systems of Construction, with numerous Formulæ and Tables giving the weight of Iron or Steel required in Bridges from 300 feet to the limiting Spans; to which are added similar Investigations and Tables relating to Short-span Railway Bridges. Second and revised edition. By B. BAKER, Assoc. Inst. C.E. *Plates*, crown 8vo, cloth, 5s.

Elementary Theory and Calculation of Iron Bridges *and Roofs.* By AUGUST RITTER, Ph.D., Professor at the Polytechnic School at Aix-la-Chapelle. Translated from the third German edition, by H. R. SANKEY, Capt. R.E. With 500 *illustrations*, 8vo, cloth, 15s.

The Elementary Principles of Carpentry. By THOMAS TREDGOLD. Revised from the original edition, and partly re-written, by JOHN THOMAS HURST. Contained in 517 pages of letter-press, and *illustrated with 48 plates and 150 wood engravings.* Sixth edition, reprinted from the third, crown 8vo, cloth, 12s. 6d.

Section I. On the Equality and Distribution of Forces—Section II. Resistance of Timber—Section III. Construction of Floors—Section IV. Construction of Roofs—Section V. Construction of Domes and Cupolas—Section VI. Construction of Partitions—Section VII. Scaffolds, Staging, and Gantries—Section VIII. Construction of Centres for Bridges—Section IX. Coffer-dams, Shoring, and Strutting—Section X. Wooden Bridges and Viaducts—Section XI. Joints, Straps, and other Fastenings—Section XII. Timber.

The Builder's Clerk: a Guide to the Management of a Builder's Business. By THOMAS BALES. Fcap. 8vo, cloth, 1s. 6d.

Practical Gold-Mining: a Comprehensive Treatise on the Origin and Occurrence of Gold-bearing Gravels, Rocks and Ores, and the methods by which the Gold is extracted. By C. G. WARNFORD LOCK, co-Editor of 'Gold, its Occurrence and Extraction.' *With 8 plates and* 271 *engravings in the text,* super-royal 8vo, cloth, 2*l.* 2*s.*

Hot Water Supply: A Practical Treatise upon the Fitting of Circulating Apparatus in connection with Kitchen Range and other Boilers, to supply Hot Water for Domestic and General Purposes. With a Chapter upon Estimating. *Fully illustrated,* crown 8vo, cloth, 3*s.*

Hot Water Apparatus: An Elementary Guide for the Fitting and Fixing of Boilers and Apparatus for the Circulation of Hot Water for Heating and for Domestic Supply, and containing a Chapter upon Boilers and Fittings for Steam Cooking. 32 *illustrations,* fcap. 8vo, cloth, 1*s.* 6*d.*

The Use and Misuse, and the Proper and Improper Fixing of a Cooking Range. Illustrated, fcap. 8vo, sewed, 6*d.*

Iron Roofs: Examples of Design, Description. *Illustrated with* 64 *Working Drawings of Executed Roofs.* By ARTHUR T. WALMISLEY, Assoc. Mem. Inst. C.E. Second edition, revised, imp. 4to, half-morocco, 3*l.* 3*s.*

A History of Electric Telegraphy, to the Year 1837. Chiefly compiled from Original Sources, and hitherto Unpublished Documents, by J. J. FAHIE, Mem. Soc. of Tel. Engineers, and of the International Society of Electricians, Paris. Crown 8vo, cloth, 9*s.*

Spons' Information for Colonial Engineers. Edited by J. T. HURST. Demy 8vo, sewed.

No. 1, Ceylon. By ABRAHAM DEANE, C.E. 2*s.* 6*d.*

CONTENTS:
Introductory Remarks — Natural Productions — Architecture and Engineering — Topography, Trade, and Natural History — Principal Stations — Weights and Measures, etc., etc.

No. 2. Southern Africa, including the Cape Colony, Natal, and the Dutch Republics. By HENRY HALL, F.R.G.S., F.R.C.I. With Map. 3*s.* 6*d.*

CONTENTS:
General Description of South Africa — Physical Geography with reference to Engineering Operations — Notes on Labour and Material in Cape Colony — Geological Notes on Rock Formation in South Africa — Engineering Instruments for Use in South Africa — Principal Public Works in Cape Colony: Railways, Mountain Roads and Passes, Harbour Works, Bridges, Gas Works, Irrigation and Water Supply, Lighthouses, Drainage and Sanitary Engineering, Public Buildings, Mines — Table of Woods in South Africa — Animals used for Draught Purposes — Statistical Notes — Table of Distances — Rates of Carriage, etc.

No. 3. India. By F. C. DANVERS, Assoc. Inst. C.E. With Map. 4*s.* 6*d.*

CONTENTS:
Physical Geography of India — Building Materials — Roads — Railways — Bridges — Irrigation — River Works — Harbours — Lighthouse Buildings — Native Labour — The Principal Trees of India — Money — Weights and Measures — Glossary of Indian Terms, etc.

Our Factories, Workshops, and Warehouses: their Sanitary and Fire-Resisting Arrangements. By B. H. THWAITE, Assoc. Mem. Inst. C.E. *With* 183 *wood engravings,* crown 8vo, cloth, 9s.

A Practical Treatise on Coal Mining. By GEORGE G. ANDRÉ, F.G.S., Assoc. Inst. C.E., Member of the Society of Engineers. *With* 82 *lithographic plates.* 2 vols., royal 4to, cloth, 3l. 12s.

A Practical Treatise on Casting and Founding, including descriptions of the modern machinery employed in the art. By N. E. SPRETSON, Engineer. Third edition, with 82 *plates* drawn to scale, 412 pp., demy 8vo, cloth, 18s.

The Depreciation of Factories and their Valuation. By EWING MATHESON, M. Inst. C.E. 8vo, cloth, 6s.

A Handbook of Electrical Testing. By H. R. KEMPE, M.S.T.E. Fourth edition, revised and enlarged, crown 8vo, cloth, 16s.

Gas Works: their Arrangement, Construction, Plant, and Machinery. By F. COLYER, M. Inst. C.E. *With* 31 *folding plates,* 8vo, cloth, 24s.

The Clerk of Works: a Vade-Mecum for all engaged in the Superintendence of Building Operations. By G. G. HOSKINS, F.R.I.B.A. Third edition, fcap. 8vo, cloth, 1s. 6d.

American Foundry Practice: Treating of Loam, Dry Sand, and Green Sand Moulding, and containing a Practical Treatise upon the Management of Cupolas, and the Melting of Iron. By T. D. WEST, Practical Iron Moulder and Foundry Foreman. Second edition, *with numerous illustrations,* crown 8vo, cloth, 10s. 6d.

The Maintenance of Macadamised Roads. By T. CODRINGTON, M.I.C.E., F.G.S., General Superintendent of County Roads for South Wales. 8vo, cloth, 6s.

Hydraulic Steam and Hand Power Lifting and Pressing Machinery. By FREDERICK COLYER, M. Inst. C.E., M. Inst. M.E. *With* 73 *plates,* 8vo, cloth, 18s.

Pumps and Pumping Machinery. By F. COLYER, M.I.C.E., M.I.M.E. *With* 23 *folding plates,* 8vo, cloth, 12s. 6d.

Pumps and Pumping Machinery. By F. COLYER. Second Part. *With* 11 *large plates,* 8vo, cloth, 12s. 6d.

A Treatise on the Origin, Progress, Prevention, and Cure of Dry Rot in Timber; with Remarks on the Means of Preserving Wood from Destruction by Sea-Worms, Beetles, Ants, etc. By THOMAS ALLEN BRITTON, late Surveyor to the Metropolitan Board of Works, etc., etc. *With* 10 *plates,* crown 8vo, cloth, 7s. 6d.

The Municipal and Sanitary Engineer's Handbook.
By H. PERCY BOULNOIS, Mem. Inst. C.E., Borough Engineer, Portsmouth. With *numerous illustrations*, demy 8vo, cloth, 12s. 6d.

CONTENTS:

The Appointment and Duties of the Town Surveyor—Traffic—Macadamised Roadways—Steam Rolling—Road Metal and Breaking—Pitched Pavements—Asphalte—Wood Pavements—Footpaths—Kerbs and Gutters—Street Naming and Numbering—Street Lighting—Sewerage—Ventilation of Sewers—Disposal of Sewage—House Drainage—Disinfection—Gas and Water Companies, etc., Breaking up Streets—Improvement of Private Streets—Borrowing Powers—Artizans' and Labourers' Dwellings—Public Conveniences—Scavenging, including Street Cleansing—Watering and the Removing of Snow—Planting Street Trees—Deposit of Plans—Dangerous Buildings—Hoardings—Obstructions—Improving Street Lines—Cellar Openings—Public Pleasure Grounds—Cemeteries—Mortuaries—Cattle and Ordinary Markets—Public Slaughter-houses, etc.—Giving numerous Forms of Notices, Specifications, and General Information upon these and other subjects of great importance to Municipal Engineers and others engaged in Sanitary Work.

Metrical Tables. By G. L. MOLESWORTH, M.I.C.E.
32mo, cloth, 1s. 6d.

CONTENTS.

General—Linear Measures—Square Measures—Cubic Measures—Measures of Capacity—Weights—Combinations—Thermometers.

Elements of Construction for Electro-Magnets. By
Count TH. DU MONCEL, Mem. de l'Institut de France. Translated from the French by C. J. WHARTON. Crown 8vo, cloth, 4s. 6d.

Practical Electrical Units Popularly Explained, with
numerous illustrations and Remarks. By JAMES SWINBURNE, late of J. W. Swan and Co., Paris, late of Brush-Swan Electric Light Company, U.S.A. 18mo, cloth, 1s. 6d.

A Treatise on the Use of Belting for the Transmission of Power.
By J. H. COOPER. Second edition, *illustrated*, 8vo, cloth, 15s.

A Pocket-Book of Useful Formulæ and Memoranda
for Civil and Mechanical Engineers. By GUILFORD L. MOLESWORTH, Mem. Inst. C.E., Consulting Engineer to the Government of India for State Railways. With *numerous illustrations*, 744 pp. Twenty-second edition, revised and enlarged, 32mo, roan, 6s.

SYNOPSIS OF CONTENTS:

Surveying, Levelling, etc.—Strength and Weight of Materials—Earthwork, Brickwork, Masonry, Arches, etc.—Struts, Columns, Beams, and Trusses—Flooring, Roofing, and Roof Trusses—Girders, Bridges, etc.—Railways and Roads—Hydraulic Formulæ—Canals, Sewers, Waterworks, Docks—Irrigation and Breakwaters—Gas, Ventilation, and Warming—Heat, Light, Colour, and Sound—Gravity: Centres, Forces, and Powers—Millwork, Teeth of Wheels, Shafting, etc.—Workshop Recipes—Sundry Machinery—Animal Power—Steam and the Steam Engine—Water-power, Water-wheels, Turbines, etc.—Wind and Windmills—Steam Navigation, Ship Building, Tonnage, etc.—Gunnery, Projectiles, etc.—Weights, Measures, and Money—Trigonometry, Conic Sections, and Curves—Telegraphy—Mensuration—Tables of Areas and Circumference, and Arcs of Circles—Logarithms, Square and Cube Roots, Powers—Reciprocals, etc.—Useful Numbers—Differential and Integral Calculus—Algebraic Signs—Telegraphic Construction and Formulæ.

Hints on Architectural Draughtsmanship. By G. W. TUXFORD HALLATT. Fcap. 8vo, cloth, 1s. 6d.

Spons' Tables and Memoranda for Engineers; selected and arranged by J. T. HURST, C.E., Author of 'Architectural Surveyors' Handbook,' 'Hurst's Tredgold's Carpentry,' etc. Ninth edition, 64mo, roan, gilt edges, 1s.; or in cloth case, 1s. 6d.

This work is printed in a pearl type, and is so small, measuring only 2½ in. by 1¾ in. by ¼ in. thick, that it may be easily carried in the waistcoat pocket.

"It is certainly an extremely rare thing for a reviewer to be called upon to notice a volume measuring but 2½ in. by 1¾ in., yet these dimensions faithfully represent the size of the handy little book before us. The volume—which contains 118 printed pages, besides a few blank pages for memoranda—is, in fact, a true pocket-book, adapted for being carried in the waistcoat pocket, and containing a far greater amount and variety of information than most people would imagine could be compressed into so small a space. . . . The little volume has been compiled with considerable care and judgment, and we can cordially recommend it to our readers as a useful little pocket companion."—*Engineering.*

A Practical Treatise on Natural and Artificial Concrete, its Varieties and Constructive Adaptations. By HENRY REID, Author of the 'Science and Art of the Manufacture of Portland Cement.' New Edition, *with* 59 *woodcuts and* 5 *plates*, 8vo, cloth, 15s.

Notes on Concrete and Works in Concrete; especially written to assist those engaged upon Public Works. By JOHN NEWMAN, Assoc. Mem. Inst. C.E., crown 8vo, cloth, 4s. 6d.

Electricity as a Motive Power. By Count TH. DU MONCEL, Membre de l'Institut de France, and FRANK GERALDY, Ingénieur des Ponts et Chaussées. Translated and Edited, with Additions, by C. J. WHARTON, Assoc. Soc. Tel. Eng. and Elec. *With* 113 *engravings and diagrams*, crown 8vo, cloth, 7s. 6d.

Treatise on Valve-Gears, with special consideration of the Link-Motions of Locomotive Engines. By Dr. GUSTAV ZEUNER, Professor of Applied Mechanics at the Confederated Polytechnikum of Zurich. Translated from the Fourth German Edition, by Professor J. F. KLEIN, Lehigh University, Bethlehem, Pa. *Illustrated*, 8vo, cloth, 12s. 6d.

The French-Polisher's Manual. By a French-Polisher; containing Timber Staining, Washing, Matching, Improving, Painting, Imitations, Directions for Staining, Sizing, Embodying, Smoothing, Spirit Varnishing, French-Polishing, Directions for Re-polishing. Third edition, royal 32mo, sewed, 6d.

Hops, their Cultivation, Commerce, and Uses in various Countries. By P. L. SIMMONDS. Crown 8vo, cloth, 4s. 6d.

The Principles of Graphic Statics. By GEORGE SYDENHAM CLARKE, Capt. Royal Engineers. *With* 112 *illustrations.* 4to, cloth, 12s. 6d.

Dynamo-Electric Machinery: A Manual for Students of Electro-technics. By SILVANUS P. THOMPSON, B.A., D.Sc., Professor of Experimental Physics in University College, Bristol, etc., etc. Third edition, *illustrated*, 8vo, cloth, 16s.

Practical Geometry, Perspective, and Engineering Drawing; a Course of Descriptive Geometry adapted to the Requirements of the Engineering Draughtsman, including the determination of cast shadows and Isometric Projection, each chapter being followed by numerous examples; to which are added rules for Shading, Shade-lining, etc., together with practical instructions as to the Lining, Colouring, Printing, and general treatment of Engineering Drawings, with a chapter on drawing Instruments. By GEORGE S. CLARKE, Capt. R.E. Second edition, *with* 21 *plates*. 2 vols., cloth, 10s. 6d.

The Elements of Graphic Statics. By Professor KARL VON OTT, translated from the German by G. S. CLARKE, Capt. R.E., Instructor in Mechanical Drawing, Royal Indian Engineering College. *With* 93 *illustrations*, crown 8vo, cloth, 5s.

A Practical Treatise on the Manufacture and Distribution of Coal Gas. By WILLIAM RICHARDS. Demy 4to, with *numerous wood engravings and* 29 *plates*, cloth, 28s.

SYNOPSIS OF CONTENTS:

Introduction — History of Gas Lighting — Chemistry of Gas Manufacture, by Lewis Thompson, Esq., M.R.C.S.—Coal, with Analyses, by J. Paterson, Lewis Thompson, and G. R. Hislop, Esqrs.—Retorts, Iron and Clay—Retort Setting—Hydraulic Main—Condensers — Exhausters — Washers and Scrubbers — Purifiers — Purification — History of Gas Holder — Tanks, Brick and Stone, Composite, Concrete, Cast-iron, Compound Annular Wrought-iron — Specifications — Gas Holders — Station Meter — Governor — Distribution— Mains—Gas Mathematics, or Formulæ for the Distribution of Gas, by Lewis Thompson, Esq.— Services—Consumers' Meters—Regulators—Burners—Fittings—Photometer—Carburization of Gas—Air Gas and Water Gas—Composition of Coal Gas, by Lewis Thompson, Esq.— Analyses of Gas—Influence of Atmospheric Pressure and Temperature on Gas—Residual Products—Appendix—Description of Retort Settings, Buildings, etc., etc.

The New Formula for Mean Velocity of Discharge of Rivers and Canals. By W. R. KUTTER. Translated from articles in the 'Cultur-Ingénieur,' by LOWIS D'A. JACKSON, Assoc. Inst. C.E. 8vo, cloth, 12s. 6d.

The Practical Millwright and Engineer's Ready Reckoner; or Tables for finding the diameter and power of cog-wheels, diameter, weight, and power of shafts, diameter and strength of bolts, etc. By THOMAS DIXON. Fourth edition, 12mo, cloth, 3s.

Tin: Describing the Chief Methods of Mining, Dressing and Smelting it abroad; with Notes upon Arsenic, Bismuth and Wolfram. By ARTHUR G. CHARLETON, Mem. American Inst. of Mining Engineers. *With plates*, 8vo, cloth, 12s. 6d.

B 3

Perspective, Explained and Illustrated. By G. S. CLARKE, Capt. R.E. *With illustrations,* 8vo, cloth, 3s. 6d.

Practical Hydraulics; a Series of Rules and Tables for the use of Engineers, etc., etc. By THOMAS BOX. Fifth edition, *numerous plates,* post 8vo, cloth, 5s.

The Essential Elements of Practical Mechanics; based on the Principle of Work, designed for Engineering Students. By OLIVER BYRNE, formerly Professor of Mathematics, College for Civil Engineers. Third edition, *with* 148 *wood engravings,* post 8vo, cloth, 7s. 6d.

CONTENTS:

Chap. 1. How Work is Measured by a Unit, both with and without reference to a Unit of Time—Chap. 2. The Work of Living Agents, the Influence of Friction, and introduces one of the most beautiful Laws of Motion—Chap. 3. The principles expounded in the first and second chapters are applied to the Motion of Bodies—Chap. 4. The Transmission of Work by simple Machines—Chap. 5. Useful Propositions and Rules.

Breweries and Maltings: their Arrangement, Construction, Machinery, and Plant. By G. SCAMELL, F.R.I.B.A. Second edition, revised, enlarged, and partly rewritten. By F. COLYER, M.I.C.E., M.I.M.E. *With* 20 *plates,* 8vo, cloth, 18s.

A Practical Treatise on the Construction of Horizontal and Vertical Waterwheels, specially designed for the use of operative mechanics. By WILLIAM CULLEN, Millwright and Engineer. *With* 11 *plates.* Second edition, revised and enlarged, small 4to, cloth, 12s. 6d.

A Practical Treatise on Mill-gearing, Wheels, Shafts, Riggers, etc.; for the use of Engineers. By THOMAS BOX. Third edition, *with* 11 *plates.* Crown 8vo, cloth, 7s. 6d.

Mining Machinery: a Descriptive Treatise on the Machinery, Tools, and other Appliances used in Mining. By G. G. ANDRÉ, F.G.S., Assoc. Inst. C.E., Mem. of the Society of Engineers. Royal 4to, uniform with the Author's Treatise on Coal Mining, containing 182 *plates,* accurately drawn to scale, with descriptive text, in 2 vols., cloth, 3l. 12s.

CONTENTS:

Machinery for Prospecting, Excavating, Hauling, and Hoisting—Ventilation—Pumping—Treatment of Mineral Products, including Gold and Silver, Copper, Tin, and Lead, Iron Coal, Sulphur, China Clay, Brick Earth, etc.

Tables for Setting out Curves for Railways, Canals, Roads, etc., varying from a radius of five chains to three miles. By A. KENNEDY and R. W. HACKWOOD. *Illustrated,* 32mo, cloth, 2s. 6d.

The Science and Art of the Manufacture of Portland Cement, with observations on some of its constructive applications. With 66 *illustrations.* By HENRY REID, C.E., Author of 'A Practical Treatise on Concrete,' etc., etc. 8vo, cloth, 18s.

The Draughtsman's Handbook of Plan and Map Drawing; including instructions for the preparation of Engineering, Architectural, and Mechanical Drawings. *With numerous illustrations in the text, and* 33 *plates* (15 *printed in colours*). By G. G. ANDRÉ, F.G.S., Assoc. Inst. C.E. 4to, cloth, 9s.

CONTENTS:

The Drawing Office and its Furnishings—Geometrical Problems—Lines, Dots, and their Combinations—Colours, Shading, Lettering, Bordering, and North Points—Scales—Plotting—Civil Engineers' and Surveyors' Plans—Map Drawing—Mechanical and Architectural Drawing—Copying and Reducing Trigonometrical Formulæ, etc., etc.

The Boiler-maker's and Iron Ship-builder's Companion, comprising a series of original and carefully calculated tables, of the utmost utility to persons interested in the iron trades. By JAMES FODEN, author of 'Mechanical Tables,' etc. Second edition revised, *with illustrations,* crown 8vo, cloth, 5s.

Rock Blasting: a Practical Treatise on the means employed in Blasting Rocks for Industrial Purposes. By G. G. ANDRÉ, F.G.S., Assoc. Inst. C.E. *With* 56 *illustrations and* 12 *plates,* 8vo, cloth, 10s. 6d.

Painting and Painters' Manual: a Book of Facts for Painters and those who Use or Deal in Paint Materials. By C. L. CONDIT and J. SCHELLER. *Illustrated,* 8vo, cloth, 10s. 6d.

A Treatise on Ropemaking as practised in public and private Rope-yards, with a Description of the Manufacture, Rules, Tables of Weights, etc., adapted to the Trade, Shipping, Mining, Railways, Builders, etc. By R. CHAPMAN, formerly foreman to Messrs. Huddart and Co., Limehouse, and late Master Ropemaker to H.M. Dockyard, Deptford. Second edition, 12mo, cloth, 3s.

Laxton's Builders' and Contractors' Tables; for the use of Engineers, Architects, Surveyors, Builders, Land Agents, and others. Bricklayer, containing 22 tables, with nearly 30,000 calculations. 4to, cloth, 5s.

Laxton's Builders' and Contractors' Tables. Excavator, Earth, Land, Water, and Gas, containing 53 tables, with nearly 24,000 calculations. 4to, cloth, 5s.

Egyptian Irrigation. By W. WILLCOCKS, M.I.C.E., Indian Public Works Department, Inspector of Irrigation, Egypt. With Introduction by Lieut.-Col. J. C. ROSS, R.E., Inspector-General of Irrigation. *With numerous lithographs and wood engravings*, royal 8vo, cloth, 1*l.* 16*s.*

Screw Cutting Tables for Engineers and Machinists, giving the values of the different trains of Wheels required to produce Screws of any pitch, calculated by Lord Lindsay, M.P., F.R.S., F.R.A.S., etc. Cloth, oblong, 2*s.*

Screw Cutting Tables, for the use of Mechanical Engineers, showing the proper arrangement of Wheels for cutting the Threads of Screws of any required pitch, with a Table for making the Universal Gas-pipe Threads and Taps. By W. A. MARTIN, Engineer. Second edition, oblong, cloth, 1*s.*, or sewed, 6*d.*

A Treatise on a Practical Method of Designing Slide-Valve Gears by Simple Geometrical Construction, based upon the principles enunciated in Euclid's Elements, and comprising the various forms of Plain Slide-Valve and Expansion Gearing; together with Stephenson's, Gooch's, and Allan's Link-Motions, as applied either to reversing or to variable expansion combinations. By EDWARD J. COWLING WELCH, Memb. Inst. Mechanical Engineers. Crown 8vo, cloth, 6*s.*

Cleaning and Scouring: a Manual for Dyers, Laundresses, and for Domestic Use. By S. CHRISTOPHER. 18mo, sewed, 6*d.*

A Glossary of Terms used in Coal Mining. By WILLIAM STUKELEY GRESLEY, Assoc. Mem. Inst. C.E., F.G.S., Member of the North of England Institute of Mining Engineers. *Illustrated with numerous woodcuts and diagrams*, crown 8vo, cloth, 5*s.*

A Pocket-Book for Boiler Makers and Steam Users, comprising a variety of useful information for Employer and Workman, Government Inspectors, Board of Trade Surveyors, Engineers in charge of Works and Ships, Foremen of Manufactories, and the general Steam-using Public. By MAURICE JOHN SEXTON. Second edition, royal 32mo, roan, gilt edges, 5*s.*

Electrolysis: a Practical Treatise on Nickeling, Coppering, Gilding, Silvering, the Refining of Metals, and the treatment of Ores by means of Electricity. By HIPPOLYTE FONTAINE, translated from the French by J. A. BERLY, C.E., Assoc. S.T.E. *With engravings.* 8vo, cloth, 9*s.*

Barlow's Tables of Squares, Cubes, Square Roots,
Cube Roots, Reciprocals of all Integer Numbers up to 10,000. Post 8vo,
cloth, 6s.

A Practical Treatise on the Steam Engine, containing Plans and Arrangements of Details for Fixed Steam Engines, with Essays on the Principles involved in Design and Construction. By ARTHUR RIGG, Engineer, Member of the Society of Engineers and of the Royal Institution of Great Britain. Demy 4to, *copiously illustrated with woodcuts and* 96 *plates,* in one Volume, half-bound morocco, 2l. 2s.; or cheaper edition, cloth, 25s.

This work is not, in any sense, an elementary treatise, or history of the steam engine, but is intended to describe examples of Fixed Steam Engines without entering into the wide domain of locomotive or marine practice. To this end illustrations will be given of the most recent arrangements of Horizontal, Vertical, Beam, Pumping, Winding, Portable, Semi-portable, Corliss, Allen, Compound, and other similar Engines, by the most eminent Firms in Great Britain and America. The laws relating to the action and precautions to be observed in the construction of the various details, such as Cylinders, Pistons, Piston-rods, Connecting-rods, Cross-heads, Motion-blocks, Eccentrics, Simple, Expansion, Balanced, and Equilibrium Slide-valves, and Valve-gearing will be minutely dealt with. In this connection will be found articles upon the Velocity of Reciprocating Parts and the Mode of Applying the Indicator, Heat and Expansion of Steam Governors, and the like. It is the writer's desire to draw illustrations from every possible source, and give only those rules that present practice deems correct.

A Practical Treatise on the Science of Land and Engineering Surveying, Levelling, Estimating Quantities, etc., with a general description of the several Instruments required for Surveying, Levelling, Plotting, etc. By H. S. MERRETT. Fourth edition, revised by G. W. USILL, Assoc. Mem. Inst. C.E. 41 *plates, with illustrations and tables,* royal 8vo, cloth, 12s. 6d.

PRINCIPAL CONTENTS:

Part 1. Introduction and the Principles of Geometry. Part 2. Land Surveying; comprising General Observations—The Chain—Offsets Surveying by the Chain only—Surveying Hilly Ground—To Survey an Estate or Parish by the Chain only—Surveying with the Theodolite—Mining and Town Surveying—Railroad Surveying—Mapping—Division and Laying out of Land—Observations on Enclosures—Plane Trigonometry. Part 3. Levelling—Simple and Compound Levelling—The Level Book—Parliamentary Plan and Section—Levelling with a Theodolite—Gradients—Wooden Curves—To Lay out a Railway Curve—Setting out Widths. Part 4. Calculating Quantities generally for Estimates—Cuttings and Embankments—Tunnels—Brickwork—Ironwork—Timber Measuring. Part 5. Description and Use of Instruments in Surveying and Plotting—The Improved Dumpy Level—Troughton's Level—The Prismatic Compass—Proportional Compass—Box Sextant—Vernier—Pantagraph—Merrett's Improved Quadrant—Improved Computation Scale—The Diagonal Scale—Straight Edge and Sector. Part 6. Logarithms of Numbers—Logarithmic Sines and Co-Sines, Tangents and Co-Tangents—Natural Sines and Co-Sines—Tables for Earthwork, for Setting out Curves, and for various Calculations, etc., etc., etc.

Health and Comfort in House Building, or Ventilation with Warm Air by Self-Acting Suction Power, with Review of the mode of Calculating the Draught in Hot-Air Flues, and with some actual Experiments. By J. DRYSDALE, M.D., and J. W. HAYWARD, M.D. Second edition, with Supplement, *with plates,* demy 8vo, cloth, 7s. 6d.

The Assayer's Manual: an Abridged Treatise on the Docimastic Examination of Ores and Furnace and other Artificial Products. By BRUNO KERL. Translated by W. T. BRANNT. *With* 65 *illustrations,* 8vo, cloth, 12s. 6d.

Dynamo - Electric Machinery: a Text-Book for Students of Electro-Technology. By SILVANUS P. THOMPSON, B.A., D.Sc., M.S.T.E. Third Edition, revised and enlarged, 8vo, cloth, 16s.

The Practice of Hand Turning in Wood, Ivory, Shell, etc., with Instructions for Turning such Work in Metal as may be required in the Practice of Turning in Wood, Ivory, etc.; also an Appendix on Ornamental Turning. (A book for beginners.) By FRANCIS CAMPIN. Third edition, *with wood engravings,* crown 8vo, cloth, 6s.

CONTENTS:

On Lathes—Turning Tools—Turning Wood—Drilling—Screw Cutting—Miscellaneous Apparatus and Processes—Turning Particular Forms—Staining—Polishing—Spinning Metals—Materials—Ornamental Turning, etc.

Treatise on Watchwork, Past and Present. By the Rev. H. L. NELTHROPP, M.A., F.S.A. *With* 32 *illustrations,* crown 8vo, cloth, 6s. 6d.

CONTENTS:

Definitions of Words and Terms used in Watchwork—Tools—Time—Historical Summary—On Calculations of the Numbers for Wheels and Pinions; their Proportional Sizes, Trains, etc.—Of Dial Wheels, or Motion Work—Length of Time of Going without Winding up—The Verge—The Horizontal—The Duplex—The Lever—The Chronometer—Repeating Watches—Keyless Watches—The Pendulum, or Spiral Spring—Compensation—Jewelling of Pivot Holes—Clerkenwell—Fallacies of the Trade—Incapacity of Workmen—How to Choose and Use a Watch, etc.

Algebra Self-Taught. By W. P. HIGGS, M.A., D.Sc., LL.D., Assoc. Inst. C.E., Author of 'A Handbook of the Differential Calculus,' etc. Second edition, crown 8vo, cloth, 2s. 6d.

CONTENTS:

Symbols and the Signs of Operation—The Equation and the Unknown Quantity—Positive and Negative Quantities—Multiplication—Involution—Exponents—Negative Exponents—Roots, and the Use of Exponents as Logarithms—Logarithms—Tables of Logarithms and Proportionate Parts—Transformation of System of Logarithms—Common Uses of Common Logarithms—Compound Multiplication and the Binomial Theorem—Division, Fractions, and Ratio—Continued Proportion—The Series and the Summation of the Series—Limit of Series—Square and Cube Roots—Equations—List of Formulæ, etc.

Spons' Dictionary of Engineering, Civil, Mechanical, Military, and Naval; with technical terms in French, German, Italian, and Spanish, 3100 pp., and *nearly* 8000 *engravings,* in super-royal 8vo, in 8 divisions, 5l. 8s. Complete in 3 vols., cloth, 5l. 5s. Bound in a superior manner, half-morocco, top edge gilt, 3 vols., 6l. 12s.

Notes in Mechanical Engineering. Compiled principally for the use of the Students attending the Classes on this subject at the City of London College. By HENRY ADAMS, Mem. Inst. M.E., Mem. Inst. C.E., Mem. Soc. of Engineers. Crown 8vo, cloth, 2s. 6d.

Canoe and Boat Building: a complete Manual for Amateurs, containing plain and comprehensive directions for the construction of Canoes, Rowing and Sailing Boats, and Hunting Craft. By W. P. STEPHENS. *With numerous illustrations and 24 plates of Working Drawings.* Crown 8vo, cloth, 7s. 6d.

Proceedings of the National Conference of Electricians, Philadelphia, October 8th to 13th, 1884. 18mo, cloth, 3s.

Dynamo - Electricity, its Generation, Application, Transmission, Storage, and Measurement. By G. B. PRESCOTT. With 545 *illustrations.* 8vo, cloth, 1l. 1s.

Domestic Electricity for Amateurs. Translated from the French of E. HOSPITALIER, Editor of "L'Electricien," by C. J. WHARTON, Assoc. Soc. Tel. Eng. *Numerous illustrations.* Demy 8vo, cloth, 9s.

CONTENTS:

1. Production of the Electric Current—2. Electric Bells—3. Automatic Alarms—4. Domestic Telephones—5. Electric Clocks—6. Electric Lighters—7. Domestic Electric Lighting—8. Domestic Application of the Electric Light—9. Electric Motors—10. Electrical Locomotion—11. Electrotyping, Plating, and Gilding—12. Electric Recreations—13. Various applications—Workshop of the Electrician.

Wrinkles in Electric Lighting. By VINCENT STEPHEN. *With illustrations.* 18mo, cloth, 2s. 6d.

CONTENTS:

1. The Electric Current and its production by Chemical means—2. Production of Electric Currents by Mechanical means—3. Dynamo-Electric Machines—4. Electric Lamps—5. Lead—6. Ship Lighting.

The Practical Flax Spinner; being a Description of the Growth, Manipulation, and Spinning of Flax and Tow. By LESLIE C. MARSHALL, of Belfast. *With illustrations.* 8vo, cloth, 15s.

Foundations and Foundation Walls for all classes of Buildings, Pile Driving, Building Stones and Bricks, Pier and Wall construction, Mortars, Limes, Cements, Concretes, Stuccos, &c. 64 *illustrations.* By G. T. POWELL and F. BAUMAN. 8vo, cloth, 10s. 6d.

Manual for Gas Engineering Students. By D. LEE.
18mo, cloth 1s.

Hydraulic Machinery, Past and Present. A Lecture delivered to the London and Suburban Railway Officials' Association. By H. ADAMS, Mem. Inst. C.E. *Folding plate.* 8vo, sewed, 1s.

Twenty Years with the Indicator. By THOMAS PRAY, Jun., C.E., M.E., Member of the American Society of Civil Engineers. 2 vols., royal 8vo, cloth, 12s. 6d.

Annual Statistical Report of the Secretary to the Members of the Iron and Steel Association on the Home and Foreign Iron and Steel Industries in 1887. Issued March 1888. 8vo, sewed, 5s.

Bad Drains, and How to Test them; with Notes on the Ventilation of Sewers, Drains, and Sanitary Fittings, and the Origin and Transmission of Zymotic Disease. By R. HARRIS REEVES. Crown 8vo, cloth, 3s. 6d.

Well Sinking. The modern practice of Sinking and Boring Wells, with geological considerations and examples of Wells. By ERNEST SPON, Assoc. Mem. Inst. C.E., Mem. Soc. Eng., and of the Franklin Inst., etc. Second edition, revised and enlarged. Crown 8vo, cloth, 10s. 6d.

The Voltaic Accumulator: an Elementary Treatise. By ÉMILE REYNIER. Translated by J. A. BERLY, Assoc. Inst. E.E. *With 62 illustrations,* 8vo, cloth, 9s.

List of Tests (Reagents), arranged in alphabetical order, according to the names of the originators. Designed especially for the convenient reference of Chemists, Pharmacists, and Scientists. By HANS M. WILDER. Crown 8vo, cloth, 4s. 6d.

Ten Years' Experience in Works of Intermittent Downward Filtration. By J. BAILEY DENTON, Mem. Inst. C.E. Second edition, with additions. Royal 8vo, sewed, 4s.

A Treatise on the Manufacture of Soap and Candles, Lubricants and Glycerin. By W. LANT CARPENTER, B.A., B.Sc. (late of Messrs. C. Thomas and Brothers, Bristol). *With illustrations.* Crown 8vo, cloth, 10s. 6d.

The Stability of Ships explained simply, and calculated by a new Graphic method. By J. C. SPENCE, M.I.N.A. 4to, sewed, 3s. 6d.

Steam Making, or Boiler Practice. By CHARLES A. SMITH, C.E. 8vo, cloth, 10s. 6d.

CONTENTS:

1. The Nature of Heat and the Properties of Steam—2. Combustion.—3. Externally Fired Stationary Boilers—4. Internally Fired Stationary Boilers—5. Internally Fired Portable Locomotive and Marine Boilers—6. Design, Construction, and Strength of Boilers—7. Proportions of Heating Surface, Economic Evaporation, Explosions—8. Miscellaneous Boilers, Choice of Boiler Fittings and Appurtenances.

The Fireman's Guide; a Handbook on the Care of Boilers. By TEKNOLOG, föreningen T. I. Stockholm. Translated from the third edition, and revised by KARL P. DAHLSTROM, M.E. Second edition. Fcap. 8vo, cloth, 2s.

A Treatise on Modern Steam Engines and Boilers, including Land Locomotive, and Marine Engines and Boilers, for the use of Students. By FREDERICK COLYER, M. Inst. C.E., Mem. Inst. M.E. *With 36 plates.* 4to, cloth, 25s.

CONTENTS:

1. Introduction—2. Original Engines—3. Boilers—4. High-Pressure Beam Engines—5. Cornish Beam Engines—6. Horizontal Engines—7. Oscillating Engines—8. Vertical High-Pressure Engines—9. Special Engines—10. Portable Engines—11. Locomotive Engines—12. Marine Engines.

Steam Engine Management; a Treatise on the Working and Management of Steam Boilers. By F. COLYER, M. Inst. C.E., Mem. Inst. M.E. 18mo, cloth, 2s.

Land Surveying on the Meridian and Perpendicular System. By WILLIAM PENMAN, C.E. 8vo, cloth, 8s. 6d.

The Topographer, his Instruments and Methods, designed for the use of Students, Amateur Photographers, Surveyors, Engineers, and all persons interested in the location and construction of works based upon Topography. *Illustrated with numerous plates, maps, and engravings.* By LEWIS M. HAUPT, A.M. 8vo, cloth, 18s.

A Text-Book of Tanning, embracing the Preparation of all kinds of Leather. By HARRY R. PROCTOR, F.C.S., of Low Lights Tanneries. *With illustrations.* Crown 8vo, cloth, 10s. 6d.

In super-royal 8vo, 1168 pp., *with 2400 illustrations*, in 3 Divisions, cloth, price 13s. 6d. each; or 1 vol., cloth, 2l.; or half-morocco, 2l. 8s.

A SUPPLEMENT

TO

SPONS' DICTIONARY OF ENGINEERING.

EDITED BY ERNEST SPON, MEMB. SOC. ENGINEERS.

Abacus, Counters, Speed Indicators, and Slide Rule.
Agricultural Implements and Machinery.
Air Compressors.
Animal Charcoal Machinery.
Antimony.
Axles and Axle-boxes.
Barn Machinery.
Belts and Belting.
Blasting. Boilers.
Brakes.
Brick Machinery.
Bridges.
Cages for Mines.
Calculus, Differential and Integral.
Canals.
Carpentry.
Cast Iron.
Cement, Concrete, Limes, and Mortar.
Chimney Shafts.
Coal Cleansing and Washing.
Coal Mining.
Coal Cutting Machines.
Coke Ovens. Copper.
Docks. Drainage.
Dredging Machinery.
Dynamo - Electric and Magneto-Electric Machines.
Dynamometers.
Electrical Engineering, Telegraphy, Electric Lighting and its practical details, Telephones
Engines, Varieties of.
Explosives. Fans.
Founding, Moulding and the practical work of the Foundry.
Gas, Manufacture of.
Hammers, Steam and other Power.
Heat. Horse Power.
Hydraulics.
Hydro-geology.
Indicators. Iron.
Lifts, Hoists, and Elevators.
Lighthouses, Buoys, and Beacons.
Machine Tools.
Materials of Construction.
Meters.
Ores, Machinery and Processes employed to Dress.
Piers.
Pile Driving.
Pneumatic Transmission.
Pumps.
Pyrometers.
Road Locomotives.
Rock Drills.
Rolling Stock.
Sanitary Engineering.
Shafting.
Steel.
Steam Navvy.
Stone Machinery.
Tramways.
Well Sinking.

London: E. & F. N. SPON, 125, Strand.
New York: 12, Cortlandt Street.

NOW COMPLETE.

With nearly 1500 *illustrations*, in super-royal 8vo, in 5 Divisions, cloth.
Divisions 1 to 4, 13*s.* 6*d.* each ; Division 5, 17*s.* 6*d.* ; or 2 vols., cloth, £3 10*s.*

SPONS' ENCYCLOPÆDIA
OF THE
INDUSTRIAL ARTS, MANUFACTURES, AND COMMERCIAL PRODUCTS.

EDITED BY C. G. WARNFORD LOCK, F.L.S.

Among the more important of the subjects treated of, are the following :—

Acids, 207 pp. 220 figs.
Alcohol, 23 pp. 16 figs.
Alcoholic Liquors, 13 pp.
Alkalies, 89 pp. 78 figs.
Alloys.   Alum.
Asphalt.   Assaying.
Beverages, 89 pp. 29 figs.
Blacks.
Bleaching Powder, 15 pp.
Bleaching, 51 pp. 48 figs.
Candles, 18 pp. 9 figs.
Carbon Bisulphide.
Celluloid, 9 pp.
Cements.   Clay.
Coal-tar Products, 44 pp. 14 figs.
Cocoa, 8 pp.
Coffee, 32 pp. 13 figs.
Cork, 8 pp. 17 figs.
Cotton Manufactures, 62 pp. 57 figs.
Drugs, 38 pp.
Dyeing and Calico Printing, 28 pp. 9 figs.
Dyestuffs, 16 pp.
Electro-Metallurgy, 13 pp.
Explosives, 22 pp. 33 figs.
Feathers.
Fibrous Substances, 92 pp. 79 figs.
Floor-cloth, 16 pp. 21 figs.
Food Preservation, 8 pp.
Fruit, 8 pp.

Fur, 5 pp.
Gas, Coal, 8 pp.
Gems.
Glass, 45 pp. 77 figs.
Graphite, 7 pp.
Hair, 7 pp.
Hair Manufactures.
Hats, 26 pp. 26 figs.
Honey.   Hops.
Horn.
Ice, 10 pp. 14 figs.
Indiarubber Manufactures, 23 pp. 17 figs.
Ink, 17 pp.
Ivory.
Jute Manufactures, 11 pp., 11 figs.
Knitted Fabrics — Hosiery, 15 pp. 13 figs.
Lace, 13 pp. 9 figs.
Leather, 28 pp. 31 figs.
Linen Manufactures, 16 pp. 6 figs.
Manures, 21 pp. 30 figs.
Matches, 17 pp. 38 figs.
Mordants, 13 pp.
Narcotics, 47 pp.
Nuts, 10 pp.
Oils and Fatty Substances, 125 pp.
Paint.
Paper, 26 pp. 23 figs.
Paraffin, 8 pp. 6 figs.
Pearl and Coral, 8 pp.
Perfumes, 10 pp.

Photography, 13 pp. 20 figs.
Pigments, 9 pp. 6 figs.
Pottery, 46 pp. 57 figs.
Printing and Engraving, 20 pp. 8 figs.
Rags.
Resinous and Gummy Substances, 75 pp. 16 figs.
Rope, 16 pp. 17 figs.
Salt, 31 pp. 23 figs.
Silk, 8 pp.
Silk Manufactures, 9 pp. 11 figs.
Skins, 5 pp.
Small Wares, 4 pp.
Soap and Glycerine, 39 pp. 45 figs.
Spices, 16 pp.
Sponge, 5 pp.
Starch, 9 pp. 10 figs.
Sugar, 155 pp. 134 figs.
Sulphur.
Tannin, 18 pp.
Tea, 12 pp.
Timber, 13 pp.
Varnish, 15 pp.
Vinegar, 5 pp.
Wax, 5 pp.
Wool, 2 pp.
Woollen Manufactures, 58 pp. 39 figs.

London: E. & F. N. SPON, 125, Strand.
New York: 12, Cortlandt Street.

Crown 8vo, cloth, with illustrations, 5s.

WORKSHOP RECEIPTS,
FIRST SERIES.
By ERNEST SPON.

Synopsis of Contents.

Bookbinding.
Bronzes and Bronzing.
Candles.
Cement.
Cleaning.
Colourwashing.
Concretes.
Dipping Acids.
Drawing Office Details.
Drying Oils.
Dynamite.
Electro - Metallurgy — (Cleaning, Dipping, Scratch-brushing, Batteries, Baths, and Deposits of every description).
Enamels.
Engraving on Wood, Copper, Gold, Silver, Steel, and Stone.
Etching and Aqua Tint.
Firework Making — (Rockets, Stars, Rains, Gerbes, Jets, Tourbillons, Candles, Fires, Lances, Lights, Wheels, Fire-balloons, and minor Fireworks).
Fluxes.
Foundry Mixtures.

Freezing.
Fulminates.
Furniture Creams, Oils, Polishes, Lacquers, and Pastes.
Gilding.
Glass Cutting, Cleaning, Frosting, Drilling, Darkening, Bending, Staining, and Painting.
Glass Making.
Glues.
Gold.
Graining.
Gums.
Gun Cotton.
Gunpowder.
Horn Working.
Indiarubber.
Japans, Japanning, and kindred processes.
Lacquers.
Lathing.
Lubricants.
Marble Working.
Matches.
Mortars.
Nitro-Glycerine.
Oils.

Paper.
Paper Hanging.
Painting in Oils, in Water Colours, as well as Fresco, House, Transparency, Sign, and Carriage Painting.
Photography.
Plastering.
Polishes.
Pottery—(Clays, Bodies, Glazes, Colours, Oils, Stains, Fluxes, Enamels, and Lustres).
Scouring.
Silvering.
Soap.
Solders.
Tanning.
Taxidermy.
Tempering Metals.
Treating Horn, Mother-o'-Pearl, and like substances.
Varnishes, Manufacture and Use of.
Veneering.
Washing.
Waterproofing.
Welding.

Besides Receipts relating to the lesser Technological matters and processes, such as the manufacture and use of Stencil Plates, Blacking, Crayons, Paste, Putty, Wax, Size, Alloys, Catgut, Tunbridge Ware, Picture Frame and Architectural Mouldings, Compos, Cameos, and others too numerous to mention.

London: E. & F. N. SPON, 125, Strand.
New York: 12, Cortlandt Street.

Crown 8vo, cloth, 485 pages, with illustrations, 5s.

WORKSHOP RECEIPTS,
SECOND SERIES.
By ROBERT HALDANE.

SYNOPSIS OF CONTENTS.

Acidimetry and Alkalimetry.	Disinfectants.	Isinglass.
Albumen.	Dyeing, Staining, and Colouring.	Ivory substitutes.
Alcohol.	Essences.	Leather.
Alkaloids.	Extracts.	Luminous bodies.
Baking-powders.	Fireproofing.	Magnesia.
Bitters.	Gelatine, Glue, and Size.	Matches.
Bleaching.	Glycerine.	Paper.
Boiler Incrustations.	Gut.	Parchment.
Cements and Lutes.	Hydrogen peroxide.	Perchloric acid.
Cleansing.	Ink.	Potassium oxalate.
Confectionery.	Iodine.	Preserving.
Copying.	Iodoform.	

Pigments, Paint, and Painting: embracing the preparation of *Pigments*, including alumina lakes, blacks (animal, bone, Frankfort, ivory, lamp, sight, soot), blues (antimony, Antwerp, cobalt, cæruleum, Egyptian, manganate, Paris, Péligot, Prussian, smalt, ultramarine), browns (bistre, hinau, sepia, sienna, umber, Vandyke), greens (baryta, Brighton, Brunswick, chrome, cobalt, Douglas, emerald, manganese, mitis, mountain, Prussian, sap, Scheele's, Schweinfurth, titanium, verdigris, zinc), reds (Brazilwood lake, carminated lake, carmine, Cassius purple, cobalt pink, cochineal lake, colcothar, Indian red, madder lake, red chalk, red lead, vermilion), whites (alum, baryta, Chinese, lead sulphate, white lead—by American, Dutch, French, German, Kremnitz, and Pattinson processes, precautions in making, and composition of commercial samples—whiting, Wilkinson's white, zinc white), yellows (chrome, gamboge, Naples, orpiment, realgar, yellow lakes); *Paint* (vehicles, testing oils, driers, grinding, storing, applying, priming, drying, filling, coats, brushes, surface, water-colours, removing smell, discoloration; miscellaneous paints—cement paint for carton-pierre, copper paint, gold paint, iron paint, lime paints, silicated paints, steatite paint, transparent paints, tungsten paints, window paint, zinc paints); *Painting* (general instructions, proportions of ingredients, measuring paint work; carriage painting—priming paint, best putty, finishing colour, cause of cracking, mixing the paints, oils, driers, and colours, varnishing, importance of washing vehicles, re-varnishing, how to dry paint; woodwork painting).

London: E. & F. N. SPON, 125, Strand.
New York: 12, Cortlandt Street.

Crown 8vo, cloth, 480 pages, with 183 illustrations, 5s.

WORKSHOP RECEIPTS,

THIRD SERIES.

BY C. G. WARNFORD LOCK.

Uniform with the First and Second Series.

SYNOPSIS OF CONTENTS.

Alloys.	Indium.	Rubidium.
Aluminium.	Iridium.	Ruthenium.
Antimony.	Iron and Steel.	Selenium.
Barium.	Lacquers and Lacquering.	Silver.
Beryllium.	Lanthanum.	Slag.
Bismuth.	Lead.	Sodium.
Cadmium.	Lithium.	Strontium.
Cæsium.	Lubricants.	Tantalum.
Calcium.	Magnesium.	Terbium.
Cerium.	Manganese.	Thallium.
Chromium.	Mercury.	Thorium.
Cobalt.	Mica.	Tin.
Copper.	Molybdenum.	Titanium.
Didymium.	Nickel.	Tungsten.
Electrics.	Niobium.	Uranium.
Enamels and Glazes.	Osmium.	Vanadium.
Erbium.	Palladium.	Yttrium.
Gallium.	Platinum.	Zinc.
Glass.	Potassium.	Zirconium.
Gold.	Rhodium.	

London: E. & F. N. SPON, 125, Strand.

New York: 12, Cortlandt Street.

WORKSHOP RECEIPTS,
FOURTH SERIES,
DEVOTED MAINLY TO HANDICRAFTS & MECHANICAL SUBJECTS.

By C. G. WARNFORD LOCK.

250 Illustrations, with Complete Index, and a General Index to the Four Series, 5s.

Waterproofing — rubber goods, cuprammonium processes, miscellaneous preparations.

Packing and Storing articles of delicate odour or colour, of a deliquescent character, liable to ignition, apt to suffer from insects or damp, or easily broken.

Embalming and Preserving anatomical specimens.

Leather Polishes.

Cooling Air and Water, producing low temperatures, making ice, cooling syrups and solutions, and separating salts from liquors by refrigeration.

Pumps and Siphons, embracing every useful contrivance for raising and supplying water on a moderate scale, and moving corrosive, tenacious, and other liquids.

Desiccating—air- and water-ovens, and other appliances for drying natural and artificial products.

Distilling—water, tinctures, extracts, pharmaceutical preparations, essences, perfumes, and alcoholic liquids.

Emulsifying as required by pharmacists and photographers.

Evaporating—saline and other solutions, and liquids demanding special precautions.

Filtering—water, and solutions of various kinds.

Percolating and Macerating.

Electrotyping.

Stereotyping by both plaster and paper processes.

Bookbinding in all its details.

Straw Plaiting and the fabrication of baskets, matting, etc.

Musical Instruments—the preservation, tuning, and repair of pianos, harmoniums, musical boxes, etc.

Clock and Watch Mending—adapted for intelligent amateurs.

Photography—recent development in rapid processes, handy apparatus, numerous recipes for sensitizing and developing solutions, and applications to modern illustrative purposes.

London: E. & F. N. SPON, 125, Strand.
New York: 12, Cortlandt Street.

JUST PUBLISHED.

In demy 8vo, cloth, 600 pages, and 1420 Illustrations, 6s.

SPONS'
MECHANICS' OWN BOOK;
A MANUAL FOR HANDICRAFTSMEN AND AMATEURS.

CONTENTS.

Mechanical Drawing—Casting and Founding in Iron, Brass, Bronze, and other Alloys—Forging and Finishing Iron—Sheetmetal Working—Soldering, Brazing, and Burning—Carpentry and Joinery, embracing descriptions of some 400 Woods, over 200 Illustrations of Tools and their uses, Explanations (with Diagrams) of 116 joints and hinges, and Details of Construction of Workshop appliances, rough furniture, Garden and Yard Erections, and House Building—Cabinet-Making and Veneering—Carving and Fretcutting—Upholstery—Painting, Graining, and Marbling—Staining Furniture, Woods, Floors, and Fittings—Gilding, dead and bright, on various grounds—Polishing Marble, Metals, and Wood—Varnishing—Mechanical movements, illustrating contrivances for transmitting motion—Turning in Wood and Metals—Masonry, embracing Stonework, Brickwork, Terracotta, and Concrete—Roofing with Thatch, Tiles, Slates, Felt, Zinc, &c.—Glazing with and without putty, and lead glazing—Plastering and Whitewashing—Paper-hanging—Gas-fitting—Bell-hanging, ordinary and electric Systems—Lighting—Warming—Ventilating—Roads, Pavements, and Bridges—Hedges, Ditches, and Drains—Water Supply and Sanitation—Hints on House Construction suited to new countries.

London: E. & F. N. SPON, 125, Strand.
New York: 12, Cortlandt Street.

www.ingramcontent.com/pod-product-compliance
Lightning Source LLC
Chambersburg PA
CBHW051725300426
44115CB00007B/465